Wireless SPECTRUM Finder

Telecommunications, Government and Scientific Radio Frequency Allocations in the U.S., 30 MHz—300 GHz

D1525545

Bennett Z. Kobb

McGraw-Hill
New York   Chicago   San Francisco   Lisbon
London   Madrid   Mexico City   Milan   New Delhi
San Juan   Seoul   Singapore   Sydney   Toronto

McGraw-Hill
A Division of The McGraw·Hill Companies

Copyright 2001 Bennett Z. Kobb.

All rights reserved. Printed in the United States of America. Except as permitted under the United States Copyright Act of 1976, no part of this publication may be reproduced or distributed in any form or by any means, or stored in a data base or retrieval system, without the prior written permission of the publisher.

1 2 3 4 5 6 7 8 9 0 AGM/AGM 0 7 6 5 4 3 2 1

ISBN 0-07-137506-6

The sponsoring editor for this book was Marjorie Spencer, and the production supervisor was Sherri Souffrance. It was set in New Century Schoolbook by MacAllister Publishing Services, LLC.

Printed and bound by Quebecor Martinsburg.

Trademarks are the property of their respective owners and may appear herein for editorial purposes only.

Visit us on the Web at http://www.spectrumfinder.net.

McGraw-Hill books are available at special quantity discounts to use as premiums and sales promotions, or for use in corporate training programs. For more information, please write to the Director of Special Sales, Professional Publishing, McGraw-Hill, Two Penn Plaza, New York, NY 10121-2298. Or contact your local bookstore.

Information contained in this work has been obtained by The McGraw-Hill Companies, Inc. ("McGraw-Hill") from sources believed to be reliable. However, neither McGraw-Hill nor its authors guarantee the accuracy or completeness of any information published herein and neither McGraw-Hill nor its authors shall be responsible for any errors, omissions, or damages arising out of use of this information. This work is published with the understanding that McGraw-Hill and its authors are supplying information but are not attempting to render engineering or other professional services. If such services are required, the assistance of an appropriate professional should be sought.

 This book is printed on recycled, acid-free paper containing a minimum of 50 percent recycled de-inked fiber.

Table of Contents

Preface by James F. Lovette	iv
Acknowledgments	v
Introduction	vi
Three Trends	1
Frequency Allocations:	
VHF (30 MHz–300 MHz)	11
UHF (300 MHz–3000 MHz)	93
SHF (3 GHz–30 GHz)	267
EHF (30 GHz–300 GHz)	359
International Footnotes	403
U.S. Footnotes	421
Government Footnotes	453
Non-Government Footnotes	459
Subject Index	469
Band Index	493
About the Author	499

Preface

One hundred and fifty years ago, the best-selling books in America were emigrant's guides describing trails and their landmarks, watering places, and potential dangers to the thousands of people planning or making their ways across our continent.

Wireless Spectrum Finder is today's essential equivalent for all of us involved in spectrum-dependent technologies and services. Momentous changes are occurring in spectrum management, and any assumption about the future is fragile. Even so, unimaginable fortunes are being risked (and some are being made) through creative spectrum acquisition and habitation.

Some nineteenth-century emigrant's guides were assembled by people who never reached St. Louis, but this guide has unassailable credibility because, when it comes to making or describing a trail through the *Federal Communications Commission* (FCC) rules, Bennett Kobb has "been there" and "done that."

To would-be users of spectrum, making one's way through the FCC process resembles the most antagonistic video-game treasure hunt ever, with unspeakable threats popping up everywhere. The format of the FCC rules is in part responsible. For any given frequency band, the applicable regulations may be strewn throughout five microscopic-print tomes of the Federal Code. Understanding a rule may also require combing the record of the proceeding in which it was developed and often reports from both the FCC and the *National Telecommunications and Information Administration* (NTIA).

Wireless Spectrum Finder does this research for us, and, along with straightforward facts and clear pointers to all the regulations for each band, the author develops especially insightful commentaries.

Anyone who hopes to exploit or bring about changes in the FCC rules needs the most effective and dependable guidebook. If *Wireless Spectrum Finder* didn't exist, each of us would have to create our own. We couldn't possibly do it as well.

JAMES F. LOVETTE

Jim Lovette is Director of Strategic Policies at Fantasma Networks, Inc.

Acknowledgments

Wireless Spectrum Finder benefited from the cooperation I have received for many years from the FCC and the NTIA.

I am very fortunate to have Jim Lovette, Dale Hatfield, Dewayne Hendricks, Mitchell Lazarus, and Craig Mathias among this book's friends and advocates.

Special thanks to Michael Peyton of Earthspan for photography and research, and to WNVC-TV, Merrifield, Virginia, for the author portrait location.

The support and expertise of Marjorie Spencer and Laura Belt were essential to making this book happen.

This book is affectionately dedicated to Marthe and Sol Kobb.

Introduction

The radio spectrum is that portion of the electromagnetic energy spectrum used by radio waves. Radio services are categories of radio use. The apportionment of frequency ranges, or bands, in the radio spectrum to radio services is a government function called spectrum allocation or frequency allocation.

The nations of the world allocate spectrum to radio services in the form of an International Table of Allocations maintained by the *International Telecommunication Union* (ITU) in Geneva, Switzerland. A United Nations agency, it develops the ITU Radio Regulations and convenes the World Radiocommunication Conferences in which the U.S. is a key participant.

ITU Member States create their own domestic tables of allocations based on the International Table. International allocations are usually, but not always, incorporated into domestic allocations.

The U.S. is within ITU Region 2, which includes the Americas and Greenland. (Region 1 includes Europe, Africa, and the northern and western portions of Asia. Region 3 includes the rest of Asia, Australia, and the South Pacific.)

About the FCC and NTIA

The Communications Act of 1934 established the Federal Communications Commission (FCC) as an independent regulatory agency with jurisdiction over non-federal spectrum use. The FCC officially defines spectrum allocation as "the entry in the Table of Frequency Allocations of a given frequency band for the purpose of its use by one or more terrestrial or space radiocommunication services or the radio astronomy service under specified conditions."

The radio spectrum is formally defined as extending in frequency from 0 to 3,000 gigahertz (GHz). This book is derived from the 30 megahertz (MHz) to 300 GHz portion of the U.S. Table of Frequency Allocations.

Actually, two official U.S. Tables exist. The FCC administers the Non-Federal Government Table. The *National Telecommunications and Information Administration* (NTIA) in the U.S. Department of Commerce administers the Federal Government Table. In a very few cases, the frequency boundaries in the two tables are different. *Wireless Spectrum Finder* is derived from both, but emphasizes FCC allocations current at the time of publication.

The book spans the *Very High Frequency* (VHF, 30–300 MHz), *Ultra High Frequency* (UHF, 300–3000 MHz), *Super High Frequency* (SHF, 3–30 GHz) and *Extremely High Frequency* (EHF, 30–300 GHz) bands. Most commercial activity is in these bands. To the official allocations data, we add descriptive narratives about spectrum uses and trends.

Wireless Spectrum Finder emphasizes bands allocated to services instead of the individual frequencies assigned to millions of stations. Some bands, especially those in the millimeter wave spectrum, may have little or no publicly known use at this time. We nevertheless include the allocation specifics for these bands.

All material in this book is derived from public sources. The existence or mission of some federal stations remain classified. Others, typically in federal law enforcement, are categorized as "sensitive," meaning that their operations or technical parameters are not releasable to the public under the Freedom of Information Act or other laws or regulations.

Regulation and Enforcement

FCC spectrum allocations usually involve a public comment process, as prescribed by the Administrative Procedures Act, which governs many regulations imposed on citizens.

All FCC actions are taken under the authority of five individuals: the Commissioners, nominated by the President and confirmed by the Senate. Aiding the Commissioners are bureaus and offices of telecommunications experts, attorneys, economists, and engineers. The FCC monitors radio services at all times through its Washington-based Watch Officer and a network of receiving stations and field agents.

Spectrum users should take special care in U.S. border areas where international agreements affect radio operations, in U.S. territories under jurisdiction of agencies other than the FCC, or in areas not under U.S. jurisdiction. The FCC's authority is limited to precisely delineated geographic boundaries, whereas ITU regulations apply globally and are enforced by its member nations.

The Commission's Enforcement Bureau, aided by the U.S. Marshals Service, penalizes rogue operators and seizes radio equipment that is made, sold, or operated in violation of FCC rules.

The federal government's internal use of radio is subject to presidential authority. This authority is exercised through the NTIA and its Administrator, the Assistant Secretary of Commerce for Communications and Information. The NTIA Administrator functions as the President's principal advisor and spokesperson on telecommunications policy.

The Communications Act limits NTIA's spectrum management authority to stations belonging to federal agencies. NTIA hosts and is advised by the *Interdepartment Radio Advisory Committee* (IRAC), representing spectrum users within the government. IRAC expertise is frequently utilized where proposed allocations or systems will impact safety-related or military frequencies or critical infrastructure.

Terminology

The FCC now uses the term *Federal Government* (FG) to refer to the use of bands by stations belonging to the U.S. The term *Non-Federal Government* (Non-FG) includes state or local government spectrum users, other non-commercial users and institutions, and all commercial users.

Wireless Spectrum Finder identifies these allocations as FG and Non-FG respectively. Do not confuse these with the *Government* (G), *Non-Government* (NG), and US designators that apply to Footnotes (see the following section).

U.S. frequency allocations consist of frequency bands, lists of the services permitted to use the bands, and Footnotes that modify the rights of a service to operate in a band.

The FCC occasionally authorizes licensees to use a given band for research purposes, even when the band is not allocated for such use. The Experimental Radio Service, regulated in Part 5 of the FCC rules, provides for such research. Also, Part 74 provides for Experimental Broadcast Stations, which are used to develop services intended for reception by the general public.

Commercial use of experimental stations is generally prohibited, though the FCC sometimes authorizes these stations to conduct market tests that involve selling products and/or services to the public. In these cases, subscribers or purchasers are supposed to be warned that their investment may become useless if the FCC changes its regulations or declines to adopt the radio service in question.

About Footnotes

Official Footnotes may add allocations, add protection, or remove protection from interference for a particular service or radio user.

Footnotes are designated G for government allocations and NG for non-government allocations. US Footnotes apply to both government and non-government allocations. The letter S now indicates International Footnotes established by the ITU. Where International Footnotes apply to U.S. allocations, the FCC includes them in the U.S. Table.

The FCC officially adopted some ITU Footnotes under a previous numbering scheme that did not use the letter S, but used a three-digit designator. At this writing, the FCC retains some International Footnotes under the older scheme.

Primary and Secondary

The categories of primary and secondary service are essential to understanding spectrum allocation. The FCC states, "Stations of a secondary service: (1) may not cause harmful interference to stations of primary services to which frequencies are already assigned or to which frequencies may be assigned at a later date; (2) cannot claim protection from harmful interference from stations of a primary service to which frequencies are already assigned or may be assigned at a later date; and (3) can claim protection, however, from harmful interference from stations of the same or other secondary service(s) to which frequencies may be assigned at a later date."

Radio services may be primary in some bands and secondary in other bands. A shared allocation is one available to two or more services, typically in a mixture of primary and secondary use. When two or more services share the same bands, they often must coordinate, or carefully select frequency, location, power, technology, and/or schedules to minimize interference. Trade associations and consulting engineers typically perform this coordination on behalf of FCC licensees.

By international convention, primary services appear in CAPITAL letters. For example, FIXED means that the fixed service (radio communications between stationary stations) is primary in a given frequency band. The FCC further resolves descriptions such as FIXED into individual radio services.

Examples of the fixed services include the Private Operational Fixed Point-to-Point Microwave Service for internal use only by commercial or state and local government licensees and the Common Carrier Fixed Point-to-Point Microwave Service, which carries telecommunications traffic for profit. The FCC designator Fixed Microwave (101) encompasses both examples and refers to FCC Rules Part 101.

Other types of fixed services include the Public Fixed Radio Service, the Rural Radiotelephone Service, the Multipoint Distribution Service, and the Broadcast Auxiliary Services. Also, fixed operations are available in services devoted mainly to mobile communications.

What's in a Name?

The Communications Act directs the FCC to classify radio services, but the FCC increasingly prefers general, nondescriptive service names in order to appear as accommodating as possible, or to avoid the appearance of favoritism toward any particular interest group.

"Static definitions of use, whether for service or technology, are doomed to fail and will need to be changed," according to former FCC chairman Reed Hundt. "In nearly every service the FCC authorizes, licensees come back to the Commission to ask permission to change something."

The FCC has stated, "We believe that, to the extent possible we should ensure that the market determines the most appropriate use of this spectrum. . In this regard, we are not inclined to allocate spectrum for particular kinds of services unless there is a clear and compelling public interest in doing so."

Thus, the FCC created the Wireless Services, the Wireless Radio Services, the Wireless Communications Service, the General Wireless Communications Service, the General Mobile Radio Service, the General Purpose Mobile Radio Service, the Commercial Wireless Radio Services, the Commercial Mobile Radio Services, the Miscellaneous Wireless Communications Services, and the Multi-Use Radio Service. These ambiguous names refer to different services and frequency allocations.

On the other hand, the FCC created a Mobile Satellite Service Above 1 GHz, a Non-Geostationary Mobile Satellite Service, and a 1.6/2.4 GHz Mobile Satellite Service. These are all different names for the same thing, which the FCC usually calls Big LEOs (Low Earth Orbits).

Here is part of a typical band listing:

148-149.9 MHz

FG: S5.218 608A US10 G30 FIXED. MOBILE. MOBILE-SATELLITE (Earth-to-space) 599B US319 US320 US323 US325

Non-FG: S5.218 608A US10 MOBILE-SATELLITE (Earth-to-space) 599B US319 US320 US323 US325

FCC: Satellite Communications (25)

In this example, S5.218 608A means that Footnotes S5.218 and 608A state requirements that apply to FG use of this band for any purpose. These are International Footnotes (identifiable by the letter S or by the three-digit number) established by the ITU, and which the U.S.

government has decided to impose domestically. US10 and G30 are U.S.-only Footnotes that apply to FG use of the band.

Whether international or domestic, Footnotes that concern any use of a band are shown first, before any particular radio services.

Footnotes 599B through US325 apply to FG use of the Mobile Satellite Service in this band, because they follow the MOBILE-SATELLITE (Earth-to-space) designator. The Non-FG listing may appear similar to the FG listing, but in this case the two are not identical.

The rule part cross-reference "FCC: Satellite Communications (25)" refers to a specific part of the FCC Rules. The rules appear in the Code of Federal Regulations and the Federal Register.

Not all bands include a rule part cross-reference. The FCC has taken special pains to point out that the rule part cross-references are for informational use only and cannot be considered an official part of frequency allocations. This precaution probably serves to nullify any claims of special rights to frequencies based on this informal cross-reference.

The capital letters used in the listing indicate that all of the authorized uses in this band are co-primary. Upper and lower case letters indicate secondary use.

Many radio services use multiple bands. Sometimes these bands can be separated by many megahertz. Land mobile radio systems, for example, may use separate bands for the base station and the mobile radios. You can usually find the principal narrative for a service with multiple bands in the listing for the lowest frequency band that it uses.

In *Wireless Spectrum Finder*, a range of allocated frequencies is a *band*. Two or more bands are *spectrum* and frequency ranges within a band are *segments*. Regulations sometimes use the terms segment and band interchangeably.

Licensed and Unlicensed

Not all radiators of radio energy are licensed and classified in radio services. Part 15 of the FCC rules governs unlicensed intentional, unintentional, and incidental radiation devices that may share spectrum with, but are not a part of, radio services.

Note that the official definition of frequency allocation, quoted earlier, does not include unlicensed devices or uses of the spectrum that are not classified as radio services.

As the FCC has stated, "The Part 15 rules specifically state that such devices have no vested or recognizable right to continued use of any frequency and must accept any interference received from other Part 15 devices. All Part 15 intentional radiators must carry a label informing the user that the device must accept any interference received, including undesired operation."

Most radio services involve individualized licensing and license documents. A few do not. It is important to distinguish between Part 15 unlicensed devices and those radio services that are "authorized by rule" but do not issue license documents, such as some aviation, marine, personal, and specialized services (Part 95 and others). Unlicensed devices and rule-authorized services are two different categories that may appear functionally identical.

Part 15 covers several types of radiating devices. Intentional radiators include such products as cordless phones or remote-controlled toys. These intentionally transmit *radio frequency* (RF) energy, usually at low power levels. By design, unintentional radiators such as television receivers use RF energy and may radiate it but are not intended to be radio transmitters. Incidental radiators do not use RF energy but may radiate it as a byproduct of operation.

THREE TRENDS

At this writing, the Federal Communications Commission (FCC) is examining the possibilities of *Ultra Wideband radio* (UWB, Docket ET 98-153) and *Software-Defined Radio* (SDR, Docket ET 00-47). These technologies are not limited to specific frequency bands but could enable communications across wide swaths of spectrum already in use by others, or on varying portions of the spectrum depending on immediate needs.

The FCC also is considering secondary markets (Docket WT 00-230) where licensees lease their spectrum usage rights to third parties. It created a type of secondary market in partitioning and disaggregation in which sublicensees provide wireless service in parts of a geographic area, or on frequencies, licensed to another party. The guard band manager licensing scheme in the 700-MHz spectrum is also a type of secondary market (see 746–764 MHz).

Paradigm Shift

UWB conveys data by varying the timing or phasing of extremely short *radio frequency* (RF) pulses. Several UWB techniques are available, but they all result in low-level emissions distributed across megahertz or gigahertz of spectrum, and therefore across the boundaries of numerous radio services. At this writing, FCC rules do not accommodate general UWB usage.

UWB research is focused on 2–6 GHz and other bands including low VHF. The exact frequency range UWB will use is critical to the issue of interference to non-UWB services.

UWB can offer privacy, low likelihood of interference, low cost, high capacity, and, when used in radar and localizer applications, the power to make extremely precise measurements of distance.

Ground penetrating radar (GPR) has been a key application for UWB. Some GPRs are wheeled, lawnmower-like instruments that reveal X-ray-like images of structures such as roads and bridges, for safety inspection.

Potential medical uses for UWB, the FCC says, "include the development of a mattress-installed breathing monitor to guard against Sudden Infant Death Syndrome, and heart monitors that act like an electrocardiogram except that they measure the heart's actual contractions instead of its electrical impulses. ...Other interesting UWB applications include liquid level sensors for everything from water conserving toilets to oil refinery tanks and the use of UWB technology to allow auto focus cameras to calculate distances more accurately." UWB also is used in experimental microphones that utilize radar instead of air to pick up sound from the vocal organs.

Defense, safety, and law-enforcement communities are also interested in the technology. Several companies have received conditional FCC waivers to manufacture and sell UWB radars and covert communication devices as long as distribution is limited and records are kept on purchasers and users.

Time Domain Corporation says that, "Ultra-wideband low power communications can prove beneficial in a host of applications that require relatively short range, large data capacity, a high degree of immunity to the adverse effects of multipath, and a low probability of unauthorized detection and interception."

The *National Telecommunications and Information Administration* (NTIA) has stated that "UWB devices...represent a radically different technology from that used in the conventional radios available when the [FCC] Part 15 [unlicensed device] rules were developed. UWB technologies use extremely narrow pulses with their concomitant ultra-wide bandwidths, high repetition frequencies, and low duty cycles. The effects of these types of signals upon conventional systems, be they analog, digital, or pulse modulated, are not well understood.

"The proliferation of UWB systems centered near 2 GHz could cause serious problems to several critical, sensitive services important to both the government and the public," the NTIA said.

FCC Acts

After extensive lobbying by UWB interests, the FCC initiated a wide-ranging *Notice of Inquiry* (NOI), followed by a *Notice of Proposed Rulemaking* (NPRM) in ET Docket 98-153, asking the industry for input on measurement techniques and methods of protecting incumbents from interference.

The NPRM tentatively concluded that, because most near-term UWB applications involve low power and short range, the FCC should regulate them under the Part 15 rules for unlicensed devices. The current Part 15 rules, however, pose two obstacles to the implementation of UWB technology. The FCC noted, "First, the wide bandwidth that is intrinsic to the operation of UWB devices can result in transmission of the fundamental emission into restricted frequency bands or into the television broadcast frequency bands, which is prohibited under the Part 15 rules.

"Second, the current emission measurement procedures specified in our Part 15 rules were developed for narrowband systems and may be inappropriate for, and pose unnecessary restrictions to, UWB technology, particularly impulse systems."

Characterizing UWB

"UWB can be viewed as an extreme form of spread spectrum," according to UWB developer XtremeSpectrum. "Consider that a spread spectrum system might have bandwidth of 20 MHz, while an UWB system might have 2 GHz. UWB systems have orders of magnitude more bandwidth than spread spectrum systems. This leads to a corresponding increase in process gain and all its attendant benefits. ...UWB systems can have orders of magnitude more simultaneous users in a cell, with the same data rate and multiuser interference level, than a conventional spread spectrum system."

Current spectrum stakeholders were unenthusiastic. "Clearly, it would be inconsistent with the fundamental tenets of the Commission's...policies regarding Part 15 devices if the Commission were to amend its rules to accommodate the provision of unlicensed UWB radio systems, knowing that such systems can cause harmful interference within restricted bands and the TV broadcast bands," said the Consumer Electronics Manufacturers Association and the National Association of Broadcasters.

Representatives of *Global Positioning System* (GPS) equipment makers and users argued that "There is no question that UWB operations would increase the background noise in a given spectrum. ...Increases in background noise in the GPS frequency bands may reduce the ability of the GPS receiver to acquire a GPS signal or even to maintain tracking of a GPS signal, or cause errors in position or time accuracy. Any of these consequences is intolerable for a safety-of-life service such as GPS."

FAA Opposition

UWB will cause "interference to critical aeronautical safety systems," according to the *Federal Aviation Administration* (FAA). The agency was especially concerned about UWB radiation in the restricted bands in which FCC Part 15 rules permit unlicensed devices to emit only very low-level emissions.

"The FAA is opposed to any authorization of licensed or unlicensed UWB systems to intentionally radiate in these bands. It is likely that authorizing even limited operation of such systems will lead to further proliferation of UWB systems as new applications for their use are developed.

"The FAA has documented cases of radio frequency interference caused to such [aeronautical] services from non-licensed low power devices such as television antenna amplifiers, baby monitors, personal computers, and UWB operations. In each case, these incidents caused disruption to air traffic flow within the United States."

We are skeptical of the FAA's claim of "documented cases" of UWB interference to aviation. UWB systems have not been widely deployed and are not easily detected and identified. For this reason, we question FAA's flat inclusion of UWB in incidents that "caused disruption to air traffic flow."

In its FCC comments, Lockheed Martin Corporation urged caution. The company "requests that the Commission limit the initial scope of this proceeding, and not seek to implement a broad technical framework for UWB, even on a provisional basis." In contrast, *Lockheed Martin Information Systems* (LMIS) said that it "strongly urges the FCC to promptly move forward on approval of the ultra-wideband technology for widespread use. In addition, LMIS urges the FCC to consider higher power authorizations to allow for extended range coverage (at least several miles) of wireless high bandwidth data systems."

Cumulative Impact

Opponents of UWB have told the FCC that the aggregate effects of many UWB transmitters will produce harmful interference into licensed services. Yet "the Commission should not permit any obstructionist tactics to delay the introduction of new, innovative UWB technologies," according to Fantasma Networks, developer of UWB technology for broadband home networking. "UWB transmissions will, as a practical matter, be invisible to other radio services. ...In metropolitan areas today, many existing sources of harmful interference, including spurious emissions from high-power VHF and UHF transmitters, intermodulation effects (including the incidental mixing of authorized transmissions in metal structures), existing unintentional radiators, and existing incidental radiators, already 'aggregate' to create an increased level of interference or a raised noise floor.

"Because the link budget and the number of and site selection for central stations of radio systems being designed today must already account for this radio noise, authorizing UWB emissions at levels comparable to these already existing non-thermal noise sources will not lead to an 'aggregation' problem," Fantasma said.

Radars Permitted

The NPRM proposed to permit GPRs to operate anywhere in the spectrum, subject to some emissions limits. A GPR would have to be within one meter of the ground and aimed directly into the ground to qualify as a permitted product.

The FCC said the situation was "less clear" with regard to UWB radars that are used to obtain images through walls or behind other surfaces. It asked whether it should treat such devices as GPRs, or restrict their transmission frequencies and power.

UWB devices can generally operate above 2 GHz without causing harmful interference to other radio services, the FCC said. It proposed to enable such usage without restriction. Because of intensive spectrum use and the presence of GPS, the FCC said it has "significant concerns" about UWB operation below 2 GHz, except for GPRs "and possibly through-wall imaging devices." It is monitoring UWB/GPS tests that NTIA and other researchers are conducting, and collecting comments on possible regulatory approaches.

Software-Defined Radio (SDR)

The FCC has proposed to define SDR as "a radio that includes a transmitter in which the operating parameters of the transmitter, including the frequency range, modulation type, and maximum radiated or conducted output power can be altered by making a change in software without making any hardware changes."

The radio can therefore be programmed to operate in different radio services with various transmission formats. The FCC suggested that a SDR unit could, for example, operate in the various cellular and *Personal Communications Service* (PCS) bands and standards used around the world.

"Today, we struggle to squeeze multiple services into spectrum, or to mandate specific standards to permit communications devices to work seamlessly," says FCC Commissioner Susan Ness. "With SDR, the software could make such decisions, not the FCC. The availability of such software also might make it easier for different users to share crowded spectrum. Of course, protection of other spectrum licensees from interference resulting from SDR devices is paramount."

SDR could improve interoperability between differing designs, services, and bands (an important feature in public safety communications), changing transmission and reception standards and frequency at the push of a button or upon over-the-air command. They could receive software upgrades directly by radio. SDR equipment could monitor the spectrum and locate, in the FCC's words, a "hole" or open frequency with sufficient bandwidth where the user could operate.

"The future vision of SDR-based end-user equipment is analogous to personal computers, where the end user can easily upgrade or reconfigure his or her computer by changing components or software," says Finnish phone giant Nokia. "Nokia believes that this exciting future vision is a long-term goal, unlikely to be reached earlier than 2010. ...We are far from reaching the goal of developing completely generic SDR equipment, where all system-specific components have been eliminated. Equipment that can handle a dynamic radio environment and wide spectrum range requires components that have yet to be developed."

SBC Wireless told the FCC that "In the near term, SDR is primarily an implementation technique, a method for radio system designs, and has no intrinsic ability to make spectrum use more efficient or flexible. ...[The FCC] discusses the use of such technology in terms of allowing access to a wide bandwidth of spectrum and enabling a handset that can operate across wide frequency bands, using multiple air interfaces. It is the belief of SBC Wireless that such use is not technically or commercially feasible at this time. Care must be taken in making sweeping assumptions about the ability of such systems."

The SDR Forum, a trade association, has said that, "Multi-band SDR can provide non-disruptive transitions to dominant air interface choices. Once SDR infrastructure is widely deployed, it will enable transitions of varying types and complexity, including overnight updates to improve waveforms and service features, as well as gradual changes as more subscribers are able to accommodate new waveforms."

According to the American Radio Relay League (ARRL), which is the national association for Amateur Radio, "The Amateur Radio Service is a fertile testing ground for SDR technology. The flexible regulations, shared frequency allocations, and the multiplicity of transmission modes in common use in the Amateur Service constitute the proper environment for the development and deployment of SDR equipment."

At this writing, the FCC proposed in Docket ET 00-47 to permit SDR across all radio services by minor changes to its equipment authorization rules. These rules will be streamlined (legally categorized as a Class III permissive change), but manufacturers would still have to submit technical documentation and test results if software downloads make substantial changes in the frequency band or radio service.

Instead of the exterior FCC ID number, a label on every Commission-authorized device, the number could appear on the display screen of an SDR and change when new software reconfigures the unit.

FCC officials emphasized that manufacturers will have to incorporate methods such as encryption and authentication into their software to prevent hacking and fraud. Such unauthorized modifications became the bane of the cellular service when it became clear that even minor changes resulted in no-charge calling.

Secondary Markets

Ultimately, SDR-based systems could identify and buy available wireless communications capacity on an open market. The FCC, concerned about a possible spectrum drought, established as a goal that spectrum become like any other commodity that flows fluidly in the marketplace.

In certain markets, the FCC said, "spectrum is becoming increasingly congested and spectrum constraints are threatening to limit the growth of new services, particularly in the more densely populated urban areas where wireless carriers have tended to concentrate their initial build-out efforts."

To remedy the problem, the Commission proposed to allow Wireless Radio Services licensees—those with exclusive rights in particular geographic areas, such as auction winners—to lease their rights in any amount of spectrum and for any period during the term of the license, as long as the lessees comply with FCC rules. Lessees could further sublease to other lessees and might be able to use their leased spectrum for purposes not contemplated in the FCC rules.

The FCC proposed to grant the expanded leasing authority to exclusive-use licensees and not to radio services in which licensees share the use of spectrum. Shared spectrum raises interference and frequency coordination issues that are more complex than for licensees that have exclusive rights to use their spectrum, the Commission said.

"In addition, where licensees do not hold spectrum on an exclusive basis, other potential spectrum users are not precluded from obtaining their own licenses, provided that appropriate sharing arrangements can be reached. This may reduce the need for leasing as an alternative to facilitate efficient spectrum use," the FCC added.

Real-Time Capacity Auction

A development-stage company, Red Bat Communications, presented to the FCC an exchange system in which the user's handset bids to make a call with wireless carriers that continuously auction capacity over the air. An independent entity collects and broadcasts carrier pricing information, and the handsets monitor this ticker for prices and terms. Red Bat said that the FCC "may want to consider licensing a small sliver of spectrum to provide an independent broadcast channel for the auction. This will help ensure the auction entity's independence, neutrality, and low transaction costs —all of which will help protect consumers and ensure the most efficient market use of spectrum.

"The single, one-way broadcast enables one limited information broadcast to support an infinite number of users. Two-way systems involving query and response, or systems that propose a negotiation between buyer and seller, will consume far greater quantities of spectrum just to establish the price and contract structure," the company said.

The user could accept different pricing levels depending on the significance of the call (voice call or e-mail download, for example), the identity of the called or calling party, the time of day, or special promotions offered by a carrier. Carriers could increase or decrease prices on the fly, depending on current demand levels or competitive conditions.

Eventually, such developments as UWB and SDR may remove technological and regulatory barriers and permit licensees to buy and sell spectrum and wireless services on the spot. The challenges of retaining public accountability and open entry for different technical approaches will remain.

Frequency Allocations

VHF

30–30.56 MHz

FG: FIXED. MOBILE

Non-FG: None

This band and adjacent bands in the low VHF spectrum are allocated for federal government use.

Communication in this spectrum can receive interference from long-distance signal propagation, especially during high sunspot activity. Higher bands are not so affected. Ambient noise levels in the low VHF bands are also higher than in other bands.

A longstanding Army portable and mobile system, operating across 30–87.975 MHz, is the *Single Channel Ground and Airborne Radio System* (SINCGARS). These radios provide frequency-hopping, secure voice, and data communications.

Experimental *ultra-wideband* (UWB) systems have been developed in the 30–50 MHz spectrum. These high data rate transmissions use ground or surface wave propagation to communicate over distances of several kilometers, through foliage and buildings and over hills. Most UWB technology uses higher bands (see Three Trends).

Other military services and federal agencies make use of these low VHF bands, especially the Air Force, Navy, and the Interior and Agriculture Departments.

30.56–32 MHz

FG: None

Non-FG: NG124 FIXED. LAND MOBILE

FCC: Private Land Mobile (90)

This and adjacent low VHF bands are available to the *Private Mobile Radio Services* (PMRS), also called *Private Land Mobile Radio Services* (PLMRS). These are categories of not-for-profit one- and two-way radio for licensees' internal use. They typically are used for voice and computer-aided dispatch of workers and vehicles, and for paging.

An essential distinction is that between mobile and fixed services. "Mobile service is one that allows the end user to communicate while moving or from different locations. Fixed service requires the end user to be at a set location," the FCC explains.

Key Definitions

The Commission has changed its classification of radio services. The new terminology— revised again in 2000—is somewhat opaque. The FCC administers a category known, if redundantly, as the Wireless Radio Services, which encompass most forms of radio communication.

Within the Wireless Radio Services are the Private Mobile Radio Services, regulated largely in Part 90 of the FCC rules and the *Commercial Wireless Radio Services* (CWRS), comprised of two types of *Commercial Mobile Radio Services* (CMRS), regulated in Part 20 of the FCC rules. Licensees of CMRS businesses provide for-profit, telephone-interconnected (mobile and portable telephone or dialup paging) service to the public.

The two CMRS components of the Commercial Wireless Radio Services are:

- The CMRS Mobile Services, including most, but not all, of the *Specialized Mobile Radio Services* (SMR, see 806–821 MHz); Public Coast Stations (see 154–156.2475 MHz); the Public Mobile Radio Services, which include the Cellular Radiotelephone Service (see 824–849 MHz); the Public Mobile Air-Ground Radiotelephone Services, which consist of the General Aviation Air-Ground Radiotelephone Service (see 454–455 MHz) and Commercial Aviation Air-Ground Systems (see 849–851 MHz); the Offshore Radiotelephone Service (see 470–512 MHz); several types of mobile satellite services; and the Broadband *Personal Communications Service* (PCS, see 1850–1990 MHz).

- The CMRS Messaging Services, which include the Private Paging and Radiotelephone Service, the Public Mobile One-Way Paging Service, the *Narrowband Personal Communications Service* (N-PCS, see 901–902 MHz), and the "220 MHz Band and Interconnected Business Radio Service" (not to be confused with 220 MHz services that are not interconnected to the telephone network. See 220–222 MHz).

Incidentally, not all forms of SMR are considered CMRS Mobile Services. Some SMR licensees, including those in the 220 MHz, 470 MHz, and higher bands, elected to remain in the PMRS ambit.

Previously, most for-profit mobile services were called *radio common carriers* (RCCs). This term has mostly disappeared from use.

PMRS usually is limited to specific missions or industries, such as police dispatch, manufacturing, or transportation. In contrast, any customer may use CMRS for any business or personal communications. PMRS licenses are traditionally issued on a shared basis, with many licensees using the same channels.

"While commercial radio providers offer communications services as their end product, private land mobile wireless licensees use radio as a tool to enhance the safety and/or efficiency of their non-communications businesses," the FCC has stated. "This difference is the foundation of the different regulatory treatments afforded to private, as opposed to commercial, wireless services."

Going forward, CMRS licenses are issued on a mutually exclusive basis, with only one licensee awarded a license for a given spectrum assignment in an area.

CMRS Niche Programming

At this writing, the FCC was considering a petition to allow CMRS licensees to provide *Limited Program Distribution Service* (LPDS) for transmission of audio to specialized audiences. Conventional scanner receivers could be used to hear LPDS. The petitioner is Matthew Edwards, president of FreePage Corporation. Edwards was one of the earliest pioneers in U.S. PCS, having filed the original petition that prompted the FCC to study and eventually create PCS.

FreePage cited precedents for the use of common carrier channels in this manner:

"For instance, the C-band satellite dish industry has become a broadcasting service for hire using common carrier frequencies in a manner not originally envisioned by the Commission. [See 3.7–4.2 GHz.] LPDS could provide program or data services to very small groups of niche audiences who otherwise could not receive programming in their language or information not otherwise economically transmitted."

Major paging carriers supported the petition, while the *National Association of Broadcasters* (NAB) opposed it. "The LPDS model assumes that consumers are both willing and able to pay for niche broadcasting of one of a few channels and assume the risk that their 'niche' does not become unprofitable," it said. "Because FreePage would have the public shoulder the financial burdens of creating LPDS and because there is no evidence of consumer demand for such a subscription-based service, the Commission should not allow FreePage or other CMRS licensees to abandon their primary service."

NAB recommended that FreePage transmit over the Internet instead: "The Internet can simultaneously serve audiences on a local, national and international basis. It is not bound by the physical constraints of the electromagnetic spectrum. [T]he barrier costs to entry on the Internet are very low. ...Internet audiocasting has the flexibility of either being a subscription-based service, a free advertising-based service or a free government subsidized service."

FreePage wryly responded, "The NAB's position on the future of the Internet as a delivery vehicle suggests that the Commission should examine the eventual take-back of AM and FM frequencies."

Providing Fixed Service

The Commission made extensive changes to the U.S. Table of Allocations in 1996 to permit fixed service in CMRS bands on a co-primary basis to mobile service. Before these changes, CMRS operators were only allowed to provide fixed service on an "ancillary," "auxiliary," or "incidental" basis.

The FCC removed those constraints. This means that the CMRS operators listed above may offer wireless local loop fixed telephone service in competition with local telephone companies. (The FCC defines "local loop" as "the wires, poles, and conduit that connect the telephone company end office to the customer's home or business.") CMRS operators need not actually offer any mobile service.

CMRS fixed service may be the means by which meaningful local exchange telephone competition will emerge. According to the FCC, "Potential fixed wireless services include not only 'wireless local loop,' i.e. fixed wireless links to connect residences, apartment buildings, office buildings, and other structures with wireline local exchange networks, but also fixed wireless architectures that can link end users to cellular switches and remote base stations."

The FCC grants PMRS licenses only to individuals, organizations or agencies that qualify under eligibility rules developed over decades of proceedings. Individuals who don't qualify for PMRS licenses may meet their needs in the Personal Radio Services (see 462.5375–462.7375 MHz); or, if they require telephone interconnection, in CMRS services such as cellular and PCS.

Frequency Coordination and Service Pools

Trade associations representing each PMRS category perform quasi-licensing functions under FCC oversight.

In their role as "frequency coordinators," these associations are authorized by the Commission to recommend frequencies for licensing to applicants, based on technical review of proposed and existing PMRS stations. The coordinators normally charge fees for this service.

The PMRS encompass a diverse array of services. There were formerly 20 principal PMRS services. But in a streamlining campaign of the mid-1990s, the FCC consolidated the 20 services into two principal "pools" or collections of PMRS frequencies: the Public Safety pool and the Industrial/Business pool.

In the 800 MHz spectrum, the FCC adds the *Specialized Mobile Radio* (SMR, see 806–821 MHz) and General pools to the Public Safety and Business pools.

Each of the frequency coordinators that provided coordination services in any of the radio services included in the Industrial/Business pool became eligible to coordinate any frequency in the Industrial/Business pool. This policy drastically increased the universe of frequencies available to each coordinator. The FCC made a few exceptions to this policy, however. Because of safety concerns, the FCC ruled that frequencies formerly allocated exclusively to or shared by the Power, Petroleum, and Railroad Radio Services would continue to be available only to coordinators historically devoted to and connected with those industries.

The Commission later added to this policy the frequencies formerly allocated to the *Automobile Emergency Radio Service* (AERS). Accordingly, coordination of such frequencies is performed only by the nationwide organization devoted to emergency road service—namely, the *American Automobile Association* (AAA).

With the explosive growth in business and public safety mobile communications, demand for channels is much greater than the supply. As a solution, an FCC "refarming" initiative sought to increase spectrum efficiency, reduce constraints on spectrum availability, and minimize bureaucratic nomenclature in the PMRS. Besides refarming and changing its regulatory structure for PMRS, the FCC will allocate some additional PMRS spectrum (see the following "New service proposed").

The FCC eliminated the identities of the individual radio services within each pool. As an aid to understanding, listed below are the radio services that comprised the two pools.

The Public Safety pool includes the (former):

- Local Government Radio Service for state, county and municipal administration
- Police and Fire Radio Services for the dispatch of these public guardians
- Highway Maintenance Radio Service used by road construction and repair crews
- Forestry-Conservation Radio Service used in game law enforcement, forest fire protection, reforestation, flood and erosion control
- Emergency Medical Radio Service (the "MED" channels) for physicians and medical personnel directly involved in life support, for example in transporting injured persons to care facilities
- *Special Emergency Radio Service* (SERS) for general assistive radio users, such as rescue teams, beach patrols, veterinarians, disabled individuals, disaster relief organizations, school buses, standby and emergency repair facilities, and establishments in isolated places. Eligibility for SERS frequencies is carefully controlled. For example, the FCC established numerical values for visual or hearing acuity necessary to qualify as a disabled licensee.

The Industrial/Business pool includes theeded (former):

- Business Radio Service, with broad eligibility for commercial, educational, philanthropic, and ecclesiastical users
- Power Radio Service for electric, gas and water utilities
- Petroleum Radio Service for oil and natural gas industries
- Forest Products Radio Service for the lumber and paper industries
- Film and Video Production Radio Service for media producers
- Relay Press Radio Service for newspapers
- Special Industrial Radio Service for construction, farming, and mining
- Manufacturers Radio Service for communications at factories
- Telephone Maintenance Radio Service for telephone repair crews
- Motor Carrier Radio Service for bus and trucking companies
- Railroad Radio Service, involving sophisticated train monitoring and control systems
- Taxicab Radio Service
- Automobile Emergency Radio Service for tow truck dispatch

The FCC retained the Radiolocation Service as a separate private service that provides for all non-federal use of radar. Most radiolocation frequencies are above 800 MHz and are available only on a secondary basis to other operations, particularly federal government operations. Frequency coordination is not required in the Radiolocation Service.

Refarming is necessary because of the extremely uneven use of the PMRS bands. Channels in some services were congested while channels in other services in the same location were lightly used. Each PMRS radio channel accommodates hundreds of licensees and thousands of transmitters.

Refarming entails technical changes as well as regulatory changes. New spectrum efficiency standards are imposed, reducing the power, height, and bandwidth of new PMRS stations.

Current PMRS channels are generally spaced at intervals of 15, 20, or 25 kHz, depending on the band, with 5 kHz spacings in 220–222 MHz and 12.5 kHz in the 821–824 MHz/866–869 MHz and 896–901 MHz/935–940 MHz bands. Under its refarming rules, the FCC specified channels every 7.5 kHz in the 150–174 MHz VHF band and every 6.25 kHz in the 421–430 MHz, 450–470 MHz, and 470–512 MHz UHF bands.

Channels may be aggregated where necessary for such purposes as high-speed data communications.

The FCC is authorizing new radios for market that are capable of operating on the narrower bandwidths. This narrower band equipment is supposed to be deployed through normal equipment replacement cycles.

Exclusive Use

A proposed, controversial element of refarming, "exclusive use overlay," was not implemented precisely as the Commission proposed. It would have permitted licensees to purchase exclusive use of their channels at auction and prevent the addition to those channels of conventional radio users beyond control of the auction winner. The exclusive licensees could then sell communications service to customers.

In 1999, the FCC terminated this proposal. It claimed (in Docket PR 92-235) that it was already providing the functional equivalent of exclusive use overlay, through a policy that forbids the addition of conventional users to channels controlled by PMRS licensees authorized to operate trunking systems on those channels.

New Service Proposed

In a November 1999 spectrum policy statement, the FCC announced its intention to allocate 1390–1395 MHz, 1427–1429 MHz, and 1432–1435 MHz to a new *Land Mobile Communications Service* (LMCS). Licenses in the new service are likely to be auctioned to "band managers," trade groups, or mega-licensees that would parcel out authorizations to actual users.

LMCS is intended to answer the insistent call of private land mobile radio users for additional spectrum, without allowing them to obtain licenses on the traditional first-come, first-serve basis that obviates the need for auctions.

32-33 MHz

FG: FIXED. MOBILE

Non-FG: None

This band, along with 34-35 MHz, 36-37 MHz, 38-39 MHz, and 40-42 MHz is for federal agency and Army, Navy, Marine, and Air Force fixed and mobile use, supporting training, test range operations, research and development, and search and rescue.

33-34 MHz

FG: None

Non-FG: NG124 FIXED. LAND MOBILE

FCC: Private Land Mobile (90)

This band is within the Public Safety frequency pool of the *Private Mobile Radio Services* (PMRS, see 30.56-32 MHz). Frequencies in this band are used principally for special emergency and fire communications.

34-35 MHz

FG: FIXED. MOBILE

Non-FG: None

See 32-33 MHz for information about this band.

35–36 MHz

FG: None

Non-FG: FIXED. LAND MOBILE

FCC: Public Mobile (22). Private Land Mobile (90)

In addition to Private Radio use (see 30.56–32 MHz), this band is used in the Paging and Radiotelephone Service, one of the Public Mobile Services, and a *Commercial Mobile Radio Service* (CMRS).

Radio paging—the use of radio to find, summon or notify someone—is a highly competitive business. Advanced paging or messaging networks support two-way communication, but paging today still is largely a one-way medium.

The two-way feature originally was employed mostly to confirm that the paged party received the message, but it is now used to convey interactive choices in wireless news services, and to compose e-mail that the paging carrier transfers to the Internet.

History of the Beep

Cleveland electronics engineer Al Gross invented radio paging in the early 1950s. Gross was a pioneer in portable VHF-UHF devices and an early proponent of "citizens radio," now called "personal radio" (see 462.5375–462.7375 MHz). He provided paging systems for use in hospitals, though doctors initially resisted the technology.

The basic paging receiver beeps when it receives a distinctive code from a transmitter controlled by a "paging terminal," connected to the public telephone network. Today's more sophisticated pagers display the telephone number of the calling party, or alphanumeric messages. The World Wide Web is now the principal data input medium for alphanumeric paging.

Pagers often are used to signal waiting voice mail messages and are increasingly connected to electronic mail services. Voice paging declined in the 1980s, but was supposed to resurge with the advent of digital voice storage in the pager. Considerable investment was made in voice paging in the Narrowband Personal Communications Service (see 901–902 MHz). The reinstituted voice paging services failed to attract enough subscribers, however.

Delighting Guests

Paging also is used to control industrial equipment, to summon animals to feeding, and—in perhaps the most entertaining use we have discovered—to activate fireworks contained in banquet floral decorations.

Today most paging is provided by FCC licensees on a commercial basis. Traditionally, the two types of for-profit paging companies were designated *radio common carriers* (RCCs), which enjoyed interference-protected service areas and, in some cases, were required to have state utility certification; and *Private Carrier Paging* (PCP) providers, who operate on channels shared with other such carriers and are supposed to be exempt from state regulation.

Federal budget legislation in 1993 requires similar or comparable treatment, or "regulatory symmetry," for substantially similar *Commercial Mobile Radio Services* (CMRS).

CMRS describes radio services offered to the public, or a large segment of the public, for profit and which interconnect to the public telephone network. Many types of CMRS services exist. PCP services were classified as CMRS in 1996. "There are no longer any real differences between private carrier and common carrier paging systems," according to the FCC.

Nevertheless, the former system continues in that separate regulations govern RCC-based paging in FCC Part 22 rules and PCP-based paging in Part 90. "Not all substantially similar services must have identical technical and operational rules," the FCC said.

Major Changes

Paging licenses were traditionally granted on a first-come, first-serve basis using careful frequency coordination to avoid interference between systems. The FCC is introducing a new regime into CMRS paging: license auctions in specified geographic areas. Incumbent licensees will retain certain interference protections.

"We noted that if an incumbent already has a significant presence in a geographic area, other potential applicants may choose not to bid for that geographic area," the FCC said. "Thus, market forces, not regulation, would determine participation in competitive bidding for geographic area licenses.

"...Even where only 30 percent of a geographic area is available to a potential new entrant, we do not believe that it has been shown that the new entrant cannot establish a viable system that serves the public as well as the incumbent." The Commission dismissed all mutually exclusive applications for paging licenses filed after July 31, 1996, in preparation for the auction regime (Docket WT 96-18).

In the 929–930 MHz and 931–932 MHz bands, the FCC established paging licenses in *Major Economic Areas* (MEAs) developed by the Commerce Department and composed of *Economic Areas* (EAs, see the following).

Special exemptions from auctions apply to certain established paging licensees in the 929–930 MHz and 931–932 MHz bands who earned exclusive licenses through extensive network construction under earlier regulations.

In other, older paging bands, such as 35–36 MHz, 43–44 MHz, 152–159 MHz, and 454–460 MHz, containing small- and medium-sized paging systems, the FCC adopted *Economic Areas* (EAs), each consisting of metropolitan or similar areas that are centers of economic activity, with surrounding economically related counties.

As is the case with most auctioned licenses, each geographic area licensee is required to provide coverage to specified portions of the population by a certain date. Alternatively, the licensee may demonstrate that it provides "substantial service," that is, "service that is sound, favorable, and substantially above a level of mediocre service." Failure to meet these requirements is supposed to automatically terminate the license.

Incumbents (who did not obtain their licenses in auctions) may make certain changes to their operations, including adding or modifying their transmission sites. To expand their existing radio contours, they must obtain the consent of the geographic licensee for the expansion area, or buy the expansion area at auction even if the area is larger than needed. If the incumbent stops operating, the FCC performs "spectrum reversion," turning the incumbent's spectrum over to the geographic licensee for that area.

Other services use the 152–159 MHz and 454–460 MHz bands allocated to paging services. These include the *Basic Exchange Telecommunications Radio Service* (BETRS), which provides rural subscribers with phone service; and pre-cellular two-way mobile telephone services, which operate mostly in the western U.S.

BETRS is licensed under the Rural Radiotelephone Service. Only local exchange telephone companies or companies having state approval to provide local phone service may provide BETRS. By definition, BETRS is a fixed service and is not CMRS.

The Commission decided to include both BETRS and the two-way mobile telephone services on paging channels in its auction scheme, over the protests of rural telephone advocates. However, it will still allow providers to obtain licenses for these services on a secondary, non-interference basis to geographic-area paging licensees.

36-37 MHz

FG: US220 FIXED. MOBILE

Non-FG: US220

See 32–33 MHz for information about this band.

The band also contains one of the low-band VHF frequencies used in oil spill cleanup and containment.

37-37.5 MHz

FG: None

Non-FG: NG124 LAND MOBILE

FCC: Private Land Mobile (90)

This band is part of the *Private Mobile Radio Services* (PMRS, see 30.56–32 MHz). It is mostly devoted to police radio and power utilities.

37.5-38 MHz

FG: S5.149 Radio Astronomy

Non-FG: S5.149 NG59 NG124 LAND MOBILE. Radio Astronomy

FCC: PRIVATE LAND MOBILE (90)

This Private Mobile Radio Services (see 30.56–32 MHz) band contains frequencies for public safety and power utility operations.

Radio astronomers observe broadband continuum emissions from space sources, as well as spectral lines associated with specific frequencies. They use 37.5–38.25 MHz for continuum observations, including observation of radiation from the Sun and Jupiter.

Contributions of Radio Astronomy

In 1931, Bell Laboratories engineer Karl Jansky, looking for the origins of noise in radiotelephone transmission, observed radio emissions that peaked every day when the center of the Milky Way galaxy passed overhead. Jansky's discovery gave birth to the science of radio astronomy.

Most radio astronomy is characterized as a "passive" or non-transmitting spectrum use. Allied to radio astronomy interests are scientists who use passive instruments, usually on satellites, to detect and measure weather and other natural phenomena.

In addition to advancing knowledge about the physical universe, radio astronomy is credited with contributions to telecommunications, defense, seismology, and medical imaging. Through radio astronomy, scientists discovered the first planets outside the solar system.

But radio astronomers are frequently at odds with the FCC's drive to permit ever-wider commercial use of radio.

"As passive users of the spectrum, radio astronomers and Earth scientists have no control over the frequencies that they must observe, or over the character of the 'transmitted' signal. These parameters are set by the laws of nature," according to the Committee on Radio Frequencies of the National Academy of Sciences.

"Furthermore, the emissions that radio astronomers observe are extremely weak—a typical radio telescope receives only about one-trillionth of a watt from even the strongest cosmic source. Because radio astronomy receivers are designed to pick up such remarkably weak signals, such facilities are therefore particularly vulnerable to interference from spurious and out-of-band emissions from users of neighboring bands, and those that produce harmonic emissions that fall into the radio astronomy bands."

This spectrum also is one of many restricted bands in which the FCC Part 15 rules permit unlicensed devices to emit only very low level emissions.

38-38.25 MHz

FG: S5.149 US81 FIXED. MOBILE. RADIO ASTRONOMY

Non-FG: S5.149 US81 RADIO ASTRONOMY

This is part of the 37.5–38.25 MHz radio astronomy spectrum. Footnote US81 provides that military transportable and mobile stations will continue to use the band (see 32–33 MHz).

The 37.5–38.25 MHz spectrum is one of many restricted bands in which FCC Part 15 rules permit unlicensed devices to emit only very low level emissions.

38.25-39 MHz

FG: FIXED. MOBILE

Non-FG: None

See 32–33 MHz for information about this federal band.

39-40 MHz

FG: None

Non-FG: NG124 LAND MOBILE

FCC: Private Land Mobile (90)

This band is available to Public Safety users (see 30.56–32 MHz). It is largely devoted to police radio.

40-40.98 MHz

FG: S5.150 US210 US220 FIXED. MOBILE

Non-FG: S5.150 US210

FCC: ISM Equipment (18). Private Land Mobile (90)

See 32–33 MHz for information about this band.

40.98-42 MHz

FG: S5.150 US210 US220 FIXED. MOBILE

Non-FG: US220

FCC: ISM Equipment (18)

See 32–33 MHz for information about this band.

42-43.69 MHz

FG: None

Non-FG: NG124 NG141 FIXED. LAND MOBILE

FCC: Public Mobile (22). Private Land Mobile (90)

This band includes public safety, industrial, and public mobile services (see 30.56–32 MHz).

This spectrum also supports meteor burst data communications, which reflects signals from the trails of micrometeors that constantly shower into the upper atmosphere.

This part of the spectrum is preferred for meteor burst use because of greater signal return from the meteor trails and the channel throughput available.

The pioneering meteor burst system is Snotel, the U.S. Department of Agriculture's network of precipitation monitoring stations in western states.

• **RECORDS through a RADIO**

Any radio, whether or not it has a connection for a phonograph pick-up, can be made to reproduce records. All you need is a turntable, and the RCA Phonograph Oscillator. To the service man and dealer, it offers a profit in itself and through the future sale of records; to the set owner, it immensely increases the enjoyability of his instrument. Simple to install. Priced at $7.57 net, less tube. Can also be used with a microphone as a public address system. To the trade: ask your RCA Parts Distributor for a free copy of the RCA Radio Parts Catalogue.

(RCA) **PARTS DIVISION**
RCA Mfg. Co., Inc., Camden, N. J.
A subsidiary of the
RADIO CORPORATION OF AMERICA

Figure 1
The phono oscillator was a transmitter for broadcasting phonograph records to nearby receivers. Such devices were the original unlicensed "intentional emitters," the progenitors of today's FCC Part 15 wireless products. (RCA advertisement from Radio News, January 1936.)

43.69–46.6 MHz

FG: None

Non-FG: NG124 NG141 LAND MOBILE

FCC: Private Land Mobile (90)

This band includes public safety, industrial, and public mobile services (see 30.56–32 MHz) and supports meteor burst data communications (see 42–43.69 MHz). It also includes cordless telephone channels. Most cordless phones in the U.S. operate under Part 15 of the FCC Rules.

Part 15 governs many aspects of device design, authorization, and marketing throughout the radio spectrum. It draws its authority not from the federal power to license stations—Part 15 devices are not licensed—but its authority to control radio interference.

The FCC has permitted low-power, unlicensed devices since the 1930s. The earliest of these so-called "intentional radiators" were "phono oscillators," used with phonographs to transmit music over a short distance to radio receivers.

Unlicensed devices do not represent a "radio service" in the conventional sense, so no formal spectrum allocations to unlicensed devices exist. Instead, the FCC allows unlicensed devices to use many bands on a shared, non-interference—sometimes called "sufferance"—basis.

Changed Status

This scheme changed somewhat when the FCC designated certain near-exclusive spectrum for these devices (see 1850–1990 MHz) and granted protections to Part 15 devices in other spectrum (see 902–928 MHz).

Historically, FCC regulation of unlicensed devices was highly restrictive, aimed more at suppressing interference than promoting commercial opportunities. Since the 1980s, however, unlicensed devices became a multibillion dollar business, with cordless phones leading the category. The lobbying resources of Part 15 manufacturers consequently increased. The FCC is now more inclined to promote development of the devices.

Proponents of unlicensed devices frequently clash with other spectrum users, who fear interference to their operations or competition with their licensed offerings.

Some commenters frequently remind the Commission of the unique status of unlicensed devices. On this topic, the *American Radio Relay League* (ARRL), the national Amateur Radio association, is firm. ARRL told the FCC: "Part 15 devices have no allocation status, and have had none, internationally or domestically. They are permitted on an 'at-sufferance' basis: they must not cause interference to licensed radio services, and they must tolerate interference received from licensed radio services in the same bands.

"The Communications Act of 1934 is devoid of any authority to accord Part 15 type devices any allocation status, or interference protection from licensed services, at all," ARRL said.

Available frequencies for cordless phones shifted in 1984 to 10 channels at 46.61–46.97 (base stations) and 49.67–49.97 MHz (handsets). The base stations previously used the noise-plagued 1.625–1.8 MHz band, now part of the AM broadcast band. Handsets used 27 MHz but were shifted to 49 MHz.

Dangerous Naivete

Millions of 49 MHz/1.7 MHz phones were sold. They became infamous for making squawking noises and ringing without being called. In FCC files, we found complaints from startled consumers whose cordless phones rang even before they were taken out of the box.

The consumer electronics industry devoted special attention to understanding the weird ability of these phones to spontaneously dial emergency 911 services. It eventually traced the problem to spurious responses of the telephone, sometimes under low battery conditions, to radio noise and signals from other co-channel phones. This resulted in random switchhook closures that the telephone network interpreted as dialing sequences.

In what FCC Commissioner Ervin Duggan called, "stunning, even dangerous naivete," the Commission declined for years to require features in cordless phones that would have reduced these problems.

The all-VHF cordless phones included required security features and experienced fewer problems than older models. The multiple channels made trunking (automated channel selection) feasible, more efficiently using the limited spectrum available.

Even so, in high-density areas, the original 46/49 MHz channels became saturated, leading to blocking (no clear channels available) and crosstalk (audible interference from units nearby). Returns for cordless phones reached nearly three times the average return rate for all consumer electronics, except for home computers and video games.

Aggravating the problem are baby monitors, which the FCC allows to use 49.82–49.90 MHz. They are supposed to transmit baby sounds to parents elsewhere in the home. But they transmit any sound within microphone range, and typically are left on continuously. Recreational monitoring of neighbor conversations broadcast by baby monitors is a popular underground pastime.

Spectrum Expansion

The perceived shortage of frequencies for cordless telephones has long been of concern to the *Telecommunications Industry Association* (TIA) which lobbies the FCC on behalf of phone manufacturers.

TIA asked the FCC to add 30 frequencies at 43.72–44.48 (base) and 48.76–49.5 MHz (handsets) to cordless phones. At the low power radiated by these products—roughly 25 microwatts—TIA believed they would not interfere with private land mobile radios operating in these bands.

The FCC granted the request in 1995. To further prevent interference with land mobile systems, phones using the new frequencies must incorporate mechanisms that prevent use of occupied frequencies. The added frequencies are not paired; any new base transmitter frequency may be used with any new handset transmitter frequency.

Use of the new frequencies was relatively short-lived. Current models of cordless telephones use digital technology in the 902–928 MHz and 2400–2483.5 MHz *Industrial, Scientific, and Medical* (ISM) bands (see 2400–2402 MHz).

The 46/49 MHz phone is informally known internationally as CT0. CT1 refers to analog UHF cordless phones sold in Europe. CT2 and its successors are European digital cordless telephone standards that improved on CT1.

High-end cordless phone systems for business use are now in unlicensed PCS spectrum at 1920–1930 MHz (see 1850–1990 MHz). Some models use licensed cellular spectrum (see 824–849 MHz). Customers typically pay a flat rate to the cellular carrier for on-premises usage.

46.6–47 MHz

FG: FIXED. MOBILE.

Non-FG: None

This band, otherwise allocated to federal use, contains cordless telephone channels (see 43.69–46.6 MHz).

47–49.6 MHz

FG: None

Non-FG: NG124 LAND MOBILE

FCC: PRIVATE LAND MOBILE (90)

This band is partially within the Public Safety frequency pool (see 30.56–32 MHz) and is used especially for state highway maintenance including self-powered vehicle detectors.

The band also includes cordless telephone handset channels (see 43.69–46.6 MHz) and the perennial, if little-used, national Red Cross frequency 47.42 MHz.

49.6–50 MHz

FG: FIXED. MOBILE

Non-FG: None

The band, otherwise allocated to federal use, contains cordless phone handset channels (see 43.69–46.6 MHz).

FCC Part 15 rules permit other unlicensed devices, including home-built devices, to use 49.82–49.90 MHz. The FCC started to migrate them to 49.82–49.90 MHz in 1976, after contention between then-licensed Citizens Band operations and unlicensed devices in the 27 MHz spectrum.

50–54 MHz

FG: None

Non-FG: AMATEUR

FCC: Amateur (97)

The U.S. authorizes the Amateur Radio Service as a voluntary, non-commercial service that provides emergency communications, contributes to the advancement of the radio art, expands the national pool of trained radio personnel, and fosters international goodwill by facilitating direct communication between people of different nations.

Radio amateurs (hams) obtain FCC licenses through qualifying examinations at several levels of proficiency. Amateur Radio is recognized by the *International Telecommunication Union* (ITU). Domestic amateur regulation is based on ITU regulations and spectrum allocations.

Amateurs are active in all portions of the spectrum, from the *Very Low Frequencies* (VLF)—where they use digital signal processing to extract information from signals buried in noise—to the highest millimeter wave bands where they experiment with construction techniques and compete for records in communication distance.

Recent FCC actions reduced the requirements for Morse Code testing (to five words per minute) and simplified the amateur licensing structure. Within the next few years, we anticipate an ITU decision to eliminate the requirement for Morse testing. At this writing, there were more than 675,000 U.S. amateur licensees.

Six Meters

Nicknamed the Magic Band, this six meter 50–54 MHz band holds a special fascination for radio amateurs. It can appear devoid of signals much of the time. It is subject, however, to sudden signal propagation by "sporadic-E," meandering clouds of metallic ions in the E-layer of the ionosphere about 70 miles above Earth.

These clouds act as mirrors, reflecting even low-power 50 MHz signals across thousands of miles and causing the band to break out in a frenzy of stations.

Propagation by auroral effects also is characteristic of these frequencies. The aurora is associated with solar activity that affects Earth's magnetic field. Tropospheric bending of 50 MHz signals also occurs, caused by atmospheric changes between air masses of different temperature and humidity. These phenomena can bring distant interferers to VHF TV and mobile radio.

Six meters is home to a subculture of amateur operators who enjoy the band's surprising behavior. Typical uses include single-sideband voice, Morse Code, data, and FM mobile radio with repeaters that retransmit signals over a citywide area. The FCC issued special permission to some amateurs for spread-spectrum experiments in this band.

Future of Amateur Radio

"It seems to me that given the increased pressure on the underlying resource from commercial and other non-commercial uses, the key issue for the amateur service is maintaining access to an adequate amount of spectrum," according to former FCC Office of Engineering and Technology chief Dale Hatfield, in an address to amateurs.

"The rapidly growing demand for spectrum coupled with the increased visibility of its economic value due to auctions makes it almost inevitable that amateurs will be under a certain amount of pressure to justify their 'free' use of this precious resource," he said.

Hatfield urged radio amateurs to continue shifting towards spectrally efficient communications techniques, especially digital. Such a shift has several benefits, he said. "First of all, it demonstrates to policymakers and regulators that you are good stewards of the public's airwaves even without direct economic incentives.

"Second, by using what you have efficiently, it strengthens your case when you need to ask for additional spectrum. Third, by allowing more users to access the available allocations simultaneously, it improves the amateur experience and ultimately increases the attractiveness of the service to new and old users alike.

"Fourth, it provides the opportunity or 'headroom' for increases in data rates to more closely match those available on wireline networks and, in the future, on commercial wireless networks as well.

"Fifth, as the rest of the telecommunications world makes the transition to digital techniques—and there are very few exceptions to that trend—the amateur service will look antiquated if it is not making progress in that direction as well."

54-72 MHz

FG: None

Non-FG: NG128 NG149 BROADCASTING

FCC: Broadcast Radio (TV) (73). Auxiliary Broadcasting (74)

This band contains TV channels 2, 3, and 4. The U.S. is making a major transition to *Digital Television* (DTV, see 470–512 MHz).

This band is one of several used in *magnetic resonance imaging* (MRI), a tool used in medical diagnosis and in guiding the application of treatment instruments.

MRI does not expose the patient to ionizing radiation. It is based on *nuclear magnetic resonance* (NMR), a phenomenon observed in atoms subjected to magnetic fields.

NMR also is used in the measurement of substance characteristics. NMR spectroscopy is used in food and textile manufacturing and analytical chemistry.

A related phenomenon is *nuclear quadrupole resonance* (NQR), wherein low-intensity radio waves cause nuclei in a substance to tip out of alignment. When they return to alignment, they emit a unique radio signal that can be received and analyzed. NQR is used to detect explosives contained in plastic land mines.

72-73 MHz

FG: None

Non-FG: NG3 NG49 NG56 FIXED. MOBILE

FCC: Public Mobile (22). Private Land Mobile (90). Personal Radio (95)

The bands 72–73 MHz, 74.6–74.8 MHz, 75.2–75.4 MHz, and 75.4–76 MHz have such diverse uses as emergency communications, narration of golf tournaments and horse races, local communications at industrial sites, the remote control of model aircraft and Public Mobile transmitters, and radio-aided education of the hearing-impaired.

Common terminology refers to the "72–76 MHz band." This term does not include the 73–74.6 MHz radio astronomy or the 74.8–75.2 MHz aeronautical marker-beacon bands.

Private Mobile Radio Services use 72–76 MHz for fixed and mobile operations, including frequencies for emergency fire call boxes, shared with the Industrial/Business pool.

Public Mobile Services stations have channels in 72–73 MHz and 75.4–76 MHz for point-to-point operation, for such uses as control of paging systems.

Auditory assistance devices use 72–73 MHz and 75.4–76 MHz to transmit teachers' speech to students who wear hearing-aid-like receivers, or to loudspeakers distributed throughout a classroom. These systems also are used in theaters, conference centers, churches, and other assembly areas.

According to the National Association of the Deaf, "the 72 to 76 MHz band, used by FM auditory assistance devices, has been inundated with interference from pagers, cellular phones, emergency dispatch vehicles, electronic equipment, CB/ham radio transmitters, walkie-talkie radios, and vehicle dispatchers. The interference from such high powered users has made these auditory assistance devices virtually unusable in a multitude of settings, the most serious of which is our schools."

Trying to find a better home for auditory devices, the FCC created a *Low Power Radio Service* (LPRS) in 216–217 MHz (see 216–220 MHz).

The 72.44–75.6 MHz spectrum contains channels used to broadcast narration of sporting events to spectators who rent special portable radios.

The *Radio Control* (R/C) Radio Service, a Personal Radio Service in Part 95 of the FCC Rules, provides 50 channels in 72–73 MHz for the remote control of model aircraft, and 30 channels in 75.4–76 MHz for model boats and cars.

73-74.6 MHz

FG: US74 RADIO ASTRONOMY

Non-FG: US74 RADIO ASTRONOMY

This exclusive radio astronomy allocation is used to study continuum emissions and to monitor interplanetary "weather" conditions of the solar wind.

It is one of many restricted bands in which the FCC Part 15 rules permit unlicensed devices to emit only very low-level emissions.

74.6-74.8 MHz

FG: US273 FIXED. MOBILE

Non-FG: US273 FIXED. MOBILE

FCC: Private Land Mobile (90)

See 72–73 MHz for the uses of this band.

74.8-75.2 MHz

FG: S5.180 AERONAUTICAL RADIONAVIGATION

Non-FG: S5.180 AERONAUTICAL RADIONAVIGATION

FCC: Aviation (87)

This band contains beacons in the Instrument Landing System (see 108–117.975 MHz). It is one of many restricted bands in which the FCC Part 15 rules permit unlicensed devices to emit only very low level emissions.

75.2-75.4 MHz

FG: US273 FIXED. MOBILE

Non-FG: US273 FIXED. MOBILE

FCC: Private Land Mobile (90)

See 72–73 MHz for the uses of this band.

75.4-76 MHz

FG: None

Non-FG: NG3 NG49 NG56 FIXED. MOBILE

FCC: Public Mobile (22). Private Land Mobile (90). Personal Radio (95)

See 72–73 MHz for the uses of this band.

76-88 MHz

FG: None

Non-FG: BROADCASTING. NG128 NG129 NG149

FCC: Broadcast Radio (TV) (73). Auxiliary Broadcasting (74)

This band contains TV channels 5 and 6.

In rare cases, the frequency 87.9 MHz is available for noncommercial FM radio broadcasting.

The U.S. is making a major transition to *Digital Television* (DTV, see 470–512 MHz).

88-108 MHz

FG: US93

Non-FG: US93 NG2 NG128 NG129 BROADCASTING

FCC: Broadcast Radio (FM) (73). Auxiliary Broadcasting (74)

U.S. FM radio broadcasting is allocated one hundred 200 kHz-wide channels numbered 201 (88.1 MHz) through 300 (107.9 MHz). The FCC distributes groups of FM channels to geographic areas in a process called allotment. Entrepreneurs often petition the FCC to allot channels to communities on the basis of population and commercial growth.

Non-commercial educational (NCE) FM stations have a reserved segment, 88.1–91.9 MHz (channels 201–220). Some older NCE stations operate in the non-reserved segment above 91.9 MHz.

Operating on the FM channels are main transmitters, as well as "translator" and "booster" stations that extend coverage to areas where terrain renders reception inadequate. No analogous facilities exist for AM stations, whose signals may propagate for great distances.

FM stations are classified by allowable transmitter power and antenna height. Detailed rules prescribe geographic separations between stations of each class.

Selecting Licensees

Historically, the FCC assigned broadcast licenses through lengthy comparative hearings that examined the relative merits of license applicants. Frequently, these hearings would award licenses based on minor distinctions between applicants, such as board members' residential histories or records of their civic involvement.

Such decisional factors made little audible difference in today's homogenized radio programming environment. Today, the FCC uses auctions to select from among competing applicants for commercial licenses.

NCE licenses in the reserved segment are awarded on a first-come, first-served basis, unless competing applicants for an available channel (mutual exclusivity) exist. In such cases, the FCC awards the license through a simplified comparative system—with applicants earning points for local ownership and technical merit, for example. To break ties, the FCC chooses the applicant with the fewest existing stations.

NCE licenses for frequencies above 91.9 MHz are subject to auctions, a policy bitterly disputed among the FCC Commissioners. Officially, the FCC claims that there are "conflicting directives" in the legislation that conferred its auction authority, which it resolved by applying auctions to competing NCE applicants outside the reserved segment.

Auxiliary Services

Many FM stations also transmit subcarriers: signals subsidiary to the main FM signal, but within its authorized bandwidth.

Subcarriers traditionally have been used to transmit commercial background music, paging, private data communications, differential *Global Positioning System* (GPS) corrections (see 1215–1240 MHz), foreign language programming, and non-profit *radio reading services* (RRS) for the visually handicapped. Special receivers pick up these signals.

An RRS association, complaining of exorbitant subcarrier lease fees, once tried and failed to persuade the FCC to grant it a separate spectrum allocation independent of FM broadcasting.

Fewer than four percent of FM stations offer RRS. Less than half of all FM stations use their subcarriers at all, and more than 90 percent of them use only one of the two subcarriers available to each station.

Some FM stations employ subcarriers to transmit short messages using the *Radio Data System* (RDS). With RDS, a station can display its callsign, program format, and even the name of the current song and artist, on the receiver or on other displays such as highway billboards. Abroad, RDS is sometimes called *Radio Broadcast Data System* (RBDS). But this system has not become popular in consumer receivers.

The broadcast industry developed *high-speed subcarriers* (HSSC) with greater capability than RDS. The *National Radio Systems Committee* (NRSC) of the *National Association of Broadcasters* (NAB) evaluated candidate HSSC systems, but did not formally adopt a standard. It appears that the drive towards HSSC has been all but arrested by the more substantial movement toward full digitalization of the broadcast signal.

Evacuate!

Federal Signal Corp. petitioned the FCC to apply RDS technology to an *Emergency Radio Data System* (ERDS). This system would cause car radios to display alerts transmitted by approaching emergency vehicles, trains, or school buses. ERDS could overcome "loud audio entertainment systems," according to the company.

Upon receiving coded information on 87.9 MHz, the ERDS receiver would display a message such as "POLICE!" "FIRE ALERT!" or "EVACUATE!" and play audio statements stored in the receiver, or live messages coming from an emergency vehicle. Some observers have suggested that ERDS could become a haven for unauthorized broadcasters who could capture the majority of car radios in a city.

ERDS has similar aims to two systems already approved: the *Safety Warning System* (SWS, see 24.05–24.25 GHz), and *Dedicated Short Range Communications* (DSRC, see 5.85–5.925 GHz).

Digital Audio Broadcasting (DAB)

Beyond ancillary data is the issue of digital improvements in the main broadcast itself. The industry is moving toward replacing FM with digital modulation and, optionally, integrating it with satellite radio.

In the U.S., the terrestrial component of digital audio broadcasting is now known as DAB while the satellite component is called *Digital Audio Radio Services* (DARS, see 2320–2345 MHz). The FCC originally considered DARS to be "an all-inclusive term that encompasses both terrestrial *digital audio broadcasting* (DAB) and digital audio provided by satellite," but later reversed itself in view of confusion over its terminology.

Unlike *digital TV* (DTV, see 470–512 MHz) the FCC has not, at this writing, established a national timetable for the introduction of terrestrial digital radio broadcasting.

USA Digital Radio Partners (USADR) filed FCC petition RM-9395 in 1998 to begin a proceeding that would launch digital transmission for AM and FM stations *"in-band and on-channel"* (IBOC)—that is, within the band already assigned for analog broadcasting.

European and Canadian broadcasters criticized IBOC and advocated Eureka, a DAB system in other spectrum (see 1435–1525 MHz). Receiver manufacturers tried, and failed, to establish IBOC in the U.S.

The USADR technology permits a station to broadcast in "hybrid" mode—both analog and digital signals in the same frequency assignment. Eventually, analog transmissions would cease, while power would increase in the digital transmissions.

"IBOC DAB provides a rational transition from analog to digital and offers immediately enhanced service without disruptions to the existing analog radio service or the need for adjustments in consumer patterns of use," USADR told the FCC. "IBOC DAB allows this to be achieved without the need for new frequency allocations, the issuance of new licenses or the creation of a new regulatory structure."

USADR later merged with Lucent Digital Radio to form Ibiquity Digital Corp., now the leading proponent of IBOC DAB and the likely source of the national IBOC standard.

FCC Answers Petition

In Docket MM 99-325, the FCC responded to the USADR petition with a historic *Notice of Proposed Rulemaking* (NPRM) to launch DAB in the U.S. DAB will "purportedly" bring near-CD quality to FM signals and FM quality to the AM band, the FCC said. "Consumer demand for improved audio fidelity is undeniable," the FCC said, announcing that it intends to deploy DAB as quickly as possible without "significant listener dislocations."

Such dislocations could occur at the time of the so-called "hard" transition from analog AM and FM to DAB, "unless everyone has acquired a digital radio, which in turn depends on the cost-effective manufacture of digital receivers and widespread consumer acceptance of these devices," the FCC said.

Though ostensibly proposing specific rules for DAB, the NPRM left key issues unresolved such as whether IBOC should really be chosen and whether the government should mandate a DAB standard or instead opt for an "open architecture" approach whereby receivers could decode transmissions in different formats.

At this writing, the NRSC is testing elements of Ibiquity's proposed unified DAB standard, with availability of commercial DAB receivers expected in 2002.

Low Power FM (LPFM)

"Micro radio," the unlicensed use of low-power broadcast transmitters, usually on a noncommercial basis, has long been illegal.

Micro radio broadcasters launched sophisticated legal challenges to the constitutionality and wisdom of the FCC policies that authorize full power stations only, even as FCC enforcement actions closed down hundreds of micro stations.

The FCC responded to the proliferation of unlicensed "pirate" micro stations, and the surging interest in micro radio, by authorizing two new classes of noncommercial *low power FM* (LPFM) stations at 10 W and 100 W maximums (Mass Media Docket MM 99-25). It placed severe eligibility and operational restrictions on LPFM stations, which are intended to serve more as community voices than as media for mass audiences.

The FCC will not confine LPFM to the 88.1–91.9 MHz segment, where most noncommercial stations are required to operate. LPFM will have the run of the entire 88–108 MHz band.

Mutually exclusive license applications will be resolved by a point system, in which FCC staff members apply merits and demerits to applicants' proposals.

Rejecting the Cube

The Commission apparently rejected an internal proposal, dubbed the "Cube," that might have placed low power broadcast stations elsewhere in the spectrum. Digital technologies and other spectrum could have accommodated far more licensees than LPFM. A special receiver or downconverter (cube) would have been necessary to receive the broadcasts.

Support for LPFM was not unanimous at the FCC. Commissioner Harold Furchtgott-Roth, a frequent dissenter to the majority's actions, scathingly condemned LPFM as irresponsible, inadequately tested, financially unsound for station operators, a source of interference, and irrelevant because, in his view, receivers will increasingly obtain audio from the Internet instead of radio transmission.

He suggested that those wishing to communicate simply buy existing stations or use Amateur Radio, Web pages, or even "bulletins and flyers" instead of LPFM.

We believed that IBOC DAB could complicate the introduction of LPFM because IBOC is premised on digital transmission by incumbent broadcasters only. It is not designed to accommodate large numbers of new entrants, even if at low power. However, IBOC technology developers did not greatly oppose LPFM on this or any other basis.

LPFM Politics

The NAB fumed that LPFM "threatens the transition to IBOC digital radio, will likely cause devastating interference to existing broadcasters, and will challenge the FCC as guardian of the spectrum."

Former FCC chairman William Kennard cited what he called "flaws" in technical studies submitted by LPFM opponents such as NAB, which predicted crippling interference: "Some of the studies…start with the premise that most existing FM radios do not provide adequate reception even today, before the creation of a low-power service. These commenters suggest that we adopt standards that bear no relation to the choices that consumers have repeatedly made in the market, and that we reject reception standards that the over one-half billion radios now in use implicitly endorse. I see no reason for the FCC to invent standards on its own, when consumers have already voted with their dollars to decide on an adequate level of performance."

"The FCC has recognized the most blatant exaggerations put forward by low-power radio opponents," said Virginia Tech Professor Theodore Rappaport, who was retained by LPFM proponents to analyze studies submitted to the FCC. "There can be no doubt that, technically speaking, the FCC has adopted an extremely cautious proposal" that could have accommodated more stations by relaxing interference protections further, he said.

National Public Radio (NPR) joined in the fervent opposition to LPFM. One of its key arguments involved the so-called "nuanced sounds" broadcast by some public stations that use minimal audio processing. Classical music listeners, a key stronghold of NPR financial support, are supposed to appreciate this quietude. NPR believed that LPFM interference would destroy the nuanced sounds, but FCC engineers demonstrated that such sounds were only detectable on the highest-quality receivers—the same equipment that is least susceptible to interference from LPFM.

NAB and NPR efforts to derail LPFM in Congress were largely successful at this writing. NAB circulated to members of Congress a compact disc, containing simulated LPFM interference.

We heard this disc and concluded that it was more of a propaganda instrument than a scientific product. In a news release, top FCC officials called it "misleading and simply wrong," "a deliberate misrepresentation," and a demonstration that "simply does not represent actual FM radio performance and therefore is meaningless."

The recording handily achieved its political objective. Shortly after the CD was circulated, legislation to suppress LPFM made progress in both houses of Congress. It ultimately was passed as a rider to appropriations legislation in December 2000.

According to the Media Access Project, the legislative wording "drastically curtails low power radio. It requires the FCC to conduct limited low power radio experiments, and requires further Congressional action to authorize the full low power radio created by the FCC. ...There will be some low power radio, but it will be cut back by 80%."

"I am troubled that several issues could not be resolved despite my Administration's best efforts during the final negotiations," President Bill Clinton said upon signing the Departments of Commerce, Justice, State, the Judiciary, and Related Agencies Appropriations Act of 2001. "This bill greatly restricts low-power FM radio broadcast. Low power radio stations are an important tool in fostering diversity on the airwaves through community-based programming. I am deeply disappointed that Congress chose to restrict the voice of our nation's churches, schools, civic organizations and community groups. I commend the FCC for giving a voice to the voiceless and I urge the Commission to go forward in licensing as many stations as possible consistent with the limitations imposed by Congress."

PUCK-FM

At this writing, the FCC is considering petition RM-9682 from the National Hockey League. It proposes an *Indoor Sports and Entertainment Radio Service* (ISERS), which would broadcast low-power FM signals confined to the interior of arenas. Ticket holders would listen to these broadcasts on portable FM receivers.

The NHL said that no radio service other than ISERS could explain the calls of referees and the rules of the game, address fans in foreign languages, provide entertainment during non-play periods, offer directions, parking, and emergency information, and accommodate the needs of the hearing and visually impaired.

Although ISERS would not be limited to hockey, it would be limited to certain kinds of events. "The fact that the broadcast signal remains indoors more critically defines the proposed radio service rather than the event itself," the NHL said, but it did offer to "limit the proposed new radio service to sports and entertainment activities as another means of controlling the potential for interference."

Emergency Broadcasting

Emergency broadcasts also are part of the digital revolution. The FCC requires AM, FM, and TV stations to adopt the *Emergency Alert System* (EAS), which replaced the *Emergency Broadcast System* (EBS).

EBS was intended to warn the public of enemy attack and natural disasters. It often is cited as the successor to the CONELRAD AM radio system of the Cold War era. Another claimed CONELRAD successor, however, was *Public Emergency Radio* (PER)—itself supplanted by NOAA Weather Radio (see 162.0125–173.2 MHz).

The irritating EBS Attention Signal (853+960 Hz) has been shortened and is accompanied by digital bursts as well as voice announcements in the new EAS. The data includes detailed information about the type and source of emergency. EAS decoders monitor multiple inputs, such as phone lines and radio receivers, for redundancy. They alert broadcast operators if they detect a genuine emergency message, although inadvertent EAS activations have been propagated through the system at this writing.

The EAS is at the President's command to disseminate a special announcement—the *Emergency Action Notification* (EAN)—representing the highest level of national crisis. The President may use the EAN to commandeer radio and TV without local intervention, at least in theory. An actual EAN had never been issued at this writing.

Local and state officials may also originate EAS alerts. Broadcasters warned the FCC that trigger-happy officials might interrupt local radio and TV with political messages; but no changes were made to the system in response to such concerns.

EAS can address discrete geographic units, including individual homes. It can activate radio and TV sets that are switched off, if the receivers contain special EAS circuitry—and provided that consumers correctly program their sets to respond to the many types of EAS alerts.

Yet the FCC's gushing prediction that EAS circuitry will ubiquitously appear in radio and TV sets has not materialized at this writing. Nothing requires consumer electronics manufacturers to implement the long-forgotten EAS celebratory pronouncements.

EAS has shown up in a few models of NOAA Weather Radio receiver. EAS was derived from a technology known as *Specific Area Message Encoding* (SAME), developed by the National Weather Service.

EAS systems did not actually function at their grand unveiling at FCC headquarters. Instead of correctly sending and receiving EAS alerts, vendors simply activated sirens and alarms by manually switching them on.

Fax Broadcasting

In this 88–108 MHz band, Non-Government Footnote NG2 permits "Facsimile Broadcasting Stations." This is a quaint relic of plans of the 1950s–60s to transmit newspapers by radio.

No such stations exist. The FCC dutifully supplied license application forms for fax stations for decades, and in 1999, it reaffirmed NG2.

The electronic fax newspaper was a typical element in "home of the future" scenarios. Criswell—the eccentric columnist and horror-film narrator portrayed in the Oscar-winning movie *Ed Wood*—predicted: "After 1980, newspapers will be fed into the home by television transmission, where hard copies of print-outs will roll out of the set each morning and afternoon, complete with photos, features, etc. And, because television is under government control, every newspaper will be under government control also."

Criswell also predicted (in 1968) that the capital of the U.S. would move to Wichita, Kansas by 1983, an alien space ray would decimate Denver in 1989, anti-gravity technology from Nebraska would power all cars by 1995, and the world would end on his birthday, August 18, 1999.

E-bombs

The 100-MHz spectrum has been cited as the low end of a wide bandwidth covered by a new class of "E-bombs" or radiofrequency energy weapons.

RF weapons come in two main types. *High Power Microwave* (HPM) weapons apply focused microwave energy on a target in order to induce heating. HPM devices are characterized by narrowband emissions. HPM energy tuned to the input frequency of a receiver—for example, an adversary's radar—can heat it sufficiently to burn out the receiver's front-end circuit (the so-called "front door" approach). Alternatively, it may be coupled to the target device via cables or, for higher-frequency weapons, even vent holes or enclosure gaps (the "back door"). The world's military forces have spent heavily to develop HPM weapons.

A more insidious weapon, the *Transient Electromagnetic Device* (TED), is causing grave concern among military experts. TEDs are bombs that use spark generators or explosive materials encased in electrical structures. They produce broadband pulses, lasting as little as 100 picoseconds, at very high peak power levels—rendering affected computers, vehicles, communications equipment, weapons, and other electronic and electrical equipment inoperative.

"The massed application of electromagnetic bombs in the opening phase of an electronic battle will allow much faster attainment of command of the electromagnetic spectrum, as it will inflict attrition upon electronic assets at a much faster rate than possible with conventional means," according to Australian defense analyst Carlo Kopp in his seminal paper, "The E-Bomb: A Weapon of Electrical Mass Destruction."

TED devices are simpler, smaller, and less expensive than HPM weapons, and have been known to produce energy a thousand times greater than the energy in a lightning strike. Building TEDs is said to be within the resources even of a nation with a 1940s technology base. Powerful TEDs have been built with a few hundred dollars worth of mail-order parts.

"Narrowband systems operate at some given frequency with a small bandwidth, and you will find them at one spot on the radio dial," according to the chilling 1998 Congressional testimony of weapons expert David Schriner. "The TEDs do not even have a definable frequency but instead, because of their short time duration, they occupy a very large spectrum space, and you will find it everywhere on every radio dial.

"When a TED pulse is generated, it will have the ability to excite responses in systems designed to receive at any frequency from as low as 100 MHz up to several GHz, from the FM band up to the lower microwave bands. A narrowband system would excite only those systems that were operating at its frequency, say 2.345 GHz, so a narrow band system must be tuned to a given target's known soft spot but a TED system would go after any soft spot of the target platform, back-door or front door."

Schriner described several types of terroristic TEDs, including a briefcase-sized unit, a vehicle-mounted unit, and "a system that could be located in one's back yard such that it could be aimed at overflying aircraft." His TED prototype, built in two weeks from $500 worth of auto parts, radiated "a very significant power level that can be compared against available vulnerability and susceptibility levels of military equipment. ...From the measurements and known signal levels, this unit is expected to be consistently deadly to many types of infrastructure items at ranges suitable for terrorist usage."

"The materials needed are nothing special," Schriner said, "and if the effort is made, advanced concepts can be made using everyday hardware such as automotive ignition systems. ...A terrorist would have little trouble developing such technology and he would have a high probability of success in the use as an RF weapon against our infrastructure elements found in any city or near facilities around the country."

108–117.975 MHz

FG: US93 AERONAUTICAL RADIONAVIGATION

Non-FG: US93 AERONAUTICAL RADIONAVIGATION

Note: The NTIA Manual (footnote G126) states that differential GPS stations may be authorized in the 108–117.975 MHz band, but the FCC has not yet addressed this footnote.

This band contains the *VHF Omnidirectional Range* (VOR) and *Instrument Landing System* (ILS), used worldwide in air navigation.

VOR beacons transmit two phase-related signals to aircraft. An aircraft receiver computes a bearing to the VOR source from the characteristics of these signals.

Often co-located with VOR but operating in 960–1215 MHz is *Distance Measuring Equipment* (DME) that provides pilots with distance to the DME station.

As an aid to landing aircraft, the ILS incorporates a localizer at 108.1–111.95 MHz, typically 1,000 feet beyond the stop end of the runway; a glide slope transmitter at 328.6–335.4 MHz, 1,000 feet from the approach end of the runway; and 75 MHz marker beacons that identify positions along the path to the runway.

Aviation users are sensitive to the location, power, and interference potential of FM radio stations because of the proximity of the 108–117.975 MHz band to the 88–108 MHz FM broadcast band.

The 112–118 MHz VOR segment also is used to transmit differential corrections that augment the *Global Positioning System* (GPS, see 1215–1240 MHz), particularly for precision approach and landing procedures. The FCC placed a special note, shown previously, regarding this use in its part of the U.S. Table of Allocations.

The FAA anticipates less demand for VOR and DME services for navigation and instrument approaches. It expects to "phase down" VOR/DME and ILS starting in 2008.

The military counterpart of VOR/DME is TACAN (see 960–1215 MHz).

The 108–121.94 MHz spectrum is one of many restricted bands in which the FCC Part 15 rules permit unlicensed devices to emit only very low level emissions.

Figure 2
The Old Executive Office Building is next to the White House and contains the Office of the Vice President. It is long rumored to harbor ground-to-air missiles. Its antenna shown here may be an aeronautical direction-finder, connected with radars on adjacent buildings. The need for such systems became evident after a plane intentionally crashed into the White House grounds in 1994. (Earthspan photo)

117.975–121.9375 MHz

FG: S5.111 S5.199 S5.200 591 US26 US28 AERONAUTICAL MOBILE (R)

Non-FG: S5.111 S5.199 S5.200 591 US26 US28 AERONAUTICAL MOBILE (R)

FCC: Aviation (87)

Allocated to the Aeronautical Mobile (Route) Service, air traffic controllers use this band to communicate with aircraft. Aeronautical utility mobile stations use it for communications among vehicles under tower direction at an airport. Federal government aeronautical mobile operations use this and the 225–400 MHz bands.

The FCC defines air traffic control as "a service operated by an appropriate authority to promote the safe, orderly, and expeditious flow of air traffic. It consists of communications between aircraft pilots and airport control towers or *Federal Aviation Administration* (FAA) aeronautical enroute stations [see 128.8125–132.0125 MHz] regarding aircraft landings and take-offs, aircraft transiting the airport traffic area on approach and departure, and aircraft traveling along domestic or international air routes." (See also 960–1215 MHz.)

A long history of congestion in the VHF aviation bands exists, according to the FAA, with frequency assignments growing about four percent a year. The FAA will replace the current analog AM radio system with the *Time Division Multiple Access* (TDMA)-based *VHF Digital Link* (VDL) system (see 136–137 MHz).

Distress Signals

Emergency Position Indicating Radio Beacons (EPIRBs) on ships and *Emergency Locator Transmitters* (ELTs) on aircraft transmit distress signals on frequencies including 121.5 MHz, 406.025 MHz, and 243 MHz. Satellites in the COSPAS/SARSAT *search and rescue* (SAR) satellite system monitor these distress signals.

The U.S. spacecraft in the system include the GOES and POES weather satellites (see 1670–1675 MHz and 1675–1700 MHz). A related system, INMARSAT E, transmits in the L-band (see 1645.5–1646.5 MHz). Both are part of the *Global Maritime Distress and Safety System* (GMDSS, see 154–156.2475 MHz) which mandates carriage of emergency beacons onboard vessels.

Ineffective, Wasteful

For aircraft, ELTs on 121.5 MHz have proven to be highly ineffective and "are a large source of wasted effort by search and rescue forces," according to U.S. SARSAT authorities. The units reportedly have a 98 percent false alarm rate and activate properly in only 12 percent of air crashes. ELTs are available on the better-supported 406.025 MHz frequency, but these are more expensive and less widely deployed.

COSPAS/SARSAT has warned all users that satellite reception of beacons on the older 121.5 MHz and 243 MHz frequencies will be withdrawn, with 406–406.1 MHz the main focus of international satellite-based SAR. The 121.5 MHz signal will still be useful for local homing (close-in direction finding) and is available as an option on some units.

The 108–121.94 MHz spectrum is one of many restricted bands in which the FCC Part 15 rules permit unlicensed devices to emit only very low-level emissions.

121.9375–123.0875 MHz

FG: 591 US30 US31 US33 US80 US102 US213

Non-FG: 591 US30 US31 US33 US80 US102 US213
AERONAUTICAL MOBILE

FCC: Aviation (87)

This band serves many types of aviation radio, including flight service communication with private and corporate aircraft, weather data, mobile units at airports and flight testing.

Aeronautical advisory stations (unicoms) in this band provide information on local conditions, fuel availability, and ground services. The FCC prescribes specific regulations governing the types and impartiality of information that unicoms transmit concerning ground services.

Automated unicoms automatically send aviation advisories and provide a simple test of aicraft radio equipment. A pilot can click a radio microphone to interrogate the unicom, which responds with a recorded or synthesized message.

Aeronautical search and rescue stations use the frequency 122.9 MHz for training missions.

The 108–121.94 MHz spectrum is one of many restricted bands in which the FCC Part 15 rules permit unlicensed devices to emit only very low level emissions.

123.0875-123.5875 MHz

FG: S5.200 591 US32 US33 US112 AERONAUTICAL MOBILE

Non-FG: S5.200 591 US32 US33 US112 AERONAUTICAL MOBILE

FCC: AVIATION (87)

Aeronautical flight testing stations use this band. Aeronautical search and rescue stations use 123.1 MHz during emergency missions.

The 123–138 MHz spectrum is one of many restricted bands in which the FCC Part 15 rules permit unlicensed devices to emit only very low-level emissions.

123.5875-128.8125 MHz

FG: 591 US26 AERONAUTICAL MOBILE (R)

Non-FG: 591 US26 AERONAUTICAL MOBILE (R)

FCC: Aviation (87)

This is one of several VHF air traffic control bands.

The 123–138 MHz spectrum is one of many restricted bands in which the FCC Part 15 rules permit unlicensed devices to emit only very low level emissions.

128.8125-132.0125 MHz

FG: 591

Non-FG: 591 AERONAUTICAL MOBILE (R)

FCC: Aviation (87)

Networked stations in the Aeronautical Enroute Service use this band in the *Aircraft Communications Addressing and Reporting System* (ACARS).

This data network is licensed to *Aeronautical Radio Inc.* (ARINC), a communications organization established by the airlines. Local aviation facilities providers typically operate these stations under contract with ARINC.

ACARS conveys critical information such as the time of aircraft gate departure, liftoff and landing, aircraft position and weight, amount of fuel, and weather conditions. It also provides alphanumeric message service for aircrews, controllers, and airline offices.

ACARS was implemented in 1977, and is planned to transition to a higher data rate standard as a part of the aviation industry's adoption of Data Link (see 1215–1240 MHz).

Non-networked, local area aeronautical enroute stations are used to communicate about terminal area matters (parking, maintenance, and clearance) with aircraft approaching or moving within the area of an airport.

Airport ground crews may use hand held radios to directly communicate on aeronautical enroute frequencies with flight crews in aircraft.

The 123–138 MHz spectrum is one of many restricted bands in which the FCC Part 15 rules permit unlicensed devices to emit only very low level emissions.

132.0125–136 MHz

FG: 591 US26 AERONAUTICAL MOBILE (R)

Non-FG: 591 US26 AERONAUTICAL MOBILE (R)

FCC: Aviation (87)

This is one of several VHF air traffic control bands.

The 123–138 MHz spectrum is one of many restricted bands in which the FCC Part 15 rules permit unlicensed devices to emit only very low-level emissions.

136-137 MHz

FG: 591 US244

Non-FG: 591 US244 AERONAUTICAL MOBILE (R)

FCC: Satellite Communications (25). Aviation (87)

This aviation band is devoted to air traffic control, *Automated Weather Observation Systems* (AWOS), and *Automatic Terminal Information Systems* (ATIS), operating in the Aeronautical Mobile (Route) Service.

AWOS systems observe weather conditions, without human involvement, and broadcast computer-generated voices that report variables such as visibility and temperature. ATIS (not to be confused with Automatic Transmitter Identification Systems, see 154–156.2475 MHz) are recorded broadcasts of airport conditions and procedures.

Few general aviation aircraft have radios that can tune 136–137 MHz. Aviation authorities have targeted this band for deployment of digital technology that eventually will spread across the aviation VHF bands.

At this writing, this band is the subject of a rulemaking proceeding (Docket WT 00-77) initiated by the *Federal Aviation Administration* (FAA) and the Small Aircraft Manufacturers Association.

They asked the FCC to accommodate *Flight Information Services* (FIS) on several frequencies in this band. FIS includes ground-based broadcasts of weather and other advisory information like AWOS and ATIS, but using data transmission.

Sometimes called "graphical weather in the cockpit," FIS generally includes weather hazard graphics, official *Notices to Airmen* (NOTAMS), and messages on the status of special-use airspace. FIS is considered a form of Data Link (see 1215–1240 MHz). Some basic FIS information will be available free; value-added information will carry a fee. The FAA authorizes private sector FIS providers, and the FCC regulates FIS stations.

The provision of weather information via Data Link is one of the highest priority elements in modernization of the U.S. airspace system. Weather accidents account for the second largest type of aviation accidents, and are the largest cause of delays in all aviation industry segments.

Beyond Weather Broadcasts

The FAA and aviation industry urged the FCC in this NPRM to adopt rules to permit use of digital voice and data technology in the next generation of aviation communications.

Based on years of technical studies, the FAA decided in 1998 to replace the current AM-based aviation radio infrastructure with *Time Division Multiple Access* (TDMA) throughout 118–137 MHz. The new systems are known as *VHF Digital Link* (VDL) Mode 3 (voice) and Mode 2 (data). The rulemaking contemplates a bandplan that would permit introduction of VDL and FIS in the currently underutilized 136–137 MHz band.

The 123–138 MHz spectrum is one of many restricted bands in which FCC Part 15 rules permit unlicensed devices to emit only very low-level emissions.

137–137.025 MHz

FG: 599A SPACE OPERATION (space-to-Earth). METEOROLOGICAL-SATELLITE (space-to-Earth). MOBILE-SATELLITE (space-to-Earth) 599B US318 US319 US320. SPACE RESEARCH (space-to-Earth)

Non-FG: 599A SPACE OPERATION (space-to-Earth). METEOROLOGICAL-SATELLITE (space-to-Earth). MOBILE-SATELLITE (space-to-Earth) 599B US318 US319 US320. SPACE RESEARCH (space-to-Earth)

The low-Earth orbit ("Little LEO") *Non-Voice Non-Geostationary Mobile Satellite Service* (NVNG MSS) uses small satellites at approximately 600 miles altitude to provide data communications.

NVNG MSS applications include vehicle tracking and monitoring, environmental data collection, maritime safety communications, electronic mail, meter reading, and security alerting among the many possibilities.

NVNG MSS can truthfully claim to be the first commercial global satellite service. It originated in the non-governmental, non-commercial small satellite programs of the Amateur Radio Service (see 144–146 MHz).

The NVNG MSS downlink bands are in 137–138 MHz and 400.15–401 MHz, with uplinks in 148–150.05 MHz and 399.9–400.05 MHz.

Little LEO Licensees

The first NVNG MSS licensee to begin revenue service is Orbcomm Global L.P., founded by rocket and satellite maker Orbital Sciences Corp. and now majority owned by Teleglobe of Canada. Its system provides two-way data services to users worldwide through 48 small satellites.

Orbcomm began operating in February 1996 with customers in transportation and environmental monitoring. The service is available through resellers in each participating country. Orbital subsidiary Magellan manufactures mobile units for vehicle tracking and telemetry, and a handheld message transceiver, for use with the Orbcomm service.

Another NVNG MSS licensee is *Volunteers in Technical Assistance* (VITA), a humanitarian organization supported largely by U.S. foreign aid and research funds. VITA supports technical and economic programs in developing countries. It earned the first Pioneer's Preference—an early assurance of license grant—from the FCC.

The Commission noted that VITA started experimenting with satellite communications in the early 1980s and applied for an experimental license for a more advanced system in 1988.

"We believe that VITA's efforts have advanced the authorization of the new service...that will provide reliable, low-cost data communications between ground stations located around the world," the FCC said. (Congress later withdrew the FCC's authority to provide preferential treatment to pioneering applicants.)

The first VITASAT satellite was destroyed in a failed launch. A replacement was part of an experimental satellite launched by *Final Analysis Communication Services* (FACS) in 1997, but it developed technical problems. VITA now leases capacity onboard satellites operated by British LEO developer Surrey Satellite Technology Ltd., and operates under the terms of the VITA FCC license. VITA subleases some of this capacity to Wavix, a company that provides oceanographic data collection.

FACS was one of the applicants licensed in the "second round" of NVNG MSS licensing. It is authorized to operate a $400 million system of 32 commercial satellites, planned for launch around the end of 2002. FACS equity investors include General Dynamics and Raytheon. Other second round licensees are E-SAT and Leo One USA Corp.

Spectrum Sharing

These firms, along with Orbcomm and VITA, agreed to a plan that permits all five licensees to share the available NVNG MSS spectrum.

Consequently, the FCC was not required to use auctions to award NVNG MSS licenses.

The plan requires specific transmission controls in order to prevent interference from NVNG MSS to the other inhabitants of the spectrum, notably the *Defense Meteorological Satellite Program* (DMSP) and the *National Oceanic and Atmospheric Administration* (NOAA), as well as satellites of other nations. The NOAA POES weather satellites output video imagery and telemetry in 137–138 MHz (see 1675–1700 MHz).

Most NVNG companies target industrial, utility, transportation, and environmental monitoring as key markets; but we expect novel uses to emerge. E-SAT will use its satellites to read utility meters and monitor energy facilities at remote locations. At this writing, the E-SAT management firm NewStar had launched ESAT-1, a demonstration "nanosatellite."

A prospective rival to NVNG MSS is SENS, under experimental development (see 2300–2305 MHz).

The 123–138 MHz spectrum is one of many restricted bands in which the FCC Part 15 rules permit unlicensed devices to emit only very low-level emissions.

137.025–137.175 MHz

FG: 599A SPACE OPERATION (space-to-Earth).

METEOROLOGICAL-SATELLITE (space-to-Earth). SPACE RESEARCH (space-to-Earth). Mobile-Satellite (space-to-Earth) 599B US318 US319 US320

Non-FG: 599A SPACE OPERATION (space-to-Earth). METEOROLOGICAL-SATELLITE (space-to-Earth). SPACE RESEARCH (space-to-Earth). Mobile-Satellite (space-to-Earth) 599B US318 US319 US320

FCC: Satellite Communications (25)

This is a downlink band in the Non-Voice Non-Geostationary Mobile Satellite Service (NVNG MSS, see 137–137.025 MHz).

The allocation in this band is identical to the adjacent 137.175–137.825 MHz band, except that the allocation to the Mobile Satellite Service is secondary in this band and co-primary in the other band.

The 123–138 MHz spectrum is one of many restricted bands in which the FCC Part 15 rules permit unlicensed devices to emit only very low level emissions.

137.175-137.825 MHz

FG: 599A SPACE OPERATION (space-to-Earth). METEOROLOGICAL-SATELLITE (space-to-Earth). SPACE RESEARCH (space-to-Earth). MOBILE-SATELLITE (space-to-Earth) 599B US318 US319 US320

Non-FG: 599A SPACE OPERATION (space-to-Earth). METEOROLOGICAL-SATELLITE (space-to-Earth). SPACE RESEARCH (space-to-Earth). MOBILE-SATELLITE (space-to-Earth) 599B US318 US319 US320

FCC: Satellite Communications (25)

This is a downlink band in the *Non-Voice Non-Geostationary Mobile Satellite Service* (NVNG MSS, see 137–137.025 MHz).

The allocation in this band is identical to the adjacent 137.025–137.175 MHz band, except that the allocation to the MSS is co-primary in this band and secondary in the other band.

The 123–138 MHz spectrum is one of many restricted bands in which the FCC Part 15 rules permit unlicensed devices to emit only very low-level emissions.

137.825-138 MHz

FG: 599A SPACE OPERATION (space-to-Earth). METEOROLOGICAL-SATELLITE (space-to-Earth). SPACE RESEARCH (space-to-Earth). Mobile-satellite (space-to-Earth) 599B US318 US319 US320

Non-FG: 599A SPACE OPERATION (space-to-Earth). METEOROLOGICAL-SATELLITE (space-to-Earth). SPACE RESEARCH (space-to-Earth). Mobile-satellite (space-to-Earth) 599B US318 US319 US320

This is a downlink band in the "Little LEO" *Non-Voice Non-Geostationary Mobile Satellite Service* (NVNG MSS, see 137–137.025 MHz).

The 123–138 MHz spectrum is one of many restricted bands in which the FCC Part 15 rules permit unlicensed devices to emit only very low-level emissions.

138-144 MHz

FG: US10 G30 FIXED. MOBILE

Non-FG: US10

The military applications in this government band include air traffic control, security and alarms, depot maintenance, special investigations, fire and medical communications, robotic handling of explosives, ground threat early warning systems, naval tracking and instrumentation data links and buoy monitoring.

The Civil Air Patrol and Coast Guard use this band for search and rescue. NASA and Energy Department land mobile operations also use 138–144 MHz.

This band is being reconfigured to narrowband 12.5 kHz FM channels, as part of a long-term initiative to recover more communications capacity in federal mobile bands.

To comply with the *Balanced Budget Act* of 1997 (BBA), 139–140.5 MHz and 141.5–143 MHz were supposed to be reallocated to the non-federal sector in January 2008. But later Defense Department authorization legislation cancelled reallocation of those two segments and 1385–1390 MHz (see 1350–1390 MHz).

The BBA required the FCC and NTIA to identify spectrum below 3 GHz to be auctioned, some from spectrum allocated to the federal government (see 2390–2400 MHz).

144-146 MHz

FG: None

Non-FG: AMATEUR 510.

AMATEUR-SATELLITE FCC: Amateur (97)

The two-meter 144–148 MHz spectrum is allocated to the Amateur Radio Service. Hams use the 144–146 MHz portion for experimental, Morse code and single-sideband communications and reflection of signals off the Moon (Earth-Moon-Earth or EME).

Amateurs pioneered small satellite communications in this band, via the *Radio Amateur Satellite Corporation* (AMSAT) and its affiliates in several countries.

Data communications via packet radio is a key use of this band. Amateur packet radio had its debut in Canada in 1978, when experimenters egged on by forward-thinking regulators transmitted the first amateur packet messages.

Inspired by that work, the *Tucson Amateur Packet Radio Corporation* (TAPR) became phenomenally successful selling kits and licensing designs for packet controllers and related devices.

A packet controller formats messages and files into "packets" or short transmissions, encapsulating them with addressing information and sending them via a connected transmitter. A form of time-division multiplexing, packet radio enables many users to efficiently share channels.

Today, amateurs use packet radio networks that integrate with satellites and automatically convey electronic-mail messages, in this and other bands. Amateur packet radio is evolving to higher speeds, better networks, and Internet connectivity.

In 1999, the FCC dramatically streamlined its amateur licensing structure, by reducing the number of license classes to three (Technician, General, and Extra Class) and by reducing the Morse Code speed test to five words per minute maximum.

The code requirement may be dropped entirely if the *International Telecommunication Union* (ITU) changes its regulations for the Amateur Service, which impose a demonstration of Morse skill at a minimum 5 WPM for high-frequency operation privileges. Increasingly, ITU member nations are reducing the code requirements that they impose on applicants.

At this writing, ham operation aboard the Space Station was slated to begin on the worldwide downlink frequency 145.8 MHz, under the aegis of the *Amateur Radio International Space Station* (ARISS) organization.

146-148 MHz

FG: None

Non-FG: AMATEUR

FCC: Amateur (97)

This is part of the amateur two-meter band (see 144–146 MHz). Unlike the worldwide 144–146 MHz band, 146–148 MHz is allocated to the Amateur Service only in ITU Region 2 (the Americas and Greenland) and 3 (Asia-Pacific).

Portable and mobile amateur communications in 146–148 MHz exploded in the 1970s, when manufacturers introduced affordably priced FM amateur radios. Wider availability of these units prompted radio clubs to install repeater stations, leading to nationwide coverage aided by volunteer frequency coordination groups.

Repeaters provide metropolitan area mobile radio and telephone service to amateur operators, for little or no outlay of fees other than club dues.

148-149.9 MHz

FG: S5.218 608A US10 G30 FIXED. MOBILE. MOBILE-SATELLITE (Earth-to-space) 599B US319 US320 US323 US325

Non-FG: S5.218 608A US10 MOBILE-SATELLITE (Earth-to-space) 599B US319 US320 US323 US325

FCC: Satellite Communications (25)

In 1993, the FCC allocated 148–150.05 MHz to "Little LEOs," the *Non-Voice Non-Geostationary Mobile Satellite Service* (NVNG MSS, see 137-137.025 MHz).

The 148–149.9 MHz band contains other satellites, government and non-government fixed, and mobile uses. POES weather satellites receive commands in this band (see 1675–1700 MHz). This is another band used by the Civil Air Patrol and Coast Guard for search and rescue.

This federal land mobile band is used for many agency missions including law enforcement, transportation, natural resource protection, emergency response, utility operation, and for tactical aviation communications.

Scientific data collection such as wildlife telemetry also is conducted in this band. Animals as exotic as gila monsters and sea turtles are outfitted with small transmitters and tracked by researchers.

The main federal land mobile bands are 162.0125–173.2 MHz, 173.4–174 MHz, and 406.1–420 MHz.

149.9–150.05 MHz

FG: S5.223 608B MOBILE-SATELLITE (Earth-to-space) 599B US319 US322. RADIONAVIGATION-SATELLITE

Non-FG: S5.223 608B MOBILE-SATELLITE (Earth-to-space) 599B US319 US322. RADIONAVIGATION-SATELLITE

FCC: Satellite Communications (25)

This band is internationally allocated to the Mobile Satellite Service for land operation only and to the *Radio Navigation Satellite Service* (RNSS).

It is available in the U.S. to uplinks for Little LEOs, the *Non-Voice Geostationary Mobile Satellite Service* (NVNG MSS, see 137–137.025 MHz), subject to coordination with the systems of other nations.

The Navy's celebrated Transit satellites used this band until 1997 to provide position-location services. Transit and related spacecraft in this spectrum were first launched in 1960 to support submarines. Their non-military users included commercial shipping and drilling platforms. The Transit program led to today's *Global Positioning System* (GPS, 1215–1240 MHz).

The former Soviet Union has operated Transit-like RNSS systems in this band for decades, including the TSYKADA system.

This is one of many restricted bands in which the FCC Part 15 rules permit unlicensed devices to emit only very low-level emissions.

150.05-150.8 MHz

FG: US216 G30 FIXED. MOBILE

Non-FG: US216

Army, Air Force, and Navy systems principally use this band, as does the non-government Emergency Medical Radio Service.

This band and adjacent bands contain frequencies used by the *Emergency Managers Weather Information Network* (EMWIN) to broadcast digital weather data and warnings directly to the public. Specialized receivers display EMWIN information on computers.

Although the EMWIN datastream originates with the National Weather Service, most of these transmissions are funded by other public and private sources.

EMWIN is a separate service from NOAA Weather Radio, a voice broadcast (see 162.0125–173.2 MHz).

150.8-152.855 MHz

FG: US216

Non-FG: US216 NG4 NG51 NG112 NG124 FIXED. LAND MOBILE

FCC: Public Mobile (22). Private Land Mobile (90)

This band contains both Public Safety pool frequencies, for such uses as highway maintenance and forestry, and frequencies in the Industrial/Business pool for general commercial use (see 30–30.56 MHz), all under Part 90 of the FCC rules.

This is one of the bands containing low power "color dot" frequencies, which have contributed to problems with unlicensed operation.

"Color dots" are inexpensive, off-the-shelf radios, pretuned to one of several frequencies and sold at low prices by mail-order merchants, discount stores and two-way radio dealers. The "dots" are colored stickers that indicate operating frequency. Choosing radios with the same color dots ensures that the units will be on the same frequency and thus able to intercommunicate.

"Many advertisements imply that these radios can be used by anybody for any purpose, whether commercial or recreational, and make no mention of the licensing requirement," the FCC said in Docket WT 98-182. "Manufacturers have informally indicated to us that it is their belief that only a small percentage of persons buying these radios actually apply for a license."

Price-conscious users widely employ color dots to avoid FCC licensing procedures, application fees, and frequency-coordination fees. This behavior directly affronts conventional mobile radio sales and coordination structures.

Creating the New CB

Realizing that neither the Commission nor the frequency coordinators have any idea of the location, operators and type of use of these radios, the FCC eliminated the licensing and coordination requirements for several of the color-dot frequencies.

In the process, the FCC reallocated these frequencies to a new, minimally regulated *Multi-Use Radio Service* (MURS) within the Citizens Band Radio Services (see 402–403 MHz and 462.5375–462.7375 MHz).

Congress permitted the FCC to exempt any Citizens Band service from licensing. As was the case with the *Family Radio Service* (FRS) in the UHF spectrum, Radio Shack was a key proponent of the MURS allocation.

The MURS frequencies are 151.820 MHz, 151.880 MHz, 151.940 MHz, 154.570 MHz, and 154.600 MHz. Manufacturers such as Radio Shack and Motorola asked the FCC to add more frequencies, but trade associations argued against it. For example, the *Personal Communications Industry Association* (PCIA) told the FCC that other frequencies in the Industrial/Business Pool should not become a "haven" in which manufacturers are allowed to promote unlicensed consumer radios.

The FCC said, "We are not persuaded that there is sufficient support in the record to justify reallocation of additional Part 90 frequencies at this time. We may, however, revisit this issue at a later date should additional support develop. We will therefore include in the new Multi-Use Radio Service only the five frequencies listed in our original proposal."

Unlike the 27-MHz CB service, the FRS, or the General Mobile Radio Service, MURS may be used to transmit data and image communications as well as voice.

What Rules Don't Say

"The significance of the rules governing the Multi-Use Radio Service is not in what they say, but in what they don't say," according to Corwin D. Moore, Jr. of the *Personal Radio Steering Group* (PRSG).

He pointed out that MURS carries no restriction on connecting to external antennas, nor on antenna height (so long as a 2 W ERP restriction is observed). Two watts at an even modest height could produce substantial coverage.

"There is no constraint on communications with other radio services, or with retransmitting signals from other stations. How soon will we see repeaters? There is no restriction on interconnection with the public telephone network. ...The available emission modes would seem to permit a wide variety of choices, openly inviting hobbyist activities such as video and packet radio."

At this writing, PRSG petitioned the FCC to reconsider certain parts of MURS, including the power measurement rules, antenna height, and permissibility of repeaters and telephone interconnection. The PRSG petition would not, however, alter the basic status of MURS as a personal radio service available directly to the general public.

Motorola, apparently concerned about impacts of MURS on its commercial radio business, also petitioned for reconsideration: "Motorola wishes to reemphasize to the Commission its position that these frequencies should be reserved for business and industrial use only," the company said. "If the frequencies are placed in the MURS category, manufacturers will likely develop radio applications that will be marketed to a broader consumer population.

"The expanded availability of the frequencies to general consumers will increase traffic congestion and interference, thereby harming business users. ...The addition of potentially millions of additional consumer users on the MURS frequencies will render it even more difficult to maintain disciplined and orderly use, which is essential for the needs of the business community. The increased interference will be unacceptable to business users, and may eventually lead to their abandonment of this popular technology."

152.855–154 MHz

FG: None

Non-FG: NG4 NG124 LAND MOBILE

FCC: Auxiliary Broadcasting (74). Private Land Mobile (90)

Frequencies in this band are used for fire protection and other public safety applications as well as Industrial/Business uses, especially power utility communications and remote pickup systems in radio and TV broadcasting.

154–156.2475 MHz

FG: S5.226

Non-FG: S5.226 NG112 NG117 NG124 NG148 FIXED. LAND MOBILE

FCC: Maritime (80). Private Land Mobile (90)

Police, fire, emergency medical, and highway maintenance are among the public safety operations that use the *Private Mobile Radio Service* (PMRS) allocations in this band.

This band contains some of the Industrial/Business pool "color dot" frequencies used by inexpensive radios from retail and mail order sources.

Faced with an epidemic of unlicensed radio operation on the color dot frequencies, the FCC eventually reallocated some of them to a *Multi-Use Radio Service* (MURS) in the Citizens Band. Two of the five MURS frequencies are in this 154 MHz band. See 150.8–152.855 MHz for more on MURS.

Transmitter ID

In an effort to combat the problem of unlicensed operation, the FCC once proposed to require an *Automatic Transmitter Identification System* (ATIS) in mobile and portable radios. The FCC later dumped the idea after trade associations howled.

The FCC decided, however, to require ATIS on satellite uplinks. The need for uplink ATIS became obvious after the 1986 escapades of "Captain Midnight," an employee of a commercial earth station who transmitted unauthorized video text messages through a satellite. FCC agents found him after comparing the messages' typography with records of purchasers of certain models of video character generation equipment.

The ATIS proceeding revealed that erroneous, multiple illuminations of satellites by uplink stations was a common occurrence, prompting the need for automatic identification. It later required ATIS in millimeter wave unlicensed devices (see 59–64 GHz).

Maritime VHF

The Radio Regulations of the *International Telecommunication Union* (ITU) prescribe maritime mobile communications worldwide in 156–174 MHz. Regulations mandate radio capability in vessels of various types and sizes.

The U.S. VHF maritime spectrum uses frequencies in 156.05–162.025 MHz, with channels devoted to safety, commercial and non-commercial vessel communications, bridge-to-bridge navigation, and federal, state, and local control of maritime traffic. Another key U.S. maritime service is provided by *Automated Maritime Telecommunications Systems* (AMTS, see 216–220 MHz).

The maritime service is introducing new, digital communications technologies. The FCC has begun to deregulate the service and stimulate business interest in it.

Chief among new technologies in maritime use is *Digital Selective Calling* (DSC) for automatic call set-up between suitably equipped stations. DSC can automatically interconnect marine radios to the public telephone network.

The FCC describes DSC as "a digital signaling system that allows ship and shore stations to call each other directly, rather than requiring a radio operator to continuously monitor a common calling channel to identify specific calls directed to the station."

The international VHF DSC distress frequency is 156.525 MHz (channel 70). The maritime community will migrate to that DSC distress frequency from the conventional distress and calling frequency 156.8 MHz, channel 16. Vessels and coast stations monitor channel 16, as does the U.S. Coast Guard's (USCG) *National Distress System* (NDS).

NDS is a network of hundreds of shore, waterway, and urban stations remotely controlled by regional USCG centers.

It receives tens of thousands of distress calls annually. The NDS was originally put into service in the 1970s and now suffers from technological obsolescence.

According to the USCG, "Much of the existing equipment is no longer commercially available off-the-shelf; and is becoming increasingly difficult to support. ...The system is not *Global Maritime Distress and Safety System* (GMDSS)-compatible, coverage gaps exist in several locations, it cannot operate on public safety channels, it has no direction finding capability, distress calls cannot be received at a high site when the site is transmitting on any channel, and the system is near the end of its useful life."

GMDSS

The GMDSS is the principal international maritime radio safety system, utilizing satellite and terrestrial communications including DSC. It essentially replaced maritime radiotelegraphy, after ceremonial sign-off messages tapped out by tearful Morse Code operators.

Years in development, GMDSS was fully activated worldwide on February 1, 1999. GMDSS mandates detailed equipment carriage requirements for vessels in various sea areas, prescribes emergency communications procedures, and requires radio officer certification. In the U.S., GMDSS certifications are issued by FCC-authorized *Commercial Operator License Examination Managers* (COLEMs) and by the Coast Guard.

Consistent with the universal adoption of GMDSS, the USCG is implementing DSC in a modernized NDS that also will be capable of direction finding and survivability in adverse conditions such as war or natural disaster. The FCC will require all new marine radios to include DSC.

According to the FCC and NTIA, "Except for the digital selective calling distress and safety channel 70, no VHF channel has been set aside for data purposes. ...Additionally, congestion has become a serious problem in many areas of the world because of the rapid increase in maritime mobile usage of VHF-FM, particularly data communications.

"This has resulted in degraded effectiveness of distress and safety calls on the calling channel. Since use of the maritime mobile service for voice and data is continuing to grow, this situation will worsen unless action is taken, intensifying the impact on critical services, including those used for safety and distress.

"In addition, administrations implementing modern vessel traffic services (VTS), using such techniques as automatic dependent surveillance and NAVTEX-like broadcasts, will need internationally compatible radio channels set aside for data transmissions."

The Coast Guard uses VTS as an advisory service to coordinate the movement of vessels in ports. Vessels report their position and conditions affecting navigation to the Coast Guard, which tracks their movements. Large ships, tow and tugboats and certain other vessels must participate in VTS.

The FCC describes NAVTEX as "an automated system that distributes maritime navigation warnings, weather forecasts and warnings, search and rescue notices, and other safety and urgent information to mariners."

The U.S. is supporting increased spectrum provisions for maritime data communications at ITU World Radio Conferences. *Automatic dependent surveillance* (ADS), being developed both in maritime and aviation communications, can provide position and status telemetry from a craft to land-based controllers and to nearby craft.

U.S. Regulatory Structure

The maritime service consists of public and private coast stations and ship stations. Public coast stations provide for-profit radiotelephone service to ships. They are similar to cellular and *Personal Communications Service* (PCS) carriers that provide wireless phone service. Private coast stations communicate with ships as part of an internal business operation, and do not derive profit from the communications directly.

Public coast stations now operate like other forms of *Commercial Mobile Radio Service* (CMRS), such as cellular and PCS. Public coast stations may serve land-based users as long as maritime communications are not compromised. Public coast station licenses are subject to auctions where mutually exclusive license applications exist.

To auction these licenses, the FCC established a geographic area scheme, with nine licensing regions near major waterways, defined as maritime *VHF Public Coast* (VPC) Areas and based on Coast Guard districts, and 33 inland VPCs based on the Economic Area scheme used in other radio services.

The FCC permits only one geographic area licensee in each of these areas. Incumbent public coast station licensees and licensees of land mobile radio systems sharing maritime spectrum in inland areas are required to afford each other protection from interference.

Public coast licensees may partition and disaggregate portions of their licensed spectrum and geographic areas to other parties. The FCC's December 1998 public coast service auction raised about $7.4 million.

156.2475-157.0375 MHz

FG: S5.226 S5.227 US77 US106 US107 US266

Non-FG: S5.226 S5.227 US77 US106 US107 US266 NG117 MARITIME MOBILE

This is part of the VHF maritime spectrum (see 154–156.2475 MHz). It includes the distress, calling, and safety channel 16 (156.8 MHz) which all VHF equipped vessels must monitor.

A type of *Emergency Position Indicating Radio Beacon* (EPIRB), for use by boats in coastal waters, transmits an alert on channel 16 to call attention to a distress situation. It then transmits a longer signal on channel 15 (156.75 MHz) as an aid to direction finding.

These "Class C" EPIRBs are being phased out in favor of other types, particularly those using 406.025 MHz.

The segments 156.52475–156.52525 MHz and 156.7–156.9 MHz are two of the restricted bands in which the FCC Part 15 rules permit unlicensed devices to emit only very low level emissions.

157.0375-157.1875 MHz

FG: S5.226 US214 US266 G109 MARITIME MOBILE

Non-FG: S5.226 US214 US266

FCC: Private Land Mobile (90)

This is an exclusive federal maritime band. The Coast Guard broadcasts marine safety information on channel 22A (157.1 MHz).

157.1875-157.45 MHz

FG: S5.226 US223 US266

Non-FG: S5.226 US223 US266 NG111 LAND MOBILE. MARITIME MOBILE

FCC: Maritime (80). Private Land Mobile (90)

This is part of the VHF maritime spectrum (see 154–156.2475 MHz). Ships transmit in this band to maritime public coast stations that provide them with telephone service. The coast stations transmit in 161.8–162.025 MHz.

Industrial/business and public safety users have access to channels in this band away from co-channel public coast stations and waterways.

157.45-161.575 MHz

FG: S5.226 US266

Non-FG: S5.226 US266 NG6 NG28 NG70 NG111 NG112 NG124 NG148 NG155 FIXED. LAND MOBILE

FCC: Public Mobile (22). Maritime (80). Private Land Mobile (90)

This band contains Industrial/Business (including Private Carrier Paging) and Public Safety Private Mobile Radio Services frequencies (see 30.56–32 MHz).

These channels also are used by Conventional Rural Radiotelephone Stations and the somewhat more modern *Basic Exchange Telephone Radio Systems* (BETRS), providing Rural Radiotelephone Service to subscribers not reached by wireline phone service. Most BETRS operations are at 454 MHz.

This is one of several Public Mobile bands cited by the FCC for possible underutilization. In a search for new spectrum for public safety radio services, the FCC observed that VHF Public Mobile channels "may be lightly used or even unused in some regions. ...We believe that this light use of frequencies may result from the discontinuation of service by commercial providers who are intended users of that spectrum."

This is a key VHF band for railroad communications, formerly in the Railroad Radio Service and now part of the Industrial/Business frequency pool. Important developments in railroad radio include the FCC's mandated refarming of channels (see 30.56–32 MHz) and the implementation of digital voice/data equipment based on the Project 25 standard of public safety radio (see 764–776 MHz). The Association of American Railroads and its Transportation Technology Center coordinate railroad spectrum use and its technological evolution.

Others in the band include power utilities, police radio, highway maintenance, forestry, and *Broadcast Auxiliary Services* (BAS, see 1990–2025 MHz). The band contains special frequencies designated for oil spill cleanup, and is part of the larger U.S. maritime spectrum (see 154–156.2475 MHz).

161.575–161.625 MHz

FG: S5.226 US77

Non-FG: S5.226 US77 NG6 NG17 MARITIME MOBILE

FCC: Public Mobile (22). Maritime (80)

This is part of the VHF maritime spectrum (see 154–156.2475 MHz).

Ships use this band to communicate with coast stations when navigating ports, locks, and waterways.

161.625–161.775 MHz

FG: S5.226

Non-FG: S5.226 NG6 LAND MOBILE

FCC: Public Mobile (22). Auxiliary Broadcasting (74)

Radio and TV remote pickup stations use this band for on-scene reporting.

161.775-162.0125 MHz

FG: S5.226 US266

Non-FG: S5.226 US266 NG6 LAND MOBILE. MARITIME MOBILE

FCC: Public Mobile (22). Maritime (80). Private Land Mobile (90)

This is part of the VHF maritime spectrum (see 154–156.2475 MHz). Public coast stations use this band to provide telephone service to ships transmitting in 157.2–157.425 MHz.

Industrial/business and public safety users have access to channels in this band away from co-channel public coast stations and waterways.

162.0125-173.2 MHz

FG: S5.226 US8 US11 US13 US216 US223 US300 US312 G5 FIXED. MOBILE

Non-FG: S5.226 US8 US11 US13 US216 US223 US300 US312

FCC: Auxiliary Broadcasting (74). Private Land Mobile (90)

The main federal VHF mobile bands are 162.0125–173.2 MHz and 173.4–174 MHz. They include some fixed, aeronautical, maritime, and non-government uses such as broadcast remote pickup links.

These bands serve many agency missions including federal law enforcement, Presidential protection, building security, transportation, emergency response, utility operation, weather broadcasts, and scientific data collection such as wildlife telemetry.

These bands have more frequency assignments than any other federal band used for land mobile radio. This band is being converted to narrower channels in order to recover more communications capacity.

The FBI, the Drug Enforcement Administration, the Customs Service, and the Bureau of Alcohol, Tobacco and Firearms use this spectrum for agent-to-agent communications, audio bugging devices, tracking beacons, body transmitters, and control of information collection equipment.

Figure 3
Sporting an Advanced Telemetry Systems GPS tracking collar with VHF transmitter, this Woodland Caribou helps biologists understand the animal's habitat use, seasonal movement, and food resources. The device drops from the animal at the end of its battery life, or upon remote command. (Christopher Kochanny photo)

LoJack Versus Hijackers

Some of these uses were impacted by a 1989 FCC and NTIA decision to allocate a key FBI channel, 173.075 MHz, to *Stolen Vehicle Recovery Systems* (SVRS). The sole SVRS business is LoJack Corporation, which sells tracking beacons to consumers. LoJack has had impressive success in vehicle recovery, with millions of beacons in service.

Installers hide the beacons inside vehicles. When a car is reported stolen, computers cause LoJack's "Sector Activation System" network to activate the beacon. The beacon reply code includes the vehicle make, model, and registration. Police cars equipped with direction finder receivers can then locate the vehicle. Once recovered, the network sends the beacon a deactivation message.

Police often discover the stolen vehicle among many others in "chop shops" that dismember them for parts. A growing business for LoJack is the recovery of stolen construction equipment.

The U.S. Attorney General and the Director of the FBI fought reallocation of 173.075 MHz. The matter even surfaced in Congressional testimony. An FBI witness told a 1990 House Telecommunications Subcommittee panel: "The value of one of our few nationwide channels, used to service body transmitters and tracking beacons, was seriously and adversely diminished, and our operational capabilities adversely affected when a private sector firm was granted nationwide use of this channel."

The SVRS should not be confused with the *Location and Monitoring Service* (LMS) at 902–928 MHz, or with covert tracking devices used by police in 30–50 MHz. The LMS uses more sophisticated methods of pinpointing vehicles and other beacon-equipped objects.

FCC Rules place certain restrictions on SVRS use of 173.075 MHz. Officially, this channel still is shared with the federal government. The frequency may not be used for "general purpose vehicle tracking or monitoring" (like LMS) but only to recover stolen vehicles. LoJack must limit transmission duration in order to avoid interference to TV channel 7 (174–180 MHz).

But in August 2000, the FCC granted LoJack some relief from the restrictions, permitting it to "uplink" or transmit from customer vehicles to the network, a feature not previously available.

Beacons will send an "early warning message" if the car is moved without permission, permitting LoJack to contact the owner to verify the theft and to then contact police. Uplink acknowledgements to the network also will reduce the large number (hundreds) of transmissions from the network that are possible during an incident with the original system.

Figure 4
PERKI the dog was the mascot of *Public Emergency Radio* (PER), a failed Cold War scheme to warn the nation of attack. In silhouette, an evacuee family briskly walks to a country shelter. (Federal Emergency Management Agency archives)

Weather and Emergency Radio

The National Oceanic and Atmospheric Administration's *National Weather Service* (NWS) operates NOAA Weather Radio on 162.400, 162.425, 162.450, 162.475, 162.500, 162.525, and 162.550 MHz. The signal may be received on scanners and on special radios sold by major electronics dealers.

Broadcasting from hundreds of transmitters across the country, NOAA Weather Radio provides forecasts, warnings, and related information to the public. These are voice broadcasts. An all-digital NOAA service, EMWIN, is now on the air via satellite (see 1675–1700 MHz) and terrestrial stations (see 150.05–150.8 MHz).

When NOAA Weather Radio was established in the 1970s, it was proclaimed to be the successor to the ill-fated *Public Emergency Radio* (PER), internally called the *Decision Information Distribution System* (DIDS).

Though not in this book's 30 MHz–300 GHz range, DIDS warrants examination as a historical oddity. Operating in 160–190 kHz, DIDS was an ambitious scheme to use longwave to warn the public of attack. Like Weather Radio, it was to use special receivers in every home.

The $2 million flagship DIDS station, WGU-20 in Chase, Md., operated from 1972 to 1984, broadcasting nothing but time checks. It had a 700-foot antenna and was the first all solid-state 50 kW AM station.

DIDS was declared the successor to *Control of Electromagnetic Radiation* (CONELRAD), a 1960s defense program that required AM stations to turn on and off and vary their schedules. They were supposed to warn the public while confusing missiles that might home in on broadcast signals.

Triangular CONELRAD symbols were printed on the dials of radio receivers of this period. CONELRAD was discontinued when missile technology improved.

DIDS' creators regarded existing warning technology as prone to human error and delay. DIDS was to be a spoof-proof supersystem that could switch on sirens and convey news to officials by radioteletype. DIDS' unique selling proposition was that it would speak to citizens in their cars and homes, even waking them from sleep.

Most Americans do not own longwave receivers, however. "The acquisition of home warning receivers would be a voluntary decision on the part of the individual citizen," said one DIDS manual. Thus, the government had to persuade manufacturers to market and the public to buy radios whose only use would be in national emergencies.

To create an approachable image for consumers, DIDS was dubbed Public Emergency Radio and given a cartoon puppy mascot named PERKI.

The U.S. anti-ballistic missile system was designed to connect directly to PER. Radar could track the flight paths of incoming enemy missiles. While the radar network dispatched ABMs to destroy the missiles at high altitudes, it would instantly furnish a prediction to PER listeners as to where an enemy warhead would land and explode if it was not successfully intercepted by an ABM. The PER broadcast would then tell commuters which areas to avoid.

Federal Emergency Management Agency (FEMA) officials told us that they discontinued PER because of the high cost to build and operate enough stations to blanket the U.S. It also would have been a one-way system, without a means to obtain status reports and damage assessments. WGU-20 was eventually dismantled.

The principal electronic connection between federal, state, and local emergency officials remains the *National Warning System* (NAWAS), a party-line telephone network funded by FEMA.

If not by NOAA Weather Radio, most Americans would probably receive military emergency warnings over the *Emergency Alert System* (EAS, see 88–108 MHz).

The 169–172 MHz segment is used by non-government entities for licensed wireless microphones. Other frequencies in this band are used for forestry and forest firefighting, and for remote control and telemetry in the Industrial/Business frequency pool.

The segments 162.0125–167.17 MHz and 167.72–173.2 MHz are restricted bands in which the FCC Part 15 rules permit unlicensed devices to emit only very low level emissions.

173.2–173.4 MHz

FG: None

Non-FG: FIXED. Land Mobile

FCC: Private Land Mobile (90)

Operations such as power utilities, forest products, industrial, petroleum, manufacturing, business, and local government services use this Industrial/Business band for telemetry and remote control. The Public Safety pool also has several frequencies in this band.

Except for frequencies near the edges, frequencies in this band are spaced 12.5 kHz apart. In designing its refarming scheme to obtain more capacity in the Private Mobile Radio Services (see 30.56–32 MHz), the FCC decided not to change the channel bandwidth arrangements in this band due to the heavy use and unique channel spacing.

173.4-174 MHz

FG: G5 FIXED. MOBILE

Non-FG: None

This is part of the federal VHF spectrum mostly used for land mobile communications, with some fixed, aeronautical, and maritime mobile uses (see 162.0125–173.2 MHz).

174-216 MHz

FG: None

Non-FG: NG115 NG128 NG149 BROADCASTING

FCC: Broadcast Radio (TV) (73). Auxiliary Broadcasting (74)

This band contains TV channels 7 through 13. The U.S. is making a major transition to *Digital Television* (DTV, see 470–512 MHz).

Wireless microphones and medical telemetry devices such as heart monitors are permitted in 174–216 MHz as well as in TV channels 14–46 (470–668 MHz).

The Alliance of Motion Picture and Television Producers petitioned the FCC (RM-9856) to permit wireless video assist devices to operate on locally vacant TV channels in 174–216 MHz and 470–746 MHz.

Such a device employs a "light tap," which conveys the image seen by the movie camera to a small video camera and transmitter. This setup enables the director and production crew to observe the scene from the movie camera's perspective, without having to physically wire the camera to remote monitors.

Figure 5
This innocuous-looking object is the two-mile-long *Naval Space Surveillance* (SPASUR) antenna, part of the world's largest continuous-wave station, at Lake Kickapoo, Tex. (P.A. Crossley photo)

216-220 MHz

FG: US210 US229 US274 US317 MARITIME MOBILE. Fixed. Radiolocation S5.241 G2. Aeronautical Mobile. Land Mobile.

Non-FG: US210 US229 US274 US317 NG152 MARITIME MOBILE. Fixed. Aeronautical Mobile. Land Mobile

FCC: Maritime (80). Private Land Mobile (90). Personal Radio (95). Amateur (97)

Note: 216–220 MHz will become a mixed-use band in January 2002.

Operations in this band include space radar, transportation, telemetry, amateur, and personal radio services.

The Balanced Budget Act of 1997 requires the reallocation of several bands used by federal systems to the private sector (see 2390–2400 MHz). The 216–220 MHz band was one of those selected for reallocation.

At this writing, the FCC proposed to allocate this band to the fixed and mobile services on a primary basis, in preparation for eventual auctions (Docket ET 00-221). It did not further specify the actual uses to which the band may be applied.

"Any new service allocated on a primary basis in this spectrum will be required to protect existing primary licensees," the Commission warned. "In addition, any new operations in the 216–220 MHz band are likely to be constrained by the need to protect TV channel 13, which occupies the subjacent 210–216 MHz band. Protection of television channel 13 was one of the factors we considered in limiting use of this band to low power applications such as Low Power Radio Service and telemetry on a secondary basis."

New mobile service licensees in this spectrum will have to use 216–218 MHz for base station transmission and 218–220 MHz for mobile unit transmission, in order to reduce interference to channel 13.

The FCC observed that there is "relatively little new capacity" in this band. "Given the significant constraints on additional use of the 216–220 MHz band...it is unclear how this band might accommodate additional services and how we might further assign licenses in this spectrum," it said.

The Fence

The *Naval Space Command* (NAVSPACECOM) operates a colossal satellite and missile-sensing radar on 216.98 MHz. Known as the *Naval Space Surveillance System* (SPASUR), or "the Fence," it has a main station at Lake Kickapoo, near Archer City, Texas—employing the world's largest *continuous wave* (CW) transmitter—and subsidiary transmitters at Jordan Lake, Alabama, and Gila River, Arizona. On the air since 1959, this system will remain in the band even after auctions (see Footnote US229).

Energy from the multibillion-watt SPASUR forms an electromagnetic fence extending 3,000 miles from California to Georgia, 1,000 miles off each coast and thousands of miles into space.

Aircraft and spacecraft passing through the Fence reflect energy back to Earth, where it is picked up at receiving stations in San Diego, Calif.; Elephant Butte, N.Mex.; Red River, Ark.; Silver Lake, Miss.; and Hawkinsville and Ft. Stewart, Ga.

The receivers measure the reflected signals—about 60,000 per day—and convey the data to the Naval Space Operations Center in Dahlgren, Virginia. The center updates a database of orbital elements and informs the U.S. Space Command Space Surveillance Network center inside Cheyenne Mountain in Colorado.

Detection by the Fence of objects that exhibited extraordinary speed and maneuvers reportedly prompted the creation of an interagency UFO working group in the U.S. government. According to the Naval Space Operations Center, SPASUR is "often the first sensor to detect a maneuver" of a space object.

The Space Surveillance Network, using the Fence and other radar and telescope facilities, can track space objects at least 30 centimeters in size (somewhat larger than a basketball). Objects of that size are a small minority of the more than 35 million man-made objects orbiting Earth, most of which are smaller than one centimeter.

Starting in 2003, the Navy plans to replace components of the Fence with modern components. But this upgrade is not expected to enhance its capabilities. To detect most near-Earth objects larger than one centimeter would require using the *Super High Frequency* (SHF) spectrum and moving the stations near the Equator.

Maritime Communications

Automated Maritime Telecommunication Systems (AMTS) are a category of maritime public coast station that is allocated frequency groups "A," 217–218 MHz and "B," 219–220 MHz. It originally was allocated group "C" at 216–217 MHz and "D" at 218–219 MHz, but due to concerns over TV channel 13 interference, C was removed and reallocated to the *Low Power Radio Service* (LPRS). Group D became the *Interactive Video and Data Service* (IVDS).

The FCC describes AMTS as "a specialized system of coast stations providing integrated and interconnected marine voice and data communications, somewhat like a cellular phone system, for tugs, barges, and other vessels on waterways."

AMTS users can dial the number of the desired vessel, mobile unit, or landline phone. The system connects the call and records billing information. AMTS is considered a *Commercial Mobile Radio Service* (CMRS) subject to auctions of mutually exclusive license applications.

The Commission requires AMTS licensees to serve either a substantial navigational area along the Pacific, Atlantic, or Gulf of Mexico coastline,

60 percent of one or more major inland waterways, or an entire inland waterway less than 240 km (150 mi.) long. In effect, this rule permits coverage of well over 100 major markets.

AMTS network control can use the 216–217 MHz *Low Power Radio Service* (LPRS, see the following).

A nationwide land mobile SMR-like wireless phone and dispatch service is available on AMTS frequencies. In 1997, the FCC permitted AMTS licensees to serve an unlimited number of land based customers as long as priority is accorded stations operating on the water.

In the long-running Docket PR 92-257, the FCC proposed liberalized technical rules and an auction scheme for AMTS. Licensees would be allowed to use any modulation or channel scheme as long as they comply with out-of-band emissions limits.

At this writing, the FCC was accepting public comment on ways to divide up geographic areas for AMTS auctions, for example, using the *VHF Public Coast* (VPC) area scheme employed in VHF maritime service (see 154–156.2475 MHz). Citing a "demand for AMTS spectrum away from large waterways," it tentatively concluded that requiring AMTS stations to serve coastlines or sizable navigable inland waterways could prevent service from being offered in some areas. It proposed to permit licensees to place stations anywhere within their service areas.

Telemetry

Telemetry is "the transmission of non-voice signals for the purpose of automatically indicating or recording measurements at a distance from the measuring instrument," according to the FCC.

Geophysical telemetry uses this band in oil and gas exploration, where acoustic sensors periodically transmit data to collection units. Another use is seismic telemetry for earthquake detection, conducted in this band by the U.S. Geological Survey in Alaska, California, and Hawaii.

Wildlife telemetry, ocean buoy telemetry, and telemetry in the Industrial/Business frequency pool of the *Private Mobile Radio Services* (PMRS) are also permitted in 216–220 MHz (see Footnote US210). An example of the latter application is telemetry in vehicle performance testing.

Air Force hazardous material emergency communications and radar missile cross-section measurement are performed in this band.

Interactive Video and Data

The 218–219 MHz Service is the new name for the former *Interactive Video and Data Service* (IVDS), allocated in Footnote US317. "Initially designated as IVDS, the service was redesignated the 218–219 MHz Service to reflect the breadth of services available in this spectrum," the FCC said.

Although the FCC classifies the 218–219 MHz Service as a Personal Radio Service (like Citizens Band), it actually is a commercial point-to-multipoint and multipoint-to-point fixed service. The FCC now also permits mobile use of this service.

This service originated with EON Corp. (formerly TV Answer), which successfully petitioned the FCC to allocate spectrum to IVDS. EON created IVDS as an interactive service for home shopping.

Its dramatic rollout of IVDS, in Washington, D.C., was marred by equipment malfunction. When the system did work, it displayed e-mail—only from other TV Answer subscribers—and elaborate on-screen graphics that obscured TV shows. It offered the dubious ability to rate jokes for "hilarity."

"Depending on what the licensee chooses to offer, IVDS could combine the functions of computers, TV sets and compact-disk players," the Commission proclaimed. "Subscribers will be able to choose the camera angle during a sporting event, pay bills, shop-until-they-drop at malls, choose endings to TV shows, check college catalogs, play video games, choose movies on demand, or order a pizza with or without toppings."

Despite those FCC claims—direct from EON marketing literature—the IVDS market simply failed to emerge. "The potential applications for the 218–219 MHz Service go far beyond the service envisioned by TV Answer when these rules were designed," the FCC said, years later.

The service earned the ire of the Federal Trade Commission. "Scam artists raise funds from investors to bid for licenses and develop IVDS systems, touting services IVDS can't perform, such as 'movies on demand.' These scammers leave consumers little capital to develop an IVDS system or complete the payments to the FCC for the licenses they were to receive," the FTC said. The agency won court orders against several IVDS investor groups.

The practice of including due diligence warnings in FCC auction announcements originated with IVDS. The standard warning reads: "The Commission makes no representations or warranties about the use of this spectrum for particular services. An FCC auction represents an opportunity to become an FCC licensee in this service, subject to certain conditions and regulations, and does not constitute an endorsement by the FCC of any particular services, technologies or products, nor does an FCC license constitute a guarantee of business success. Applicants for an auction of FCC licenses should perform their individual due diligence before proceeding as they would with any new business venture."

Licensing and Structure

The FCC grants two 218–219 MHz Service licenses at 500 kHz each (license A, 218.0–218.5 MHz and B, 218.5–219.0 MHz), for each of 734 service areas consisting of 306 *Metropolitan Statistical Areas* (MSAs) and 428 *Rural Service Areas* (RSAs).

A 1993 lottery awarded two licenses in each of nine MSAs: New York City, Los Angeles, Chicago, Philadelphia, Boston, San Francisco, Washington D.C., Dallas, and Houston. After receiving authority from Congress, the FCC auctioned licenses in the remaining 297 MSAs in July 1994, for more than $213 million, but had not auctioned the RSA licenses at this writing.

The FCC tinkered with 218–219 MHz Service auction schedules and licensee payment obligations several times (Docket WT 98-169). It offered licensees several financing options, including amnesty from payments upon return of the licenses. Ownership of both of the available channel blocks in a market is now permitted.

The service involves a user device, the *Response Transmitter Unit* (RTU), which forwards the user's identity and data payload to the nearest base station.

At EON's request, the FCC decided to allow mobile RTUs. "This new [mobile] technology will allow IVDS licensees to leverage existing platforms such as the new generation of portable game players, portable televisions and personal organizers in order to provide customers and program providers greater options to customize and better organize the data that they transmit over the IVDS system," EON said.

"At price points similar to basic local paging services, parents will be able to confirm their child's safety or send an electronic 'come home now,'" EON added. "Broadcasters will be able to provide sports updates to people on the road and let them know to tune in at 7 PM when the evening game begins." (Like the earlier wireless pizza-ordering claims, these predictions did not come true.)

An EON patent (5,388,101) states that its RTUs may be "put into a car or van for movement within or across cell boundaries with good digital synchronous communication contact within the nationwide network of cells."

We believed that the broad claims in the EON patent appear to describe any digital cellular mobile communication system. But EON maintained that others may use technology that complies with FCC rules without infringing the patent.

Initially rejecting the idea of mobile use, the FCC said that "The primary use of the IVDS system must be to provide subscribers at fixed locations with the capability to interact with video, data, or other service providers. The offering to subscribers of mobile service only, such as paging or dispatch services, would not be permitted."

In what is now known as the "Mobility Report and Order," the FCC reversed itself after receiving a more substantive proposal from EON. "Commenters favoring mobility have stated that IVDS spectrum will be useful for TV-based consumer messaging, as well as for messaging in the business and commercial data context, with applications for meter reading, stock transactions or quotations on demand, and automatic teller machine and credit card verifications. ...

"To the extent that there is public demand for mobile IVDS, our action here [allowing mobile IVDS] will permit licensees to respond to this demand. This flexibility should also encourage a broader range of service offerings."

Separately, however, the Commission said that the "218–219 MHz band is insufficient for the transmission of conventional full-motion video. Although the new rules are designed to allow licensees the maximum flexibility to structure services to meet market demand, 218–219 MHz Service channels may be unable to support proposed operations that require large amounts of spectrum, including certain video, voice and advanced data applications."

To protect against interference to TV channel 13, the FCC strictly limits the duration of transmission from each RTU in areas served by TV stations on that channel.

More Changes

A September 1999 FCC proceeding made major changes to these and other restrictions in order to rescue the moribund service. Among the changes, the FCC liberalized construction benchmarks, allowed licensees to partition, and/or disaggregate to others, portions of their licensed service areas and spectrum, and allowed the 218–219 MHz Service to be offered as either a *Commercial Mobile Radio Service* (CMRS) or *Private Mobile Radio Service* (PMRS). The FCC also allowed the same entity to own or control both licenses in a market.

As a CMRS carrier, a 218–219 MHz Service licensee could interconnect to the public switched network and provide cellular or *Personal Communications Services* (PCS) or paging. Even RTU-to-RTU communications are allowed; but to our knowledge, no one has proposed a business model for this interesting capability.

After losing hundreds of millions of dollars in interactive TV, EON Corp. eventually turned its attention to the remote monitoring of soft-drink vending machines. And at this writing, U.S. Telemetry Corp. had acquired access to a substantial amount of 218–219 MHz Service spectrum and was developing fixed and mobile telemetry applications ("T-commerce") for this service.

Amateur Use

The Amateur Radio Service was allocated 219–220 MHz on a secondary basis for digital network links. The *American Radio Relay League* (ARRL) requested the allocation as compensation for the loss of amateur access to 220–222 MHz.

Amateur use in 219–220 MHz is limited to directional fixed stations so as not to interfere with AMTS. Operations must be coordinated with AMTS licensees.

High data rate links in this band allow amateurs to send audio and video data as well as text messages and software. Mobile and multipoint operations are not allowed in 219–220 MHz. The amateur allocation resumes at 222–225 MHz.

At this writing, the FCC requested comment (in Docket PR 92-257) about its tentative conclusion that it should retain the secondary 219–220 MHz allocation to Amateur Radio.

Beacon Bucks

This band also is home to another valuable application: tracking of beacons hidden in stolen currency and merchandise. The sole provider of this service is ProNet Inc., a company acquired by Metrocall.

ProNet's electronic tracking subsidiary ETS, acquired from Texas Instruments in 1988, operated for more than 20 years under experimental authority—apparently one of the longest running experiments the FCC ever licensed.

Banks and retailers hide the ETS "TracPac" transmitter in paper money or other valuables. "Once the TracPacs are activated by a crime, highly sensitive directional receivers, which are located in law enforcement vehicles, helicopters, or at a few fixed sites, track and locate the crime while it is still in progress," the company said. "Law enforcement directly monitors for those signals in its own dispatch center, and when triggered, previously deployed trackers scramble to locate the unlawful incident, protect the public, and apprehend the felon. On average, all this occurs in less than 15 minutes."

ETS has an impressive record of arrests and convictions in armed robberies, theft of automatic teller machines, and other serious crimes. Tens of thousands of TracPacs are in service, in more than 100 cities.

The FCC eventually granted ETS a spectrum home in the *Low Power Radio Service* (LPRS) in 216–217 MHz. The LPRS is legally one of the Citizens Band Radio Services in Part 95 of the FCC rules. LPRS radios are not considered to be unlicensed devices, but they are "authorized by rule" (no license documents are issued).

Notwithstanding the establishment of the LPRS, ProNet said, "retrofitting the many existing ETS systems, which now operate on its experimental frequency, to accommodate the new LPRS frequencies will entail enormous capital expenditure that cannot now be cost justified. ...Accordingly, ProNet will remain dependent on its experimental license for the foreseeable future—i.e., until such time as a substantial increase in market demand or technological innovation compels retrofitting existing equipment, at which time LPRS frequencies can be introduced into existing systems."

Health Care

Other LPRS channels are devoted to auditory assistance devices for the hearing impaired and a broad category of "health care assistance devices." Auditory assistance devices consist of a short range transmitter and headset receivers for use in classrooms, churches, or theaters by persons with hearing disabilities (see 72–73 MHz). Interference from other uses of the 72 MHz spectrum prompted manufacturers to request spectrum in the 216–217 MHz band for these products.

"Unfortunately, there is no way to predict what explosions of technology may occur and lead to unexpected interference sources in this band, in a manner that is similar to what occurred in the 72–76 MHz band with the advent of hundreds of paging companies," according to the National Association of the Deaf.

The FCC is similarly concerned. "The auditory assistance devices, as well as currency tracking devices authorized under LPRS, provide valuable services to the public," the Commission said when it proposed reallocation of this band. "There are over 23 million Americans that are hearing impaired. The assisted listening devices, which operate in this band, are being increasingly relied upon by persons with special auditory needs, particularly in schools by children.

"LPRS users are licensed by rule, and thus we maintain no records on the number or location of users, which would complicate coordination with new service providers. Ocean buoy and wildlife tracking, as well as being valuable services, complicate the use of the 216–220 MHz band because of their itinerant nature. Finally, we note that telemetry is heavily used in both urban and rural areas."

The FCC also allows AMTS coast stations to use the LPRS for network control.

At this writing, Docket ET 00-221 proposes to modify the allocations in this band as follows. In the *Federal Government* (FG) allocation, only Footnote US229 would remain. All other Footnotes and entries would be removed. In the Non-FG allocation, Footnotes US229 and NG152 would remain; and the band would be allocated only to FIXED and MOBILE except aeronautical mobile services.

220-222 MHz

FG: US335 FIXED. LAND MOBILE. Radiolocation S5.241 G2

Non-FG: US335 FIXED. LAND MOBILE

FCC: Private Land Mobile (90)

In 1988, the FCC reallocated 220–222 MHz to the Private Mobile Radio Services from the 220–225 MHz band used by Amateur Radio. The principal requester was *United Parcel Service* (UPS) which needed radios for its vehicles.

The FCC received tens of thousands of 220 MHz license applications in 1991. After Congressional hearings in to and years of litigation over the 220 MHz allocation and licensing process, the Commission awarded some licenses by lottery in 1992 and 1993 in Phase I of its plan for the band.

Some licenses were devoted to local commercial use and others to nationwide coverage. Fifteen channels were made available to public safety eligibles, including emergency medical service providers. Some of these public safety channels are shared and others are assigned on an exclusive basis to licensees for particular areas. Some segments are available to government users as Footnote US335 provides.

The band gradually began to develop new narrowband voice and data services. These services essentially provide dispatch of vehicle fleets, and business and maintenance communications.

Through management agreements with license holders, several firms built 220–222 MHz regional and nationwide networks to serve those markets. Relatively few manufacturers offer equipment for this band, however, constraining development.

Phase II of the 220–222 MHz band plan, finalized in March 1997, concluded that mutually exclusive commercial applications for licenses must be auctioned. The FCC believes that this band will mostly be used to provide *Commercial Mobile Radio Service* (CMRS)—that is, telephone-interconnected mobile service to the public—and thus will essentially be similar to other forms of CMRS such as cellular, *Personal Communications Services* (PCS, see 1850–1990 MHz) and *Specialized Mobile Radio* (SMR, see 806–821 MHz).

The FCC considers some 220 MHz operations to be SMR even though not interconnected to the telephone network. Other such operations are interconnected, but instead of being called SMR, they apparently were dubbed the "220 MHz Band and Interconnected Business Radio Service."

The FCC established three types of Phase II 220 MHz licenses. There are three nationwide licenses of ten channels each and five 15-channel licenses in each of six *Regional Economic Area Groupings* (REAGs), for a total of 30 REAG licenses, and five 10-channel licenses in each of 175 *Economic Areas* (EAs) for a total of 875 EA licenses.

The 5-kHz channels in this band occupy less bandwidth than conventional mobile radio channels. Licensees may aggregate channels to take advantage of non-narrowband technologies such as *Time Division Multiple Access* (TDMA). Phase II licensees may partition or assign portions of their operating area to others.

Licensees also may offer fixed and paging services in addition to, or instead of, mobile services. Each type of 220 MHz licensee is subject to construction requirements or "benchmarks" to ensure that licensed stations are actually built and placed into service.

However, in light of its liberal benchmark policy in the Wireless Communications Service (see 2305–2310 MHz) where the FCC found that "construction requirements in some cases may be unnecessary, ineffective, and potentially harmful," the agency said that it "may choose to reassess the nature of construction requirements in the 220 MHz band at some time in the future."

A September 1998 220 MHz Phase II license auction raised approximately $21.6 million. A June 1999 second Phase II auction raised more than $1.9 million, for licenses covering areas "with small populations that were overlooked in the last auction," according to former FCC chairman William Kennard.

Some frequencies in this band are administered by the Federal Highway Administration for experiments in *Intelligent Transportation Systems* (ITS). National hailing and mayday frequencies exist, as well as frequencies for vehicle traffic probes and general two-way communications.

PELTS Rejected

But for an acrimonious defeat, the *Personal Emergency Locator Transmitter Service* (PELTS), proposed by a cellular engineer, would have operated in this 220 MHz band. PELTS was a *Personal Locator Beacon* (PLB) rescue radio system for use in remote areas.

The FCC saw PELTS as an alternative to the illegal operation on land of emergency radio beacons intended for ships and aircraft (see 406–406.1 MHz). PELTS would have had channels for user contact with watch centers and for one-way broadcasts, alerting and homing, and short-distance voice.

As is often the case with novel spectrum proposals, worried stakeholders attacked PELTS. Emergency rescue groups argued that PELTS would expose them to lawsuits in responding, or failing to respond, to emergency calls.

The FCC's PELTS licensing scheme required consumers to rent the radios from search and rescue teams. Consumers could not own the radios.

Radio amateurs warned that PELTS would become another Citizens Band service, filled with chaotic traffic and false alarms. Instead of a life-saving technology, most commenters painted PELTS as a complex bureaucratic mistake.

"Based on the comments it does not appear that a watch and response system so critical to the success of PELTS will materialize," the FCC conceded. "Further, this lack of support to promote and develop the technology for PELTS would result in uneven usage and a lack of uniformity of emergency communications service. ...The unsettled nature of the liability question and the legal trend of pursuing large damage suits would likely preclude knowledgeable search and rescue individuals and organizations from participating in PELTS."

The FCC's rejection of PELTS appeared to leave Mt. Hood, Oregon's Mountain Signal system in a kind of regulatory limbo. Its users rent *Mt. Hood Locator Units* (MLUs) from outdoor outfitters. In case of emergency, they can use these transmitters to signal to receivers around the area.

MLUs are manufactured for, monitored by, and licensed experimentally to a not-for-profit corporation. Medical, legal, and wilderness experts established the corporation after mountaineering accidents occurred. State legislation gave the corporation legal immunity.

The FCC never issued Mountain Signal permanent rules or regular license and it is not legal to use anywhere else.

At this writing, the FCC is considering whether and how to permit general public use of PLBs in 406–406.1 MHz.

222–225 MHz

FG: Radiolocation S5.241 G2

Non-FG: AMATEUR

FCC: Amateur (97)

The Amateur Radio 1.25-meter band occupies 222–225 MHz. It originally extended from 220 to 225 MHz, but the FCC reallocated the lower 2 MHz to create the commercial mobile allocation in 220–222 MHz.

This band is planned in detail to accommodate a wide variety of amateur interests. Within its subbands are frequencies for weak signal modes, FM voice direct and repeater operation, Internet protocol digital communications, wide-area linking, and other applications. The amateur band at 219–220 MHz is limited to digital fixed links (see 216–220 MHz).

Signal propagation in this spectrum is sometimes affected by tropospheric ducting.

On June 1, 1945, the FCC authorized the first FM portable transceiver, operating on 224 MHz. This invention by Al Gross evolved into today's cellular, personal communications, and personal radio devices (see 35–36 MHz and 462.5375–462.7375 MHz).

225-235 MHz

FG: G27 FIXED. MOBILE

Non-FG: None

The 225–400 MHz spectrum is devoted to military aircraft, tactical and training communications, satellite links for ground, air, surface and subsurface users, rocket test and telemetry, position location networks, and Presidential communications. Other nations use portions of it for non-military purposes (see 335.4–399.9 MHz).

The FCC called on the government to offer some of this spectrum to the private sector; but this has not occurred at this writing. "Spectrum reallocated in the 225–400 MHz band could greatly facilitate our efforts by providing 'green space' in which to begin implementing spectrum-efficient systems," the FCC said. "This spectrum could also provide greater flexibility for accommodating applications that require wider bandwidth, such as transmission of data or images."

The Defense Department takes a dim view of such assertions. It regards the 225–400 MHz band as the single most critical spectrum resource of the military tactical forces, both nationally and within the *North Atlantic Treaty Organization* (NATO). We don't believe that military authorities consider reallocation of any portion of this spectrum to be a practical option.

235-267 MHz

FG: S5.111 S5.199 S5.256 G27 G100 FIXED. MOBILE

Non-FG: S5.111 S5.199 S5.256

This is part of the 225–400 MHz military spectrum (see 225–235 MHz).

NASA uses frequencies in this part of the spectrum for Space Shuttle spacesuit-orbiter, orbiter-ground, and emergency crew return communications.

Downlinking in 235–322 MHz is FLTSATCOM, the Navy's Fleet Satellite Communications System connecting aircraft, ships, submarines, and command elements (see 7.9–8.025 GHz). The Air Force AFSATCOM system and the Presidential Command Network have channels on these satellites.

The Navy is replacing FLTSATCOM and other satellites with the $1.8 billion *UHF Follow-On* (UFO) series. These nuclear-hardened satellites use the same spectrum but provide more channels.

Some of the UFOs contain packages for the *Global Broadcast System* (GBS), the Pentagon's version of *Direct Broadcast Satellite* (DBS) television for military users.

Each UFO GBS package includes transponders for spot (localized) and broad-area coverage beams, serving small receiving dishes. Field users can request video, still images, or other information and receive it via GBS in real time. Typical uses of GBS include distribution of tasking orders, targeting data, and imagery received by unmanned aerial vehicles.

The newer UFO satellites include *Extremely High Frequency* (EHF) communications packages compatible with Milstar ground terminals. Milstar is the Pentagon's UHF/SHF/EHF satellite program. It has UHF uplinks and downlinks for compatibility with older systems (see 20.2–21.2 GHz).

Emergency Beacons

The signaling devices known as *Emergency Position Indicating Radio Beacons* (EPIRBs) are carried on ships, and *Emergency Locator Transmitters* (ELTs) are carried on aircraft. Several types of EPIRBs exist, with different characteristics such as manual activation, automatic activation, survival craft use, and coastal use.

These devices transmit distress signals on 121.5 MHz and 243 MHz and the more modern 406.025 MHz in the international COSPAS/SARSAT search and rescue satellite system. U.S. satellites in this system include the GOES and POES weather satellites (see 1670–1675 MHz).

The spectrum 240–285 MHz is one of many restricted bands in which the FCC Part 15 rules permit unlicensed devices to emit only very low level emissions.

Frequency Allocations

UHF

267-322 MHz

FG: G27 G100 FIXED. MOBILE

Non-FG: None

This is part of the 225–400-MHz military spectrum (see 225–235 MHz). Car alarm and remote control manufacturers often choose this band because this—spectrum otherwise contains only federal users. Commonly used frequencies for these devices are 303.8 MHz, 315 MHz, and 318 MHz.

This approach is not foolproof. High-power transmissions in this spectrum from presidential and military aircraft are thought to be responsible for interference that causes openings of home garage doors in the flight path.

The spectrum 240–285 MHz is one of many restricted bands in which the FCC Part 15 rules permit unlicensed devices to emit only very low level emissions.

Figure 6
The *Very Large Array* (VLA), operated by the National Radio Astronomy Observatory in Socorro, New Mexico, consists of 27 antennas electrically combined into one enormous antenna. VLA observations span 300 MHz–50 GHz. The crane moved this 81-foot (25 m) diameter antenna into a maintenance building. (Earthspan photo)

322-328.6 MHz

FG: S5.149 G27 FIXED. MOBILE

Non-FG: S5.149

This is part of the 225–400 MHz military spectrum (see 225–235 MHz).

The 322–328.6 MHz band, although not allocated to passive (receive-only) services in the U.S., is widely used by radio astronomers for pulsar observations. The spinning pulsars remain after stars have expended all of their fuel and collapsed into dense spheres a few miles in diameter.

Pulsars emit rotating beams of radio energy. The extreme regularity of pulsar emissions appeared so unusual when discovered in 1967 that scientists suspected that they were transmissions from intelligent beings.

The study of pulsars is stimulating interest again in the extraterrestrial hypothesis. A January 2000 American Astronomical Society paper by Paul LaViolette, Ph.D. concludes that "Pulsar sky positions are nonrandomly distributed in a pattern that is not easily attributed to natural causes." The author suggests that pulsars could comprise an intelligently designed beacon network.

The 322–335.4 MHz spectrum is one of many restricted bands in which the FCC Part 15 rules permit unlicensed devices to emit only very low level emissions.

328.6-335.4 MHz

FG: S5.258 AERONAUTICAL RADIONAVIGATION

Non-FG: S5.258 AERONAUTICAL RADIONAVIGATION

This band is devoted to the aircraft *Instrument Landing System* (ILS) worldwide, which explains its identical allocation to federal and non-federal use (see 108–117.975 MHz).

The 322–335.4 MHz spectrum is one of many restricted bands in which the FCC Part 15 rules permit unlicensed devices to emit only very low level emissions.

335.4-399.9 MHz

FG: G27 G100 FIXED. MOBILE

Non-FG: None

This is part of the 225–400 MHz military spectrum (see 225–235 MHz).

The FCC has identified this band as a possible home for future Public Safety Radio Services. An agreement between the *North Atlantic Treaty Organization* (NATO) and European government bodies provides for European use of this band for public safety services.

"This low UHF segment, therefore, appears to be a good candidate for reallocation for public safety uses," the FCC once observed. "Its proximity to the 74 channels in the 450–470 MHz band that are currently available for public safety communications may make this segment ideal for public safety voice, data and facsimile channels." To date, the FCC has had no luck in promoting the idea of reallocation of this spectrum for Non-FG use.

A variety of illegal radiotelephone operations are active in this band in major cities. They probably choose this band because it does not normally carry civilian traffic and is therefore quiet.

399.9-400.05 MHz

FG: S5.260 MOBILE-SATELLITE (Earth-to-space) US319 US322. RADIONAVIGATION-SATELLITE

Non-FG: S5.260 MOBILE-SATELLITE (Earth-to-space) US319 US322. RADIONAVIGATION-SATELLITE

The *Non-Voice Non-Geostationary Mobile Satellite Service* (NVNG MSS) UHF uplink is allocated this band (see 137–137.025 MHz). However, the NVNG MSS companies do not appear to have expressed interest in using this band, according to the FCC.

The 399.9–410 MHz spectrum is one of many restricted bands in which the FCC Part 15 rules permit unlicensed devices to emit only very low level emissions.

400.05-400.15 MHz

FG: S5.261 STANDARD FREQUENCY AND TIME SIGNAL-SATELLITE (400.1 MHz)

Non-FG: S5.261 STANDARD FREQUENCY AND TIME SIGNAL-SATELLITE (400.1 MHz)

The frequency 400.1 MHz was allocated for satellites that would replace some high-frequency (HF) time and frequency services. The U.S. does not operate satellites in this band.

The 399.9–410 MHz spectrum is one of many restricted bands in which the FCC Part 15 rules permit unlicensed devices to emit only very low level emissions.

400.15-401 MHz

FG: 647B US70 METEOROLOGICAL AIDS (radiosonde). METEOROLOGICAL-SATELLITE (space-to-Earth). MOBILE-SATELLITE (space-to-Earth) 599B US319 US320 US324 SPACE RESEARCH (space-to-Earth) S5.263 Space Operation (space-to-Earth)

Non-FG: 647B US70 METEOROLOGICAL AIDS (radiosonde). MOBILE-SATELLITE (space-to-Earth) 599B US319 US320 US324 SPACE RESEARCH (space-to-Earth) S5.263 Space Operation (space-to-Earth)

FCC: Satellite Communications (25)

In 1993, the FCC allocated this band to Little LEO (Low Earth Orbit) Non-Voice Non-Geostationary Mobile Satellites (NVNG MSS, see 137–137.025 MHz).

The 400.15–406 MHz meteorological aids (Met Aids) spectrum is used for weather and water data collection by governments and researchers. The *Defense Meteorological Satellite Program* (DMSP) also uses this band (see 1675–1700 MHz).

NVNG MSS licensees that share this spectrum are required to avoid interference to DMSP or risk serious penalties. This avoidance will require the Department of Defense to share ephemeris (satellite orbital data) information with the licensees and to alert them to sudden shifts of DMSP operations within the band.

The 399.9–410 MHz spectrum is one of many restricted bands in which the FCC Part 15 rules permit unlicensed devices to emit only very low level emissions.

401–402 MHz

FG: US70 METEOROLOGICAL AIDS (radiosonde). SPACE OPERATION (space-to-Earth). Earth Exploration-Satellite (Earth-to-space). Meteorological-Satellite (Earth-to-space)

Non-FG: US70 METEOROLOGICAL AIDS (radiosonde). SPACE OPERATION (space-to-Earth). Earth Exploration-Satellite (Earth-to-space). Meteorological-Satellite (Earth-to-space)

FCC: Satellite Communications (25)

Radiosonde balloons, rocketsondes, and other so-called data platforms transmit in the 401–406 MHz Meteorological Aids spectrum to the NOAA *Geostationary Operational Environmental Satellites* (GOES) and *Polar-orbiting Operational Environmental Satellites* (POES) (see 1670–1675 MHz and 1675–1700 MHz).

Radiosondes worldwide are launched at the same time to provide *soundings*, data that contributes to global reports of air movement, pressure, and temperature. Computer models use this information to produce weather forecasts.

Akin to radiosondes are dropsondes, instrument packages with *Global Positioning System* (GPS) receivers or other tracking receivers that are parachuted from aircraft. Dropsondes measure temperature and humidity, and transmit the data to the aircraft.

Data platforms collect meteorological data from sensors in remote locations. They measure stream height, snow depth, and temperatures in mountain areas. This data is used to predict droughts and floods. Other sensors on ships and buoys measure ocean temperature and ice floe velocity and direction.

Animals and Adventurers

Included among these scientific uses are wildlife telemetry, the monitoring of unstable sites such as volcanoes, and emergency beacons for expeditions. These platform applications use Service ARGOS, operated by the French/U.S. CLS-Argos Inc. organization at 401.65 MHz. It enables the location of a *Platform Transmitter Terminal* (PTT) anywhere in the world to within 150–1000 meters.

The ARGOS data collection system is on board the NOAA POES satellites. It receives the incoming signal, measures its frequency and time characteristics, and retransmits this data over the POES VHF beacons to collection networks accessible to subscribing scientists.

Conservation biologists and engineers have developed ingenious PTTs that attach to birds and fish. The "bird backpack" is an aerodynamic PTT package.

Some models incorporate tiny GPS receivers to enhance location accuracy. Fish can be equipped with PTTs that release from the fish at a preprogrammed time by electrically corroding a metal linkage in the seawater. The device then floats to the surface and uploads recorded data to the satellite.

Solo adventures across great distances typically include PTTs for tracking and distress signaling. Celebrated ARGOS users include the Swiss expeditioner Mike Horn who made a nonmotorized solo trek around the Equator in 2000; and Tori Murden who rowed across the Atlantic in 1999.

CLS also operates the DORIS system at 401.25 MHz and 2036.25 MHz, providing precise orbital and geodetic measurements. At this writing, some 50 continuously operating DORIS orbitography beacons are available around the world, with portable beacons used at specific sites of interest.

Satellites with DORIS receivers use the Doppler effect to compute beacon coordinates with centimeter accuracy for applications such as the precise location of construction and engineering sites, determination of sea level, and measurement of satellite orbits.

The 401–406 MHz spectrum is a candidate for expansion of the Little LEO Non-Voice Non-Geostationary Mobile Satellite Service (NVNG MSS, see 137–137.025 MHz). The U.S. advocates the sharing of some of this spectrum between Met Aids and NVNG MSS uses. Continued and well-supported meteorological use of these bands complicates the matter.

The 399.9–410 MHz spectrum is one of many restricted bands in which the FCC Part 15 rules permit unlicensed devices to emit only very low level emissions.

402–403 MHz

FG: US345 METEOROLOGICAL AIDS (radiosonde) US70. Earth Exploration-Satellite (Earth-to-space). Meteorological-Satellite (Earth-to-space)

Non-FG: US345 METEOROLOGICAL AIDS (radiosonde) US70. Earth Exploration-Satellite (Earth-to-space). Meteorological-Satellite (Earth-to-space)

FCC: Personal Radio (95)

See 400.15–401 MHz and 401–402 MHz for uses of this Meteorological Aids band.

At the request of Medtronic, a medical device manufacturer, the FCC in Docket WT 99-66 allocated 402–405 MHz to a *Medical Implant Communications Service* (MICS), "an ultra low power radio service for the transmission of non-voice data for the purpose of facilitating diagnostic and/or therapeutic functions involving implanted medical devices."

Devices such as pacemakers implanted in the body use MICS to communicate a short distance to external programming and monitoring equipment. This is an improvement over conventional inductive pickups, which cannot support high data rates and are interrupted when the patient or device moves.

The FCC was persuaded to select 402–405 MHz because of favorable signal propagation through the human body, constraints on size and power consumption of implants, issues of noise in the radio environment, and international frequency compatibility. MICS devices are not allowed to interfere with the scientific services allocated in this spectrum.

No MICS license documents are issued, but MICS is not considered an unlicensed device. Instead, MICS is "authorized by rule" as one of the Citizens Band Radio Services, regulated under Subpart I of Part 95 of the FCC Rules.

The CB Radio Services are now a catchall category for services that do not have license documents—an exception to the Communications Act's directive that no station shall transmit without a license (see also 462.5375–462.7375 MHz). Congress granted the loophole after the Commission was deluged with CB license applications in the 1970s.

Only "duly authorized health care professionals" who do not represent foreign governments may operate MICS transmitters. MICS should not be confused with health care assistance devices permitted in another CB service, the little-known *Low Power Radio Service* (LPRS, see 216–220 MHz).

The 399.9–410 MHz spectrum is one of many restricted bands in which the FCC Part 15 rules permit unlicensed devices to emit only very low level emissions.

403–406 MHz

FG: US345 G6 METEOROLOGICAL AIDS (radiosonde) US70

Non-FG: US345 METEOROLOGICAL AIDS (radiosonde) US70

FCC: Personal Radio (95)

See 400.15–401 MHz, 401–402 MHz and 402–403 MHz for uses of this Met Aids band devoted to radiosondes.

The 403–405 MHz segment is available to the *Medical Implant Communications Service* or MICS (Subpart I) within the CB Radio Service, one of the Personal Radio Services in Part 95 (see 402–403 MHz).

International negotiations may eventually phase out meteorological aids from 405–406 MHz and make that segment available for Little LEO satellites in the *Non-Voice Non-Geostationary Mobile Satellite Service* (NVNG MSS, see 137–137.025 MHz).

After 2002, more spectrum-efficient radiosondes are supposed to work in the more limited bandwidth available at 401–405 MHz.

The 399.9–410 MHz spectrum is one of many restricted bands in which the FCC Part 15 rules permit unlicensed devices to emit only very low level emissions.

406–406.1 MHz

FG: S5.266 S5.267 MOBILE-SATELLITE (Earth-to-space)

Non-FG: S5.266 S5.267 MOBILE-SATELLITE (Earth-to-space)

Emergency Position Indicating Radio Beacons (EPIRBs) on ships and *Emergency Locator Transmitters* (ELTs) on aircraft transmit signals on 406.025 MHz to locate and identify craft and persons in distress. This band is protected for this purpose throughout the world. Newer models will use 406.028 MHz.

COSPAS/SARSAT-equipped satellites monitor EPIRB and ELT signals, and alert control centers to launch search and rescue missions. COSPAS is an acronym for a Russian phrase that means *Space System for Search of Vessels in Distress*. SARSAT stands for *Search and Rescue Satellite-Aided Tracking*. Twenty-seven nations fund the London-based COSPAS/SARSAT program.

Older EPIRBs and ELTs operate on 121.5 MHz and 243 MHz. Transmissions on the older frequencies provide distress alerting and homing assistance but not identification.

Some 406.025 MHz EPIRBs activate when removed from a bracket; others activate when they contact water. The beacon then floats free of the vessel and contacts the satellite. All 121.5 MHz EPIRBs are manual activation units.

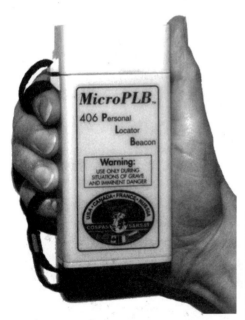

Figure 7
The MicroPLB 406 Personal Locator Beacon is a COSPAS/SARSAT transmitter, which, when activated, emits a 406 MHz distress signal to an internationally funded and operated satellite system. (Microwave Monolithics photo)

False Alarms

For aircraft, ELTs on 121.5 MHz have proven to be highly ineffective and "are a large source of wasted effort by search and rescue forces," according to U.S. SARSAT authorities. The units reportedly have a 98 percent false alarm rate and activate properly in only 12 percent of air crashes.

The 121.5 MHz signal can only be detected and relayed when a satellite is in range of the EPIRB or ELT and a ground station called a *Local User*

Terminal (LUT). NOAA operates LUTs in Alaska, California, Guam, Hawaii, Maryland, Puerto Rico, and Texas.

The 406.025 MHz units transmit identification data that must be registered with the *National Oceanic and Atmospheric Administration* (NOAA). It maintains a database of identifications and provides the database to rescue authorities.

Newer, advanced emergency beacons transmit their locations derived from the Global Positioning System (GPS, see 1215–1240 MHz).

The 406.025 MHz signal is immediately processed on board the satellite, stored, and transmitted to the ground on 1544.5 MHz from a continuous memory dump. Ground stations acquire the data and beacon identification, and forward the alerts to control stations and rescue centers.

COSPAS/SARSAT has warned all users that satellite reception of beacons on the older 121.5 MHz and 243 MHz frequencies will be withdrawn, with 406–406.1 MHz as the main focus of international satellite-based SAR. The 121.5 MHz signal will still be useful for close-in direction-finding and is available as an option on some beacon models.

U.S. satellites in the system include the GOES and POES weather satellites (1670–1675 MHz and 1675–1700 MHz). NOAA's SARSAT Operations Division maintains the U.S. control center for the system in Suitland, Maryland.

A related EPIRB system is *International Maritime Satellite Organization* (INMARSAT) E (see 1645.5–1646.5 MHz). INMARSAT E and COSPAS/SARSAT are part of the *Global Maritime Distress and Safety System* (GMDSS), which incorporates both terrestrial and space networks. GMDSS regulations mandate the carriage of satellite emergency beacons on various types of vessels. Russia also operates satellites with COSPAS/SARSAT payloads, and several additional nations will launch satellites with this capability.

Land Personal Use

COSPAS/SARSAT *personal locating beacons* (PLBs) are carried on the person instead of on board a ship or aircraft.

PLBs are only in limited use, mostly for government employees and experiments. Many rescues have been attributed to PLBs, but the general public is not officially offered access to these devices.

Misuse of EPIRBs and ELTs on land prompted the FCC to propose a new type of PLB in the *Personal Emergency Locator Transmitter Service* (PELTS). Rescue groups denounced PELTS and refused to participate, killing the proposal (see 220–222 MHz).

At this writing, the FCC was considering how to authorize 406.025 MHz PLBs "to satisfy the individual distress alerting needs of the general public" (Docket WT 99-366). Officially, the PLBs would be considered Citizens Band units in this burgeoning catchall category (see 462.5375–462.7375 MHz).

The CB PLBs would be required to contain a 121.5 MHz homing beacon. CB devices are not unlicensed, but are "licensed by rule" under special statutory exemption. The FCC said, "Because of the proposed broad eligibility and operational provisions for PLBs, we recognize that there are millions of potential users. We believe that individually licensing each one would be unnecessarily burdensome on the Commission without concomitant public interest benefit. Notably, on October 18, 1996, the Commission decided to license EPIRBs and ELTs by rule, which eliminated individual licenses. We note, however, that the current ELT/EPIRB system has been designed specifically to handle aircraft and ships in distress rather than to accommodate the general public. We are concerned that the addition of a large number of users, especially users unfamiliar with the use of radio, could hamper the present system."

The Commission tentatively proposed that PLBs be administered by the states, and asked whether the states themselves, or possibly state officials, could receive the FCC licenses and then (in effect) sublicense individual PLB users.

Pandora's Box

"The codes will not help the folks out there rescuing," said the Colorado Search and Rescue Board. "Finding the beacons is our problem and we are already faced with a 97 % false reporting network of Emergency Locator Beacons and do not wish to see more. As the state organization faced with tracking these down, we feel there is significant cost involved with these items being used in the continental United States and no one, including the FCC, has offered to pay any of these costs. ...We strongly ask the FCC to table this plan and remove any further considerations from these dangerous and ineffective tools."

"Certain agencies that would logically assume certain SAR obligations have clearly stated their concern about the potential misuse of expanded U.S. domestic PLB operations," according to the Personal Radio Steering

Group. "These agencies question how circuit discipline could be maintained in what they might euphemistically want to call a *Personal Alerting Network Deployed Only in Remote Areas* (PANDORA)."

"The Commission expects that there are millions of potential users of PLBs," observed the National Oceanic and Atmospheric Administration. "NOAA believes that the PLB market may be limited by the widespread use of cellular phones and the implementation of world-wide mobile satellite communication systems. PLBs should only be used as a last resort. In that capacity their usefulness as a general communication tool is limited."

Another type of portable emergency beacon, not generally known to the public, is part of Service ARGOS (see 401–402 MHz).

The 399.9–410 MHz spectrum is one of many restricted bands in which the FCC Part 15 rules permit unlicensed devices to emit only very low level emissions.

406.1–410 MHz

FG: US13 US117 G5 G6 FIXED. MOBILE. RADIO ASTRONOMY US74

Non-FG: US13 US117 RADIO ASTRONOMY US74

The 406.1–410 MHz and 410–420 MHz bands are changing from mostly fixed use to become the main general-purpose federal UHF land mobile bands.

Their diverse uses include mobile networks as well as weather, hydrologic, and seismic sensors, communications for law enforcement and protection of the President, other government personnel, and foreign dignitaries, flood and wind-shear warning, and telemetry for electric transmission.

Air Force, Army, and Navy uses include radar, paging, and range control, disaster preparedness, training, shipboard operations, fire protection, transportation, and security. The 410–420 MHz portion is exclusive to the government.

Radio astronomy is the exclusive non-government user in 406.1–410 MHz, one of several bands used for continuum observations, particularly of pulsars (see 322–328.6 MHz).

Trends in this spectrum include the migration of fixed systems to 932–935 MHz and 941–944 MHz, and increased use of multi-agency trunking (automated channel selection) systems. This will result in fewer single-channel, single-user systems. A private corporation operates trunked systems for the government in this spectrum in major cities.

The technology in the 406–420 MHz federal bands will change as the *National Telecommunications and Information Administration* (NTIA) implements narrowband radios here (see 162.0125–173.2 MHz).

The 399.9–410 MHz spectrum is one of many restricted bands in which the FCC Part 15 rules permit unlicensed devices to emit only very low level emissions.

410-420 MHz

FG: US13 G5 FIXED. MOBILE

Non-FG: US13

See 406.1–410 MHz for the fixed and mobile uses of this government band.

Internationally, 410–420 MHz is available to the *Space Research Service* (SRS) on a secondary basis for space-to-space links within five kilometers of an orbiting manned vehicle.

The SRS is defined as "a radiocommunication service in which spacecraft or other objects in space are used for scientific or technological research purposes." The SRS is one of the three services—the others are the Space Operation Service and the *Earth Exploration-Satellite Service* (EESS)—now collectively known as the Space Science Services.

Using these services, the U.S. civil space program performs missions of collecting satellite-derived weather data, fighting forest fires, monitoring volcanic activity, ozone depletion and pollution; atmospheric and solar measurements, and exploration of the moon; and three-dimensional imaging techniques, to name a few.

Airborne synthetic aperture radar, for uses such as target identification, operates in 410–450 MHz among its bands.

Person-overboard marine alarms use 418 MHz under FCC Part 15 rules that provide for low-power control devices in several VHF and UHF bands.

420–450 MHz

FG: S5.286 US7 US87 US217 US228 US230 G8
RADIOLOCATION G2

Non-FG: S5.282 S5.286 US7 US87 US217 US228 US230
NG135 Amateur

FCC: Private Land Mobile (90). Amateur (97)

This band is used for telemetry, telecommand, and long-range radar surveillance by land, ship, and airborne stations, andfor early missile warning, detection of low-observable military targets, and tracking of all objects in Earth's orbit.

The band also is used for development of anti-stealth and foliage penetration radars.

The *Air Force Ballistic Missile Early Warning System* (BMEWS) radars in Alaska, Great Britain, and Greenland operate in this band. BMEWS is the backbone of the U.S. missile defense system.

Established to track satellites and submarine-launched ballistic missiles, the *Precision Acquisition Vehicle Entry Phased Array Warning System* (PAVE PAWS) radars are in this band. PAVE PAWS can simultaneously detect and discriminate among many objects and provide data on launch, position, velocity, and impact.

Non-Defense Uses

The *National Oceanic and Atmospheric Administration* (NOAA) operates a network of 449 MHz wind profilers across the U.S. These upward-looking radars warn of severe wind conditions for forecasts and aviation, and help in the analysis of pollutants such as volcanic ash and acid rain. Wind profilers are considered an improvement over radiosondes lofted into the air. Profilers also operate in 902–928 MHz.

The Arecibo Observatory conducts radar and astronomy observations in this band (see 2360–2385 MHz).

NASA's National Scientific Balloon Facility uses this band for its primary balloon command frequency.

The 420–450 MHz, 70-cm band is a popular Amateur Radio allocation, especially for FM repeater use in the 440–450 MHz segment. The amateur satellite program relies on 435–438 MHz for links with spacecraft. *Amateur television* (ATV) is also growing in popularity in 420–430 MHz. We anticipate wider use of spread spectrum technology in the band in the years to come.

Sharing of 421–430 MHz with *Private Mobile Radio Services* (PMRS) is permitted under unique conditions in Buffalo, Detroit, and Cleveland. The *Land Mobile Communications Council* (LMCC), a trade association for the land mobile radio industry, petitioned the FCC to expand PMRS operations in 420–430 MHz and 440–450 MHz nationwide.

LMCC told the FCC, "A reduction in military use of this band is foreseen and it could be that most PMRS services could co-exist in most significant geographical areas of the U.S., with perhaps PAVE PAWS geographical restrictions in parts of California, Georgia, Massachusetts and Texas."

The petition evoked angry protests from radio amateurs who concluded that sharing with commercial and industrial users in this spectrum would be difficult or impossible. The *American Radio Relay League* (ARRL) argued that the petition "stands to render valueless a large portion of the huge investment of radio amateurs in equipment for public-service communications."

450-454 MHz

FG: S5.286 US87

Non-FG: S5.286 US87 NG112 NG124 LAND MOBILE

FCC: Auxiliary Broadcasting (74). Private Land Mobile (90)

The FCC consolidated 450–451 MHz and 451–454 MHz into the 450–454 MHz band in 1999.

The 450–451 MHz segment is used for base stations and 455–456 MHz for mobile units by TV and radio stations in the *Broadcast Auxiliary Service* (BAS), which includes wireless microphones, transmitter telemetry, and mobile communications for electronic news gathering (see 1990–2025 MHz).

The 451–454 MHz base station segment, which has channels for mobile units at 456–460 MHz, marks the start of the principal UHF *Private Mobile Radio Services* (PMRS) allocations, which extend to 470 MHz (512 MHz in some areas). The band includes both Public Safety and Industrial/Business frequencies (see 30.56–32 MHz) and is used especially by power utilities and railroads.

Modern mobile radio expanded beyond the 450 MHz UHF bands to the 800 MHz and 900 MHz bands, where trunking and digital technologies are used. There, the *Specialized Mobile Radio* (SMR) services have become *Commercial Mobile Radio Services* (CMRS) and are competing with cellular and other for-profit mobile radio services.

The 450–470 MHz spectrum remains the heart of conventional mobile radio in the U.S.

PMRS frequencies in 450–470 MHz are currently spaced 25 kHz apart with secondary, low-power operations allowed on frequencies offset 12.5 kHz from assigned channels. In most of these frequencies, the FCC's refarming rules establish 6.25 kHz wide channels.

454–455 MHz

FG: None

Non-FG: NG12 NG112 NG148 FIXED. LAND MOBILE

FCC: Public Mobile (22). Maritime (80)

This band contains uplinks in the General Aviation Air-Ground Radiotelephone Service, a Public Mobile Service authorized in 1969 for phones on private aircraft (sometimes also called the *Air-Ground Radiotelephone Automated Service* or AGRAS).

It is divided into 12, 25 kHz channels at 454.700–454.975 MHz. The ground signaling channel is 454.675 MHz. Downlinks are at 456–460 MHz.

The FCC recognized that the allocations at 450 MHz to this service were too small to provide capacity for consumer use. It allocated 849–851 MHz and 894–896 MHz to *Commercial Aviation Air-Ground Systems,* where several licensees offer services on scheduled airlines.

Later, a controversial service known as AirCell emerged to compete for private aircraft business (see 849–851 MHz).

The 454 MHz band also contains channels for the *Public Mobile Rural Radiotelephone Service* and the *Paging and Radiotelephone Service,* including mobile phones using pre-cellular technology (see 35–36 MHz). These uses are paired with 456–460 MHz and they also have channels at 152 MHz and 157 MHz.

455–456 MHz

FG: None

Non-FG: LAND MOBILE FCC: Auxiliary Broadcasting (74).

See 450–454 MHz for uses of this band.

This band and 459–460 MHz were allocated to ITU Region 2 (the Americas and Greenland) for uplinks in the Little LEO *Non-Voice Non-Geostationary Mobile Satellite Service* (NVNG MSS, see 137–137.025 MHz).

Based on this international action, the FCC proposed to allocate 455–456 MHz and 459–460 MHz to U.S. NVNG MSS uplinks. The ITU footnotes associated with these bands state that these MSS stations are not to constrain the development and use of this spectrum by the Fixed and Mobile Services.

"Therefore," the FCC observed, "even though Little LEO operations have a Region 2 primary allocation in these bands, their operations are effectively secondary to fixed and mobile services."

The FCC found more than 25,000 incumbent *Broadcast Auxiliary Service* (BAS) transmitters in 455–456 MHz. These remote pickup systems are used to dispatch and cue personnel, and convey program material in radio and TV operations, especially when on location, and to relay telemetry from the transmission site to the studio.

The FCC claimed, "Since many auxiliary broadcast remote pickup channels in the 455–456 MHz band tend to be used only intermittently and Little LEO transmissions are currently limited to a short duration of only 450 milliseconds in the 148–149.9 MHz band, Little LEO systems may be able to search the spectrum for unused channels and accomplish their communications without hindering incumbent use."

Little LEO satellite operators have emphasized the ability of their scanning and dynamic channel assignment technologies to avoid interfering with incumbent users of these bands.

456-460 MHz

FG: S5.288 669

Non-FG: S5.288 669 NG112 NG124 NG148 FIXED. LAND MOBILE

FCC: Public Mobile (22). Maritime (80). Private Land Mobile (90)

The 456-459 MHz segment of this band is associated with 451-454 MHz for mobile radio services (see 450-454 MHz).

The 459-460 MHz segment is the downlink band in the *General Aviation Air-Ground Radiotelephone Service* (see 454-455 MHz).

The band also is used by the *Rural Radiotelephone Service* and mobile units in the *Paging and Radiotelephone Service* (see 35-36 MHz).

This band and 455-456 MHz were allocated to ITU Region 2 (the Americas and Greenland) in 1995 for uplinks in the *Non-Voice Non-Geostationary Mobile Satellite Service* (NVNG MSS, see 137-137.025 MHz and 455-456 MHz).

This band also is available for space operations. NASA's Mars Pathfinder mission, for example, used 459.7 MHz for video and telemetry between the lander and rover on the surface of Mars.

460-462.5375 MHz

FG: S5.288 S5.289 669 US201 US209 US216 Meteorological-Satellite (space-to-Earth)

Non-FG: S5.289 US201 US209 NG124 FIXED. LAND MOBILE

FCC: Private Land Mobile (90)

This band for base stations and the 462.7375-467.5375 MHz band for mobile units encompass much of the UHF *Private Mobile Radio Services* (PMRS) spectrum and their Industrial/Business and Public Safety frequency pools (see 30.56-32 MHz).

Some channels in this band are designated for medical telemetry, alarm systems, and airport operations. Other channels are for use only at locations distant from airports.

The 460-461 MHz and 462.525-467.475 MHz segments are available for experimental use by students and schools.

462.5375–462.7375 MHz

FG: S5.288 S5.289 669 US201 US209 US216 Meteorological-Satellite (space-to-Earth)

Non-FG: S5.289 US201 LAND MOBILE

FCC: Personal Radio (95)

This is one of the UHF bands allocated to two of the Personal Radio Services: the *General Mobile Radio Service* (GMRS) and the *Family Radio Service* (FRS).

GMRS unit-to-unit and repeater output transmissions use eight primary channels from 462.550 to 462.725 MHz. Repeaters, which extend communications range over a wide area, use eight input channels from 467.550 to 467.725 MHz.

Overlapping the 462 MHz primary GMRS channels are seven low-power offset channels at 462.5625–462.7125 MHz, which are shared with the FRS.

GMRS also formerly shared the 31–31.3 GHz band with other services. No GMRS licensing occurred in that millimeter wave band. It eventually found a popular niche in controlling timed traffic lights as required by pollution control regulations.

However, the FCC considered such usage to be merely a "simplified communication function," and reallocated 31–31.3 GHz to the *Local Multipoint Distribution Service* (LMDS, see 27.5–29.5 GHz). The FCC auctioned 31 GHz licenses.

GMRS and Personal Radio History

From the pioneering work of Cleveland engineer Al Gross (see 35–36 MHz), the FCC developed the Citizens Radiocommunications Service in 1945–47.

Commissioner Ewell K. "Jack" Jett championed Citizens Radio at the FCC. Writing in *The Saturday Evening Post* of July 28, 1945, Jett described how "any American citizen, firm, group or community unit may privately transmit and receive short-range messages over certain wavelengths. From mere listeners or spectators as they are now, people in homes and offices throughout the country will become active participants."

Citizens Radio had three different classes. Class A is the remaining UHF version. Class B was a low-power UHF version that was eliminated in 1966. Class C is the *Radio Control Radio Service* (R/C) today (see 72–73 MHz).

After manufacturers pressed the FCC for a Citizens Radio allocation lower in the spectrum, to reduce costs, it created Class D in 1958 at 27 MHz over the objections of radio amateurs who lost access to the band.

The FCC later renamed Class D Citizens Band (CB), officially blessing the service's common nickname. CB became enormously popular with retail radio sales reaching $1 billion in 1975 and widespread use of illegal power amplifiers ("afterburners" or "foot warmers").

Appalled by the chaotic CB subculture and determined to avoid another loss of precious spectrum, radio amateurs foiled an FCC proposal to establish a 220 MHz Class E CB.

The Class A service survived, however. A persistent myth is that GMRS was once UHF CB or Class A Citizens Band. In fact, the FCC renamed Class A Citizens Radio to GMRS in 1976, skipping the CB name entirely.

GMRS licensees successfully lobbied Congress to block the FCC's attempts to lump GMRS in with CB. Today, CB is a catchall category for services that are "licensed by rule" and therefore do not issue individual license documents. This arrangement is an exception to the Communications Act's requirement that no station shall transmit without a license.

After the Commission was deluged with 27 MHz CB license applications, Congress granted the statutory loophole of licensing by rule. It applies to anything that the FCC officially designates as a type of Citizens Band.

Citizens Band became a crazy quilt of unrelated services, all without end user licenses. The CB Services are 27 MHz CB, the FRS, the 218–219 MHz Service, the *Low Power Radio Service* (LPRS, 216–220 MHz), the *Medical Implant Communications Service* (MICS, 402–403 MHz), the *Wireless Medical Telemetry Service* (WMTS, 608–614 MHz) and the *Multi-Use Radio Service* (MURS, 150.8–152.855 MHz) with *Personal Locator Beacons* (PLBs, see 406–406.1 MHz) likely to join this exclusive club.

GMRS Development

Starting in 1968 with the first repeater system in Chicago on 462.675 MHz, GMRS developed a nationwide not-for-profit mutual aid network of repeaters, volunteer monitors, public service teams, and user groups.

GMRS traditionally used expensive FM radios identical to those used by commercial licensees. The licensed-by-rule and less capable FRS attracted more manufacturer interest and quickly became a high-volume consumer product (see below).

The FCC *Personal Use Radio Advisory Committee* (PURAC) of the 1970s envisioned a 900 MHz personal radio service as an improvement over 27 MHz CB and a possible adjunct or successor to GMRS. This concept became the ill-fated *Citizens Band Land Mobile Private Radio Communications Service* (PRCS) on which General Electric lost millions (PRCS, 896–901 MHz).

Petitioned by Mura Corp., once a dominant maker of cordless phones, the FCC later sought to transform GMRS into a *Consumer Radio Service* (CRS). In the FCC's words, CRS involved the "abolition of repeaters, reduction of maximum authorized power, and the adoption of complex standardized algorithms and protocols."

The proposal met with widespread derision among GMRS users. The FCC closed the Consumer Radio Service effort after declaring that it had "calibrated user sentiment."

Eventually the FCC updated the GMRS rules to permit roamer usage, more channels, and other features recommended by PURAC. Later, in its Universal Licensing System proceeding, the FCC stripped most of the licensing, frequency assignment, and operational rules from GMRS. Licenses are still required in GMRS, but only a perfunctory application form is used.

At one point, the Commission confined one of the GMRS primary channels to emergency and traveler assistance use, banning all other message traffic. The agency later granted a rare reconsideration of this rule.

Only an individual person may obtain a new GMRS license or modify an existing one. A licensee's immediate family members are eligible to operate under his or her license. Before these 1989 changes, any legal entity (a corporation, association, a partnership, or a governmental agency) was also eligible to license in GMRS.

In some places, especially in the larger urban areas, these non-personal licensees still dominate the GMRS channels. The usurpation of the limited GMRS allocation by business users who are eligible in other services is a continuing problem for the service. The accountability introduced by individual-only licensing helped to somewhat slow the problem.

Many licensees belong to local cooperatives or teams that provide repeater service at little or no cost. The GMRS National Repeater Guide lists repeaters and their sponsors, and is published by the Personal Radio Steering Group (PRSG), the successor to PURAC's GMRS task force.

FRS History

In July 1994, Radio Shack proposed a Family Radio Service that would use spectrum in the GMRS band. The FCC responded with a Notice of Proposed Rulemaking—its fourth attempt to establish a new personal radio service in the VHF-UHF spectrum.

The majority of formal comments vehemently opposed FRS, mostly because of concern over interference from FRS to GMRS.

Seemingly ignoring most of the response, the FCC proclaimed "overwhelming" support for FRS and established the service on July 8, 1996. FRS operates on 14 offset channels at 462 MHz and 467 MHz (GMRS does not use the offset channels at 467 MHz).

"The FRS will enable families, friends, and associates to communicate among themselves within neighborhoods and while on group outings," the FCC said. FRS "would be useful to hunters, campers, bikers, bicyclists, and other outdoor activity enthusiasts who need to communicate with other members of their party who are out of speaking range or sight but still in the same general area."

FRS Plus GPS

The most recent FRS regulatory developments at this writing included the FCC's authorization of two manufacturers for *Global Positioning System* (GPS)-connected FRS equipment.

The Commission permitted Garmin International to test market a type of FRS radio that can transmit its GPS-derived location and can display a map of the locations of other, GPS-equipped radios in the area.

The FRS rules essentially forbid data transmission, except when used to establish voice contact with another unit.

The FCC placed important restrictions on the market trial. It only lasts one year, after which Garmin must petition for a permanent change in the FRS rules. The GPS data burst can last no longer than one second every 10 seconds. Units may not automatically poll other units to request location.

In a twist to its cherished policy of declining to regulate radio operator behavior or message content, the FCC requires Garmin to warn all customers that any GPS-derived location information is "provided for personal and public safety purposes" and any other use is "not authorized by the Commission."

The FCC also authorized Motorola to experiment with a FRS-based device that locates a car in a parking lot. The user presses a LOCATE button on the handheld radio, which requests a position advisory from a radio in the vehicle. The vehicle unit queries a GPS receiver and reports its position back to the handheld unit, which compares the received information with its own GPS-derived position. It then displays the direction and distance to the vehicle.

The 462.525–467.475 MHz segment is available for experimental use by students and schools.

462.7375–467.5375 MHz

FG: S5.288 S5.289 669 US201 US209 US216 Meteorological-Satellite (space-to-Earth)

Non-FG: S5.289 669 US201 US209 US216 NG124 FIXED. LAND MOBILE

FCC: Private Land Mobile (90)

This band includes mobile units for the Industrial/Business systems in 460–462.5375 MHz. Private Carrier Paging Systems are in this band (see 35–36 MHz).

In addition to police and fire channels, the band also contains the 462 MHz and 463 MHz MED channels for base units in medical telemetry and voice communications during patient transport, part of the former *Emergency Medical Radio Service* (EMRS) and now part of the Public Safety frequency pool (see 30.56–32 MHz). The mobile MED channels are in 467.7375–470 MHz. Currently, 40 MED channels are available.

This band and upper adjacent bands are used in *Vessel Traffic Services* (VTS) that report the positions of ships to waterway management facilities.

The 462.525–467.475 MHz segment is available for experimental use by students and schools.

467.5375–467.7375 MHz

FG: S5.288 S5.289 669 US201 US209 US216 Meteorological-Satellite (space-to-Earth)

Non-FG: S5.289 669 US201 LAND MOBILE

FCC: Personal Radio (95)

This band is used for repeater inputs in the *General Mobile Radio Service* (GMRS). The non-repeater *Family Radio Service* (FRS) also has channels in this band (see 462.5375–462.7375 MHz).

467.7375–470 MHz

FG: S5.288 S5.289 669 US201 US209 US216 Meteorological-Satellite (space-to-Earth)

Non-FG: S5.288 S5.289 US201 US216 NG124 FIXED. LAND MOBILE

FCC: Private Land Mobile (90)

This band is used by the Industrial/Business frequency pool and by Public Safety mobile units on the MED channels (see 462.7375–467.5375 MHz).

It is one of the bands containing low-power "color dot" frequencies, which have contributed to problems with unlicensed operation (see 154–156.2475 MHz).

The *Geostationary Operational Environmental Satellite* (GOES) transmits in this band among its downlink frequencies (see 1670–1675 MHz).

470–512 MHz

FG: None

Non-FG: NG66 NG114 NG127 NG128 NG149 FIXED. BROADCASTING. LAND MOBILE

FCC: Public Mobile (22). Broadcast Radio (TV) (73). Auxiliary Broadcasting (74). Private Land Mobile (90)

This band contains UHF television channels 14–20.

The American television system uses the analog NTSC standard, developed by the *National Television Systems Committee* (NTSC) for black-and-white TV in 1940–41 and color in 1950–53.

NTSC was nicknamed "never twice the same color" for its technical infirmities. It prompted generations of viewers to festoon TV antennas with kitchen foil and guaranteed the cable industry a lucrative market in improving reception.

Figure 8
The Digital Television Certification Logo of the Advanced Television Systems Committee (ATSC) signifies that the product can receive and display all ATSC video formats. The logo will appear on TV sets, computers, and other consumer devices. The Consumer Electronics Association (CEA) administers the certification program.

Digital Television

U.S. television will see fundamental transformation over the next decade. The FCC has directed a national conversion from NTSC to *Digital Television* (DTV), which it defines as "any technology that uses digital techniques to provide advanced television services such as *High Definition TV* (HDTV), *multiple standard definition TV* (SDTV) and other advanced features and services."

Complete conversion to DTV is supposed to occur by December 31, 2006. DTV will be broadcast on a core spectrum of channels 2 through 51.

The FCC's DTV Table of Allotments assigns a second 6 MHz channel—a "digital channel"—to each TV station. This table only accommodates existing TV licensees, holders of TV station construction permits, and pending applicants. The FCC has deleted the vacant NTSC channel allotments.

The FCC reallocated TV channels 52–69 (698–806 MHz) to other services. Broadcasters are supposed to use their newly allocated digital channels and return their vacated analog channels to the FCC, which will auction them for other, generally unspecified uses. Intriguing possibilities in broadband multimedia exist for some of this spectrum (see 746–764 MHz).

Licenses for the vacated channels, regardless of when they might actually be vacated, are supposed to be auctioned such that the receipts would be deposited in the Treasury by September 30, 2002.

Originally, the main attraction of DTV was its ability to deliver HDTV, which offers widescreen images at twice the resolution of NTSC, approaching the picture quality of 35mm movies.

TV broadcasters complained of the FCC's requirement to shift from NTSC to DTV, which will cost billions in new plant construction and programming. But the U.S. HDTV and DTV campaign was an elaborate project of trade associations representing the broadcasting and receiver manufacturing industries. "Nobody forced them to go digital," according to the Media Access Project, a nonprofit law firm. "They lobbied for, and received, the gift of exclusive access to public airwaves for digital TV."

Sharing with Mobile Services

The largesse began in 1987 with an industry petition to the FCC asking it to investigate advanced TV. At the time, the FCC was seeking to increase the opportunities for sharing of TV spectrum by *Private Mobile Radio Services* (PMRS).

The PMRS share 470–512 MHz (TV channels 14–20) in 13 urbanized areas and Hawaii under power and siting restrictions designed to protect TV from interference.

(Also sharing this band are the Gulf of Mexico operations licensed in the *Offshore Radiotelephone Service* (ORS), one of the *Public Mobile Services* within the *CMRS Mobile Services,* itself one of the *Commercial Wireless Radio Services* (see 30.56–32 MHz). The ORS is used in oil exploration and production off the Louisiana and Texas coasts.)

Broadcasters were alarmed at the prospect of more PMRS incursion into TV spectrum. Sharing with the mobile service dilutes, at least in theory, the value of TV licenses and invites interference. The FCC later acknowledged that the real cause of most TV interference is in the design and construction of the susceptible TV receiver itself.

Broadcasters believed that they might stave off further sharing of their spectrum by invoking the exotic promise of HDTV, which could require all of the UHF TV spectrum and could not tolerate more sharing with mobile radio. They portrayed HDTV as crucial to American dominance in consumer electronics. Japan and Europe had already launched HDTV development with a lot of fanfare. This point was not lost on Congress, which encouraged the FCC to develop HDTV regulations.

The lobbying campaign was an outrageous success. It secured free spectrum for broadcasters while other wireless players were paying billions at license auctions. "Basically, the broadcast networks were the beneficiaries of the biggest government giveaway since Peter Stuyvesant bought Manhattan from the Indians for $24," said FCC Chairman William Kennard. "You can barely buy lunch in Manhattan for that now."

The FCC convened the *Advisory Committee on Advanced Television Service* (ACATS). The broadcast, manufacturing, media, and cable industries opened HDTV testing laboratories and established a joint *Advanced Television Systems Committee* (ATSC).

Merged Systems

It did not seem that an all-digital advanced TV system was feasible, but manufacturers and labs whose competing systems were undergoing independent tests formed the Digital HDTV Grand Alliance in 1993. The Grand Alliance members merged their designs into a single DTV system. The ATSC recommended this system to the FCC, which adopted it in 1996 as the U.S. DTV transmission standard.

"Use of the ATSC DTV standard represents a rare opportunity to increase significantly the efficient use of broadcast spectrum," the FCC said. "The packetized structure of the data transport ensures a flexibility that will permit the DTV licensee to provide, for instance, several standard definition programs, or one high-definition program, or some standard definition programming together with data transfer or electronic publishing on the remaining bit streams, and to switch instantaneously between such applications. This means a wide array of innovations can be introduced without Commission action.

"Digital TV stands a risk of failing unless it is rolled out quickly," the FCC added. It required TV stations affiliated with the ABC, CBS, Fox, and NBC networks in the ten largest markets to build DTV transmission facilities by May 1, 1999. Other stations affiliated with those networks in the top 30 markets had to build DTV facilities by November 1, 1999. All other commercial stations must build DTV facilities by May 1, 2002; all noncommercial stations must build DTV facilities by May 1, 2003.

TV stations will transmit analog NTSC and DTV simultaneously on separate channels for years, in order to ensure that popular programs continue to be available on the DTV channel and to encourage viewers to switch to DTV.

FCC rules require the programming on the analog and digital channels to be identical by April 2005, making the NTSC channel superfluous and preparing the public for the supposed abandonment of NTSC in 2006. Converter devices should be available for holdouts unwilling to jettison their NTSC receivers.

Delaying Transition

Legislative loopholes in the 1997 Budget Act permit the FCC to delay the 2006 transition date if poor sales of digital sets make extensions necessary. Analog TV broadcasting is supposed to stop at the end of 2006 in markets where 85 percent of TV households can receive DTV.

A September 1999 Congressional Budget Office report said "It now appears likely that the transition will extend beyond 2006 in most markets, with its ultimate end date uncertain. ...

"The transition will almost certainly continue beyond 2006 in any television market in which less than 85 percent of television households — the legally mandated goal for ending the transition — are considered DTV households," the CBO continued. "Currently available evidence suggests two reasons the 85 percent goal is unlikely to be met by 2006. First, the availability of DTV programming on cable systems, which is crucial to meeting the goal, is an unsettled question.

"Second, the adoption of digital TV by households that do not subscribe to a *multichannel video programming distributor* (MVPD) such as a cable or satellite service is uncertain. In choosing not to pay for their TV viewing, those households would appear to have a relatively low demand for video programming services. Yet in order to reach the goal of 85 percent market penetration signaling the end of the transition, some of those households will almost certainly have to adopt digital TV. Ironically, households that value television the least may be critical in determining when all viewers and society in general receive the full benefits of the new digital broadcast technology."

DTV stations will be required to transmit one free digital video programming service, the resolution of which is comparable to or better than today's service. It must air during the same time periods that their analog channel is broadcasting.

They will not be required to broadcast HDTV. To mandate HDTV broadcasting might impose "significant burdens" on stations, according to the FCC. In theory, TV stations may find it more profitable to broadcast multiple, NTSC-like standard definition programs than to utilize the entire channel for HDTV. At the time, former FCC Chairman Reed Hundt described HDTV as "very marginal," "unnecessary," "possibly superfluous" and even "defunct" in view of the non-HDTV uses of DTV channels.

TV stations are broadcasting HDTV programming anyway. Improved resolution is, at this writing, the principal attraction for consumers to invest in DTV receivers.

Additional Digital Services

Broadcasters will be permitted to offer on the digital channel "ancillary and supplementary" services, which are considered to be any service other than free services.

These ancillary and supplementary services need not be broadcast-related. They could include, the FCC said, "subscription television programming, computer software distribution, data transmissions, teletext, interactive services, audio signals, and any other services that do not interfere with the required free service."

At this writing, several broadcast groups have emerged to develop the market for DTV data distribution. For example, iBlast Networks announced a strategy of signing carriage agreements with coverage of 95 percent of the nation's homes. IBlast data will be transmitted by as many as four TV stations in each market, increasing available capacity. iBlast stations will contribute spectrum, cash investments, and marketing commitments in exchange for equity in the datacast business. Another entrant, SpectraRep, describes itself as a "spectrum aggregation service" and will initially target business customers.

The 1996 Act requires the FCC to charge fees for ancillary and supplementary services provided by DTV licensees. Broadcasters may be able to avoid paying these fees by exploiting loopholes in the Act. For example, ad-supported ancillary services are not subject to fees unless the broadcaster charges users to receive the services.

Congress directed that the fees "avoid unjust enrichment" of broadcasters and that they recover an amount that, to the extent feasible, equals but does not exceed the amount that would have been recovered if the service been licensed by auction. The broadcasting industry lobbied heavily to set the fees at one or two percent of revenues, but at this writing, the FCC had set the fee at five percent of gross revenues obtained from the new digital services.

The wireless industry is suspicious that broadcasters will use DTV to offer ancillary, competitive mobile services. "The fact that broadcasters continue to lust greedily after additional free spectrum to use for any services they choose is certainly misguided, at best, and highway or spectrum robbery, in fact," according to Jay Kitchen, president of the Personal Communications Industry Association.

The ATSC Conditional Access Standard for DTV will permit DTV stations to broadcast pay-per-view or subscription programming, at least in standard definition.

Low-Power TV

At this writing, about 2200 *Low-Power TV* (LPTV) stations are operating. LPTV stations are secondary; that is, they may continue operating unless they interfere with full-power DTV stations. LPTV stations that face interference from or to DTV may apply to move to other channels. The FCC is supposed to process those applications quickly and will not entertain competing applications for the channel.

LPTV stations are allowed up to 3 kW on VHF channels 2–13 and 150 kW on UHF channels 14–69, although in practice both types usually radiate less power than these limits. Full-power VHF stations may radiate up to 316 kW and UHF stations up to 5 MW (5000 kW).

Pursuant to congressional legislation, some LPTV stations can obtain Class A primary status, which confers increased incumbency and interference protection from full power stations. The policy reversed somewhat the original premise of the service, namely, that these stations could be supplanted by other, primary broadcast operations or other services if warranted. Class A status might help LPTV stations attract financial support, advertising and cable carriage, and could assist them in planning programming, expansion, hiring, and investment in new equipment for eventual digital operations, according to the FCC. Class A stations have callsigns with the *-CA* suffix; LP stations identify using *-LP*.

Current TV Data Transmission Modes

The FCC currently allows data transmission on the *Vertical Blanking Interval* (VBI) portion of the NTSC TV signal.

No picture information is transmitted during the VBI when the electron beam retraces from the bottom to the top of the screen. The VBI transmits closed captioning and data that broadcasters use internally to track ads and programs.

Other TV data methods are the overscan method, which replaces video with data in a part of the picture that is usually invisible to viewers (it is hidden by the frame around the screen), and the sub-video method that distributes the data signal at low levels throughout the visible picture.

"Sub-video systems are capable of relatively high data rates, on the order of 300–500 kbps," according to the FCC. "Such high-speed uses include digital magazines, newspapers, books, catalogues, and audio materials, downloading computer software or financial data, and distributing business data to branch offices or clients." One application transmits data from the TV to stuffed animals, causing them to talk.

No-See TV

The U.S. International Broadcasting Bureau broadcasts video propaganda to Havana over TV Marti. According to the General Accounting Office, the TV Marti broadcasts are constantly and effectively jammed.

The U.S. Advisory Commission on Public Diplomacy reported that TV Marti is not cost-effective and has repeatedly recommended that it be closed, a finding consistent with the views of the majority of the station's own advisory board.

The Wall Street Journal called working at TV Marti "one of the most frustrating jobs in journalism" because the station has few viewers. Reportedly, the government's own photos of TV Marti show only a blank screen.

The station began work to convert to UHF channels 18, 50, and 64, but its board concluded that "UHF transmission is not a viable means" to overcome Cuban jamming.

Officially, TV Marti states that "international reception is possible in some outlying areas of the city and in other parts of the Havana province. As a result, the TV Marti signal is now shifted to include areas east and west of Havana randomly during the broadcast in an effort to lessen the effects of jamming."

Medical Telemetry

Wireless, unlicensed medical telemetry devices in health care facilities operate under FCC Part 15 rules in TV channels 14–46 (470–668 MHz) as well as channels 7–13 (174–216 MHz). Part 90 rules permit this equipment to operate on a secondary basis to land mobile users in the 450–470 MHz spectrum. Medical telemetry typically monitors blood pressure, cardiac, and respiration status.

Medical telemetry devices must avoid interfering with licensed services such as TV broadcasting and must accept any interference they may receive from licensed services. The FCC requires that specific distances separate medical telemetry installations from transmitters in licensed services. Installations are supposed to be carefully engineered and frequencies should be selected for each location.

In March 1998, test transmissions from digital TV broadcasters interfered with medical telemetry at a nearby hospital. The incident prompted serious concern at the FCC, leading it to seek a more stable and interference-free environment for medical telemetry.

The FCC achieved this objective in a new *Wireless Medical Telemetry Service* (WMTS) in which these devices could operate on a co-primary basis (see 608–614 MHz).

512-608 MHz

FG: None

Non-FG: NG128 NG149 BROADCASTING

FCC: Broadcast Radio (TV) (73). Auxiliary Broadcasting (74)

This band contains UHF TV channels 21–36. The U.S. is making a major transition to DTV (see 470–512 MHz).

608-614 MHz

FG: US246 LAND MOBILE US350 RADIO ASTRONOMY US74

Non-FG: US246 LAND MOBILE US350 RADIO ASTRONOMY US74

FCC: Personal (95)

This is TV channel 37 on which no TV transmissions are permitted.

The channel is used for radio astronomy continuum observations. It has very low background noise.

The University of Illinois obtained Navy funding to build a radio telescope on this channel in the late 1950s, prompting the FCC to entrench the channel for scientific uses and disallow its use for TV.

Today, radio astronomers use channel 37, among other bands, for *Very Long Baseline Interferometry* (VLBI): the connection of telescopes far apart for observing such sources as pulsars (see 4.99–5 GHz).

The *National Radio Astronomy Observatory* (NRAO) operates the Very Long Baseline Array of telescopes at ten sites in the U.S. and its territories. Data from the antennas are combined, synthesizing a single telescope 5000 miles in diameter.

Medical Telemetry

This is one of three bands allocated to the new *Wireless Medical Telemetry Service* (WMTS). The others are 1395–1400 MHz and 1429–1432 MHz. WMTS devices operate in this service on a "licensed by rule" basis as one of the growing number of Citizens Band Radio Services (see 462.5375–462.7375 MHz).

The FCC defines WMTS as "a private, short distance data communication service for the transmission of patient medical information to a central monitoring location in a hospital or other medical facility. Voice and video communications are prohibited. Waveforms such as electrocardiograms (EKG) are not considered video."

The FCC intends for WMTS to be the modern replacement for conventional medical telemetry systems that operate on an unlicensed basis in FCC Part 15, or licensed and secondary under Part 90 rules (see 470–512 MHz). When it created WMTS, the Commission said that it does not anticipate making any more allocations to medical telemetry.

WMTS is unique among the CB Services as it is the first to require frequency coordination by FCC-selected entities. These coordinators record and monitor the locations and frequencies of WMTS equipment for interference control purposes. At this writing, the FCC had invited interested organizations to apply to become WMTS frequency coordinators.

WMTS equipment will only be permitted in health care facilities, and eligibility to operate the equipment is strictly limited to health care personnel. WMTS cannot legally be used in a moving vehicle, even an ambulance.

The 608–614 MHz spectrum is one of many restricted bands in which the FCC's Part 15 rules permit unlicensed devices to emit only very low level emissions. The band will not change its status in the national transition to *Digital Television* (DTV, see 470–512 MHz).

614-698 MHz

FG: None

Non-FG: NG128 NG149 BROADCASTING

FCC: Broadcast Radio (TV) (73). Auxiliary Broadcast (74)

This part of the UHF TV spectrum contains channels 38-51.

The U.S. is making a major transition to *Digital Television* (DTV, see 470-512 MHz). Under this transition, virtually all television broadcasting will shift to TV channels 2-51.

698-746 MHz

FG: None

Non-FG: NG128 NG149 BROADCASTING

FCC: Broadcast Radio (TV) (73). Auxiliary Broadcast (74)

Note: Band to be reallocated and auctioned by Sept. 30, 2002.

This part of the UHF TV spectrum contains channels 52-59. The U.S. is making a major transition to *Digital Television* (DTV). Under this transition, virtually all TV broadcasting will shift to channels 2-51.

In a November 1999 spectrum policy statement, the FCC announced its intention to auction licenses in 698-746 MHz for Fixed, Mobile, and new Broadcasting Services for commercial uses, possibly including some type of *International Mobile Telecommunications-2000* (IMT-2000) Services (see 1850-1990 MHz).

We believe that this 48 MHz of spectrum could have immense value, depending especially on whether developers expose the Commission to persuasive service and product ideas in time for them to be considered in the reallotment and assignment process.

Other TV channels, namely 60-62 and 65-67, are in a similar position, but at a more advanced stage of regulatory development at this writing (see 746-764 MHz and 776-794 MHz).

Any new operations will have to avoid interference to TV stations during the transition to DTV service and the eventual exit of these stations from this band.

The FCC warns entrepreneurs that "TV broadcasters will continue to use UHF TV Channels 52–69 at least through 2006, and they may petition the Commission to continue to use their channel until 85 percent of the households in their Grade B contour have at least one digital TV set." This is the congressionally mandated loophole described in our DTV commentary (see 470–512 MHz).

The Grade B contour marks the far boundary of a TV station's coverage. It provides a lower signal level than available closer to the transmitter in the Grade A contour.

746–764 MHz

FG: None

Non-FG: NG128 NG159 FIXED. MOBILE. BROADCASTING

FCC: Wireless Communications (27). Broadcast Radio (TV) (73). Auxiliary Broadcasting (74). Private Land Mobile (90).

This part of the UHF TV spectrum contains channels 60, 61, and 62. However, the U.S. is making a major transition to *Digital Television* (DTV, see 470–512 MHz). The FCC will not accept any new applications for analog TV licenses in channels 60–69.

The FCC reallocated 746–764 MHz and 776–794 MHz (channels 65, 66, 67) to the *Miscellaneous Wireless Communications Services* (MWCS) in Part 27 of its rules. MWCS already was allocated 2300 MHz spectrum (see 2305–2310 MHz).

This service was formerly called *Wireless Communications Service* (WCS). The January 7, 2000 name change went virtually unnoticed, and the industry still refers to this service as WCS. Few WCS operations exist at this writing.

A 700 MHz MWCS licensee could offer a "point-to-multipoint datacast service that would distribute data such as financial or market reports or video or music streams to the general public, and intend to recoup its costs and profit by inserting commercial messages or some other non-subscription mechanism," the FCC gave as one example.

Incumbent broadcasters will be able to offer some of the same services—certainly one-way, and possibly two-way—on DTV channels. Broadcast groups have already formed to launch these services over DTV (see 470–512 MHz).

New Forms of Broadcasting

The allocation to MWCS added the fixed and mobile services and retained the allocation to broadcasting. Conventional TV stations will be exiting this spectrum, so in effect this action created a new allocation to broadcasting that "will not necessarily resemble current radio and television broadcast services...but could still meet the statutory definition of 'broadcasting'," according to the FCC.

These services definitely will not be subject to the conventional radio or TV technical rules. The Commission cautioned, however, that statutory rules (imposed by the Communications Act) will apply to broadcasting in this spectrum. MWCS licensees must disclose to the FCC at least the general character of the service they plan to offer by selecting from a limited list of service types.

The sole public proponent of broadcasting in this spectrum was the *Consumer Electronics Association* (CEA). It called on the FCC to devote the spectrum to "free, high-quality multi-channel digital audio, information and high capacity data services for a new broadcast service to mobile receivers." CEA called this the *Mobile Multimedia Broadcast Service* (MMBS).

MMBS data services could include electronic program guides; stock information; map images, traffic, and navigation information; electronic books and newspapers; advertising, games and software updates; and animated video. MMBS receivers could even be addressable by ZIP code.

CEA did not continue to develop MMBS. "We still support MMBS. We believe it has lots of opportunity, but we're not optimistic that the FCC will designate freed-up spectrum for that service," a CEA spokesperson told us.

At this writing, it appears that fixed and mobile services, not broadcasting, will probably dominate MWCS. Nevertheless, new allocations to broadcasting are very rare, and MWCS is a virtual *carte blanche* for broadcasting.

(The Communications Act defines broadcasting as "the dissemination of radio communications intended to be received by the public, directly or by the intermediary of relay stations." The FCC historically has interpreted this definition to mean free services, and to exclude such broadcast-like services as wireless cable TV and *Direct Broadcast Satellite* (DBS). This is why, for example, scrambled pay-TV is not allowed on conventional VHF or UHF broadcast channels.)

Frequency Allocations

WT Docket 99-168 adopted MWCS allocations as follows (FCC Rule 27.5):

Block A: Two paired channels of 1 MHz each in 746–747 MHz and 776–777 MHz

Block B: Two paired channels of 2 MHz each in 762–764 MHz and 792–794 MHz

Block C: Two paired channels of 5 MHz each in 747–752 MHz and 777–782 MHz

Block D: Two paired channels of 10 MHz each in 752–762 MHz and 782–792 MHz

The FCC claimed that the C block will enable some data services, including Internet access, and if used as a 10-MHz segment, it could enable a single wideband CDMA channel.

The wider, 10-MHz segments could enable Internet access at higher speeds or "substantial augmentation of existing *Commercial Mobile Radio Service* (CMRS) systems," the FCC said. The blocks could be aggregated to form a 30-MHz band.

In its *Third Generation* (3G) Notice of Proposed Rulemaking (Docket ET 00-258; see 1850–1990 MHz), the FCC noted that the C and D blocks "may be used for 3G services" while the A and B block spectrum "is not available for 3G services" because cellular-type systems are prohibited there (see below).

License Scope

The FCC will auction one C and one D 700 MHz MWCS license in each of six regions called *Economic Area Groupings* (EAGs): the Northeast, Mid-Atlantic, Southeast, Great Lakes, Central/Mountain, and Pacific, for a total of 12 licenses. Licensing the spectrum in smaller geographic areas could apparently complicate the protection of incumbent TV stations. Licensees could aggregate EAG authorizations to form a nationwide service area, or partition these licensed coverage areas into smaller service areas.

License terms will run for eight years after 2006, the year that analog TV broadcasting is supposed to cease nationwide and TV stations are to have moved to a consolidated channel group. MWCS operators who use their licenses for broadcasting will receive an eight-year term to commence on the date that they begin broadcasting even if that date is before 2006. The Communications Act limits broadcast licenses to eight years.

Performance requirements (benchmarks) for this spectrum are based on the type of service. For example, by the end of its license term, a MWCS licensee providing fixed service must show four permanent links for each one million people in its service area. Mobile or multipoint operators must show coverage of 20 percent of the population in the area.

With blocks *C* and *D*, the FCC has proposed to permit "combinatorial," or package bidding in which participants are not restricted to placing bids on individual licenses, but may bid on collections of licenses. These could be a package of all licenses in the auction, of all licenses of a particular type, or of all licenses in a region, for example. The FCC may define the permissible packages in advance of the auction.

Guard Bands

Blocks *A* and *B* are guard bands subject to special rules to prevent interference to public safety use in the adjoining 764–776 MHz and 794–806 MHz segments. Traditionally, guard bands are unavailable for transmission. Here, the FCC opened them for commercial use with a flat ban on cellular architecture and stringent emissions limits, a requirement to closely coordinate frequency usage with public safety authorities and under license to an entirely new type of entity.

Even after heavy lobbying by FreeSpace, a prospective provider of fixed cellular Internet service, the Commission forbade cellular architecture in these blocks. It defined cellular architecture as one where "large geographic service areas are segmented into many smaller areas or cells, each of which uses its own base station, to enable frequencies to be reused at relatively short distances."

The FCC called coordinating frequencies for every base station in an area a "complex, uncertain and resource-intensive task for both commercial and public safety users." It dismissed FreeSpace's proposal for sophisticated, dynamically controllable base station networks as still too likely to expose public safety radios to interference, especially in and around private residences, even if it could correct interference after the fact.

The FCC also said that it is "reluctant to adopt procedures that would be tailored to a particular entity's technology." The Commission has in fact adopted such procedures in several services. In this case, a congressional conference report directed the FCC to prevent commercial users from interfering with public safety users in this specific spectrum, increasing the agency's caution.

Band Managers

Licensees on the *A* and *B* blocks will not transmit signals themselves. These "guard band managers" who win license auctions serve as spectrum brokers, "engaged solely in the business of leasing for value spectrum to third parties on a for-profit basis," advancing the goal of enabling a "free market in spectrum to develop," according to the FCC.

The FCC will hold the managers responsible for frequency coordination and interference control. But if guard band managers wish to provide communications services, they must do so through a separate affiliate company.

The band manager concept is consistent with current FCC trends to offload much of its duties to private, non-governmental parties—especially trade groups and frequency coordinators.

Commissioner Harold Furchtgott-Roth, a frequent dissenter to FCC actions, called the separate affiliate requirement an "employment program for corporate lawyers." Commissioner Michael Powell described the guard band manager as a Frankenstein creature apparently designed to ensure that spectrum is devoted to private users instead of commercial users.

Each of the 52 *Major Economic Areas* (MEAs) has one *A* and one *B* license, making a total of 104 licenses. The FCC's September 2000 guard band auction raised more than $519 million. It sold 96 licenses to nine bidders. Unsold licenses will be part of a future auction. The three largest winners were Nextel, Pegasus Guard Band, and Access Spectrum LLC.

Access Spectrum, partly owned by the Industrial Telecommunications Association, will offer capacity to the private business and industrial users among ITA's longtime constituents.

MWCS and TV

Implementation of fixed, mobile, and new broadcasting services in these segments depends on the eventual exodus of TV stations, including DTV stations, from channels 60–69. This transition, which will consolidate all DTV in other channels, is supposed to end in 2006.

Channels 60–69 are "relatively lightly used for full service television operations," the FCC said. Any new operations in channels 60–69 will have to protect TV stations on those channels until the TV stations vacate them. Private auction and clearinghouse schemes are in discussion at this writing to fund clearing of the 700 MHz channels.

The channels already have been officially recovered in geographic areas where there are no TV stations on those channels, the FCC announced in November 1999. Nevertheless, no new broadcast or non-broadcast operations can occur on those channels until license auctions are held.

The FCC later announced its intention to make additional channels, namely 52–59, available to new fixed, mobile, and broadcasting services for commercial uses (see 698–746 MHz).

Federal law requires the FCC to deny the renewal of analog TV broadcast licenses after December 31, 2006. But broadcasters persuaded Congress to allow the FCC to extend this deadline in particular geographic markets depending on the extent of market penetration of digital TV receivers and the extent of DTV broadcasting underway in that market.

Wireless audio systems used in broadcasting operate in the TV channels as FCC-licensed *Low-Power Auxiliary Stations* (LPAS) on a secondary, non-interference basis. According to manufacturers, 746–806 MHz is desirable for LPAS because of the relatively low number of full-power TV stations in that segment.

BLASTing Zone

This is one of several bands in which Lucent Technologies' Bell Laboratories is authorized to experiment with *Bell Labs Layered Space-Time* (BLAST), an exotic technology that the company says is "capable of enormous theoretical capacities" that are "unprecedented," "unattainable using traditional techniques," "paradigm-shifting," and "unthinkable" with conventional approaches.

"Initial experiments in a laboratory facility have indicated that spectral efficiencies on the order of 20-40 bps/Hz can be realized at realistic SNRs [*signal-to-noise ratios*] and error rates," Lucent said. Typical efficiencies in fixed microwave systems are 10–12 bps/Hz.

The BLAST method splits a user's data stream into multiple substreams and transmits them in parallel over multiple antennas.

The receiving station processes the multipath-scattered signals it receives, providing what the company calls a "very useful spatial parallelism that is used to greatly improve data transmission. ...The ability to separate the substreams depends on the slight differences in the way the different substreams propagate through the environment."

764-776 MHz

FG: None

Non-FG: NG128 NG158 NG159 FIXED. MOBILE

FCC: Auxiliary Broadcasting (74). Private Land Mobile (90)

This part of the UHF TV spectrum contains channels 63 and 64. The U.S. is making a major transition to *Digital Television* (DTV, see 470–512 MHz).

As Congress directed in 1997, the FCC reallocated 24 MHz from 764–776 MHz and 794–806 MHz (channels 68 and 69) to state and local public safety agencies for communications such as police, fire, and emergency medical radio (Dockets WT 96-86 and ET 97-157).

Until the U.S. completes its transition from analog to DTV service, which is supposed to end December 31, 2006, public safety operations will have to share these 24 MHz with analog and digital TV stations. The FCC has prescribed detailed signal strength ratios that licensees must observe to fulfill the sharing requirement.

The 700-MHz band is the largest allocation the FCC has ever made for public safety communications.

Eligibility in this band also is available to federal agencies and to non-commercial, non-governmental entities such as Neighborhood Watch groups if authorized by public safety agencies. A national coordinating committee and regional planning committees of public safety users develop procedures and technology for use of this band.

The FCC called interoperability the "crowning achievement" of this allocation. Interoperability essentially means that radios from two or more government agencies can interact with one another and exchange information.

Some mandated public safety interoperability channels already exist (see 821–824 MHz), but have not been sufficient.

"The inability of public safety agencies to efficiently communicate with one another is a glaring deficiency in present day public safety communications," the Commission said. "As a result of the interaction of numerous political, technological, financial, and regulatory obstacles that work to inhibit attempts to establish universal public safety interoperability, this deficiency has persisted despite many years of efforts to eradicate it."

The FCC devoted 2.6 MHz of the available 700 MHz channels for nationwide interoperable communications. It designated 12.6 MHz for general local, regional, or state public safety use and retained 8.8 MHz as reserve spectrum, seeking comment on possible uses and licensing schemes for this portion.

The FCC further divided this spectrum into sets of narrowband (6.25 kHz) and wideband (50 kHz) channels that may be aggregated to form larger channels in order to accommodate wider bandwidth uses such as image communications. Aggregated bandwidths are limited to 25 kHz for narrowband channels and 150 kHz maximum for wideband.

"Applications such as real-time video, full-motion video, and high speed data transfers, which require larger channel aggregates than currently provided for, are inappropriate for the 700 MHz band," according to the *Federal Law Enforcement Wireless Users Group* (FLEWUG). "Additional allocations of public safety spectrum, such as above 3 GHz, are required to support such advanced data applications," it told the FCC, but so far such allocations have not been made.

Digital Radio

Public safety radios at 700 MHz must use digital modulation primarily, and may have analog modulation as a secondary mode. The FCC did not prescribe standards for digital modulation, trunking, or receiver performance, but it charged its 700-MHz national coordinating committee and regional planning committees with investigating the issues and recommending standards where necessary. All 700 MHz public safety radios must be able to operate on the narrowband interoperability channels.

Standards for digital public safety radio are the focus of Project 25 of the *Associated Public Safety Communications Officers* (APCO) and the *Telecommunications Industry Association* (TIA). Project 25 defines digital radio systems with encryption and other advanced features.

A new Project 25/34 will evolve the standard toward high data rates (155 Mb/s or higher), handoffs between base stations, GPS capability, in-building access, and "a long-term wireless subscriber unit and network strategy that includes a transparent migration path to the widespread use of mobile, notebook and handheld computers and PDAs, with both on-net and off-net priority communications, and point-to-point and point-to-multipoint video," according to briefing documents.

In addition to needs for video and position location, a key driver for improved public safety radio is the national trend toward community-based policing, which emphasizes out-of-vehicle duty. This approach places greater stress on police radio networks, which must have enough base stations and coverage to adequately serve portable, low-power radios.

The FCC's aggressive efforts to engender interoperability at 700 MHz and 800 MHz are initially of limited value because most public safety systems do not operate above 512 MHz. Even the attractive 700 MHz allocation will not be available for public safety use in many of the largest cities until after the ostensible end of the DTV transition in 2006.

The Commission proposed to add more nationwide interoperability channels within VHF and UHF bands including 150.8–152.855 MHz, 154–156.2475 MHz, 157.45–161.575 MHz, 450–454 MHz, and 456–460 MHz; it also sought comments on establishing another interoperability band in 138–144 MHz.

776-794 MHz

FG: None

Non-FG: NG128 NG159 FIXED. MOBILE. BROADCASTING

FCC: Wireless Communications (27). Broadcast Radio (TV) (73). Auxiliary Broadcast (74). Private Land Mobile (90)

This band contains TV channels 65, 66, and 67. The U.S. is making a major transition to *Digital Television* (DTV, see 470–512 MHz).

This band was reallocated to the *Miscellaneous Wireless Communications Services* (MWCS, also called WCS, see 746–764 MHz).

It is one of several bands in which Lucent Technologies' Bell Laboratories is authorized to experiment with the unusual, high-capacity *Bell Labs Layered Space-Time* (BLAST) technology (see 746–764 MHz).

794-806 MHz

FG: None

Non-FG: NG128 NG158 NG159 FIXED. MOBILE

FCC: Auxiliary Broadcasting (74). Private Land Mobile (90)

This band, as part of the UHF TV spectrum, contains channels 68 and 69. The U.S. is making a major transition to *Digital Television* (DTV, see 470-512 MHz).

This band has been reallocated to Public Safety Services, however (see 764-776 MHz).

The UHF TV spectrum originally reached to channels 70-83 (806-890 MHz). The FCC later reallocated this segment to mobile services and required the segment's TV stations to move to other channels.

Older TV sets that tune channels 70-83 can receive mobile services in this spectrum such as cellular telephone and 800-MHz mobile radio.

806-821 MHz

FG: None

Non-FG: NG30 NG31 NG43 NG63 FIXED. LAND MOBILE

FCC: Public Mobile (22). Private Land Mobile (90)

This band contains *Private Mobile Radio Services* (PMRS) systems and the mobile-to-base frequencies for the 800-MHz portion of *Specialized Mobile Radio* (SMR), a modern dispatch and mobile telephone service that is one of the *Commercial Mobile Radio Services* (CMRS). The companion base-to-mobile band is 851-866 MHz.

Together with the 900-MHz SMR spectrum (896-901 MHz and 935-940 MHz), these bands contain services that compete with cellular (824-849 MHz) and the *Personal Communications Services* (PCS, 1850-1990 MHz) as well as with conventional mobile radio (see 450-454 MHz).

This 800-MHz spectrum is divided into three "pools:" the General Category, SMR Category, and the Public Safety, Industrial/Land Transportation, and Business pool.

Trunking Technology

The bands accommodate both conventional and "trunked" radio systems. Conventional systems typically use one channel (a single frequency or pair of frequencies in two separate bands). A user who needs to transmit must listen and wait until the channel is available for use. This arrangement is inefficient and counterintuitive. A nearby channel may be unoccupied but unavailable for use because it is licensed to others.

As a solution, trunked systems automatically manage groups of channels, locating and switching user radios to an available channel without user effort. This technique allows more users to be served per radio channel, as long as the licensee has access to more than one channel.

A centralized trunking system uses a central switch that selects channels for all users. FCC rules also permit decentralized trunking in which mobile units monitor the system's assigned channels until an unused channel is found, and then they use this channel for communications.

Decentralized trunking is used in smaller, less sophisticated radio systems and was not under consideration in 1974 when the FCC decided to release 800 MHz spectrum to private mobile radio.

The FCC wanted to use centralized trunking on the new channels. Trunked central equipment, however, is ordinarily too complex and expensive to expect individual users to purchase it. (Some very large users, such as multi-state power utilities, have invested in such equipment.)

SMR History

To promote the adoption of trunking, the FCC established SMR, a novel type of business at the time in which licensed operators could sell trunked communications service and mobile radios to users eligible to hold radio licenses in the private services.

State utility commissions wanted regulatory authority over SMR and fought the FCC's creation of it, as did paging and mobile phone radio common carriers that had been the only parties allowed to derive subscriber revenue from mobile radio transmissions. These opposition campaigns were unsuccessful.

SMR's profit potential was the needed incentive for entrepreneurs to invest in trunked switches, base stations, and radio sales and service facilities in the new bands. Traditional SMR customers are in the construction, service, transportation, and utility industries.

SMR demonstrated the value of trunking, prompting the introduction of this technique into other services including public safety, cellular and federal mobile radio.

Another advantage of trunking is selective calling. The computer-based channel management of a trunked system permits dispatchers to communicate with specific users or groups of users without disturbing other users or requiring them to listen for calls. This feature is critical to the operation of modern police radio networks.

Trunking also made it easier to offer users reliable telephone interconnection: the ability to make telephone calls from mobile and portable radios.

The FCC ultimately allocated 816–821 MHz for trunked SMR mobile-to-base use and 861–866 MHz for trunked SMR base-to-mobile use. It allocated 806–810 MHz and 851–855 MHz for conventional systems.

The 810–816 MHz and 855–861 MHz segments were devoted to trunked or conventional systems in the former Business, Industrial, and Land Transportation, and Public Safety services as well as SMR. The FCC granted licenses on a site- and frequency-specific basis.

Digital, Enhanced SMR

The process of introducing high-capacity digital technologies into SMR began when Fleet Call, a large SMR owner founded by former FCC officials, launched what it called the *Enhanced SMR* (ESMR) or wide-area SMR using a digital architecture from Motorola.

"Wide-area digital SMR providers have sometimes been referred to in the industry as enhanced SMRs to reflect their ability to offer an expanded array of service capabilities (for example, paging, data transmission, positioning) made possible by the carrier's adoption of digital network systems," the FCC has stated.

Fleet Call changed its name to Nextel to symbolize its transition from dispatching to next-generation mobile telecommunications. Nextel acquired SMR licenses in the largest U.S. markets.

"ESMR involves the accumulation of SMR stations, the application of digital technology, and the operation of multiple low-power base stations with significant channel reuse," the company said. "These innovations, introduced at a cost of over one billion dollars to Nextel, make possible an advanced mobile communications system capable of providing mobile telephone service comparable to that currently provided by the cellular industry, as well as private network dispatch, paging and mobile data services."

Legal Restrictions

These developments prompted the cellular industry to lobby for laws purporting to "level the playing field" and subject cellular and ESMR to the same regulations. This principle is known as "regulatory parity," the similar treatment of like services.

As a result, Congress directed the FCC to treat cellular, ESMR, PCS, paging and other telephone-interconnected, subscriber revenue-based services on exclusive frequency, and market slots similarly—categorizing services such as *Commercial Mobile Radio Services* (CMRS). This led to additional nomenclature that distinguishes or confuses the services (see 30.56–32 MHz).

CMRS licenses are subject to auction, whereas the FCC ordinarily grants PMRS licenses on a first-come, first-served, shared-frequency basis. CMRS licensees are now permitted to offer fixed services; in fact they may legally offer only fixed services and no mobile services at all.

Desiring to promote digital ESMR, the FCC designated 10 MHz (816–821 MHz and 861–866 MHz) for SMR systems "that are comparable to and compete with cellular and Broadband PCS systems." The 10 MHz is divided into three blocks of 120, 60, and 20 channels, which are known as the "upper 200" 800 MHz SMR channels. For some $96 million, the FCC in the winter of 1997 auctioned a license for each of the three blocks in 175 *Economic Areas* (EAs), geographic zones defined by the Department of Commerce with additions by the FCC. Nextel was the leading bidder and remains the dominant ESMR provider.

Incumbent licensees in the EAs are subject to voluntary and if necessary, involuntary relocation to other SMR spectrum at the expense of the EA auction winner if the EA licensee can find other channels in which to relocate them.

All geographic area SMR licensees in 800 MHz and 900 MHz are allowed to partition, or assign portions of their operating areas to others, and to disaggregate, or assign portions of their licensed spectrum to others.

In a decision that frustrated many users, the FCC also decided to auction geographic area licenses in the congested "lower 230" 800 MHz channels. These include both SMR channels and General Category channels available for licensees' internal operations.

"While non-SMR operators may not require geographic licenses to operate systems designed for internal communications," the FCC said, "geographic area licensing remains the most efficient and logical licensing approach for the majority of licensees in the band. We are not persuaded that we should forego the benefits of geographic licensing to accommodate the interests of a small minority of systems. In any event, systems that are not SMR systems will remain fully protected under our geographic licensing rules. In addition, non-SMRs can obtain spectrum to suit their internal communications needs by forming joint bidding consortia or by entering into partitioning and disaggregation agreements with EA licensees."

Auction winners in the lower 230 channels will have to avoid interference to SMR or non-SMR operators already using that spectrum, and will not be able to compel these incumbents to relocate "because there is no identifiable alternative spectrum to accommodate such migration," according to the FCC.

"In addition, it is likely that many of the incumbents who will operate on these channels will have relocated from the upper 200 channels, and we have already determined that such relocatees should not be required to relocate more than once," the Commission added. "Therefore, EA licensees on the lower 230 channels will not have the right to move incumbents off of their spectrum blocks unless the incumbent voluntarily agrees to move." The lower 230 channel incumbents are allowed to combine their site-specific licenses into single geographic area licenses.

The FCC auctioned General Category channels and some upper band licenses in September 2000. Licenses for six contiguous 25 channel blocks in the 806–809 MHz/851–854 MHz sub-bands, designated *D*, *DD*, *E*, *EE*, *F*, and *FF* were auctioned in each of 172 EAs plus three EA-like areas in U.S. island territory. Three additional licenses in the upper SMR pool, *A*, *B*, and *C*, were offered for auction in U.S. island areas for a total of 1,053 licenses. The auction raised more than $319 million.

Another group of channels, the "SMR Lower 80," were auctioned as 16 licenses in each of 172 EAs plus three EA-like areas in U.S. island territory for a total of 2,800 licenses. Each license, designated *G* through *V*, covers five non-contiguous channels. The December 2000 Lower 80 auction pulled in about $29 million for the Treasury.

SMR, ESMR, and PMRS systems in these bands carry data communications as well as voice. Other services dedicated to portable mobile data devices also emerged. The former ARDIS network, founded by Motorola and IBM in the early life of the 800 MHz band, is now owned by Motient (see 1525–1530 MHz) and operates on nationwide and local 800 MHz business radio and SMR frequencies. It provides data communications for field service, transportation, security, and industrial customers. Similarly, the Palm.net wireless handheld service uses SMR frequencies.

Third Generation Systems

The *World Radiocommunication Conference-2000* (WRC-2000) included 806–960 MHz among the bands available for *International Mobile Telecommunications-2000* (IMT-2000), an advanced, global mobile and portable service often called "3G" for Third Generation (see 1850–1990 MHz).

However, the FCC stated in Docket ET 00-258 that it does not intend to allocate for advanced wireless services any additional spectrum in the portions of 806–960 MHz already used by cellular and SMR. And "reallocation of other parts of the 806–960 MHz band for advanced wireless systems would not further facilitate worldwide roaming or economies of scale in equipment manufacturing. Further, it appears that the remaining parts of the 806–960 MHz band are heavily occupied by existing services. Accordingly, we do not believe that any additional reallocation of spectrum for advanced Wireless Services is appropriate in the 806–960 MHz band," it said, seeking comment on this tentative conclusion.

Portions of 806–824 MHz are used for high-power Navy shipborne, long-range search radars.

821–824 MHz

FG: None

Non-FG: NG30 NG43 NG63 LAND MOBILE

FCC: Private Land Mobile (90)

To promote technology, interoperability, and local control in public safety communications, such as police and fire radio, in 1986 the FCC allocated the 821–824 MHz (mobile-to-base) and 866–869 MHz (base-to-mobile) bands nationwide for public safety use.

These bands provide 230 channels for general assignment and five channels for mutual aid, all at 12.5 kHz channel spacing. The mutual aid channels are intended to enable different jurisdictions to readily intercommunicate.

The plan established a *National Public Safety Planning Advisory Committee* (NPSPAC) to oversee committees in national regions that coordinate frequency use in these bands.

The public safety community was not satisfied with the capacity and limited interoperability afforded by these bands. It successfully lobbied Congress for additional spectrum. As Congress directed, the FCC reallocated to public safety use the bands 764–776 MHz (UHF TV channels 63 and 64) for base to mobile operation and 794–806 MHz (channels 68 and 69) for mobile to base operation.

"Over the past decade, police, fire, emergency medical, and other public safety providers have been confronted by a number of problems that threaten their ability to fulfill their mission of protecting the public," the FCC stated. "Public safety communications continue to be plagued by inefficient spectrum use, by the absence of a competitive market for public safety communications equipment and services that meet public safety agency needs, and by difficulties in building a structure for interoperable communications among public safety agencies."

According to the FCC's *Public Safety Wireless Advisory Committee* (PSWAC), interoperability is "hampered by the use of multiple frequency bands, incompatible radio equipment, and a lack of standardization in repeater spacing and transmission formats." The Commission improved the prospects for interoperability in the reallocated 764–776 MHz and 794–806 MHz spectrum by imposing detailed technical and oversight requirements.

Portions of 806–824 MHz are used for high-power Navy ship-borne, long-range search radars.

824–849 MHz

FG: None

Non-FG: NG30 NG43 NG63 NG151 FIXED. LAND MOBILE

FCC: Public Mobile (22)

The *Cellular Radiotelephone Service* expanded mobile telephony beyond its small initial audience of the professional elite and into the mass consumer market.

Key technical factors in this achievement include the generous spectrum allocation afforded to cellular carriers; the inventions of trunking and frequency reuse, which increased spectrum efficiency and capacity; the development of comprehensive technical standards; and progress in miniaturization and cost reduction in microwave components.

The U.S. cellular scheme provides for two carriers, A and B, in each geographic market, which is known as the cellular "duopoly." Originally, the B, or wireline carrier was affiliated with the local telephone company, and the A, or "non-wireline" carrier was not so affiliated. After 1986 the wireline companies were allowed to buy A carriers.

Cellular History

In 1970, AT&T proposed a high-capacity cellular phone system based on the reuse of frequencies at multiple base station sites (cells or cell sites). Mobile units communicate with cells, which connect by microwave or landline to a *Mobile Telephone Switching Office* (MTSO).

Together with hardware and software in the cells and mobile units, the MTSO manages signaling, telephone interconnection, channel and power-level selection, and transfer of traffic between base stations.

AT&T called the system *Advanced Mobile Phone Service* (AMPS); this term describes the FM-based analog standard used in U.S. cellular today. The FCC issued rules for the service in 1981 after years of proceedings and experimental operations in the Washington/Baltimore area and Chicago.

The FCC licensed cellular systems in 734 areas, comprised of 306 *Metropolitan Statistical Areas* (MSAs), the nation's major cities and towns, which encompass 75 percent of the population, and 428 *Rural Statistical Areas* (RSAs), which comprise about 80 percent of the land in the U.S.

Commercial cellular service in the U.S began in a Chicago parking lot on October 13, 1983 when technician Jeff Benuzzi beat 13 others in a race to complete a customer's car phone installation.

Cellular Band Plan

The cellular channels begin with 824.04–834.99 MHz, the A segment devoted to mobile-to-base transmissions.

Within the block A cellular mobile-to-base segment 824.04–834.99 MHz, control channels that manage the calling process are at 834.39–834.99 MHz.

Cellular base-to-mobile segments are 45 MHz higher in frequency than the mobile-to-base segments. For example, the 824.04–834.99 MHz A mobile-to-base segment corresponds to the A base-to-mobile segment at 869.04–879.99 MHz. Within that base-to-mobile segment, the control channels are at 879.39–879.99 MHz.

The newer mobile-to-base *A* segment is 845.01–846.48 MHz, corresponding to the newer base-to-mobile *A* segment at 890.01–891.48 MHz.

The original *B* mobile-to-base segment is 835.02–844.98 MHz, which has control channels occupying part of the segment at 835.02–835.62 MHz. The corresponding *B* base-to-mobile segment is 880.02–889.98 MHz, which has *B* base-to-mobile control channels at 880.02–880.62 MHz.

The newer mobile-to-base *B* segment is 846.51–848.97 MHz, corresponding to the newer base-to-mobile *B* segment at 891.51–893.97 MHz.

To accommodate the increasing demand for service, the FCC added the newer segments just mentioned in the mid-1980s. In that proceeding, hotly contested by other claimants to the spectrum, the cellular industry persuaded the FCC that cell-splitting (increased re-use of spectrum) was too expensive a substitute for additional spectrum.

The cellular industry now widely employs cell-splitting to increase capacity and is now largely based on digital technology. The FCC in 1988 allowed cellular carriers to adopt new technologies as long as they continued to support AMPS users. It declined to endorse any particular digital cellular standard.

The FCC granted most cellular licenses by hearing procedures, not competitive bidding. It later used lotteries and auctions to assign certain cellular licenses in geographic areas that were not served by existing cellular carriers. Its January 1997 auction of 14 cellular unserved area licenses raised more than $1.8 million.

The U.S. cellular digital standards are essentially *Global System for Mobile Communications* (GSM), based on *Time Division Multiple Access* (TDMA), which divides the channel into time slots; and the *Code Division Multiple Access* (CDMA) standard cdmaOne. It distinguishes between users by means of unique codes. Carriers are implementing higher-speed data and enhanced services, and evolving these standards toward *Third Generation (*3G) versions (see 1850–1990 MHz).

Cellular Mobile Data

Cellular serves wireless data users with *Cellular Digital Packet Data* (CDPD), an interoperable specification based on TCP/IP and developed by a consortium of cellular carriers. CDPD radios operate through analog cellular systems to convey data for vertical applications such as remote monitoring, credit card verification, and Internet access.

The cellular industry experienced dramatic competitive impact from the advent of *Personal Communications Services* (PCS, 1850–1990 MHz). PCS carriers compete with cellular on price, features, and convenience of purchase and use. PCS phones are somewhat less susceptible to fraudulent use, which extensively compromised analog cellular service and caused enormous losses.

Analog transmissions are not permitted in PCS. Therefore, PCS signals are inherently more secure from eavesdropping than analog AMPS cellular signals are, which can be received on unmodified television sets as well as specialized receivers.

The 824–849 MHz band is one of several in which Lucent Technologies' Bell Laboratories is authorized to experiment with the unusual, high-capacity BLAST technology (see 746–764 MHz).

849-851 MHz

FG: None

Non-FG: NG30 NG63 AERONAUTICAL MOBILE

FCC: Public Mobile (22)

This is the uplink band of *Commercial Aviation Air-Ground Systems* (often simply called *Air-to-Ground* or ATG), part of the *Public Mobile Air-Ground Radiotelephone Service*: public phones aboard scheduled airlines. Radios connected to handsets in the cabin link passengers to the public phone network through ground stations installed across the country.

Invented by communications entrepreneur and MCI founder Jack Goeken, the Airfone service operated experimentally for years only to have its request for permanent spectrum rebuffed twice by the FCC in favor of other services. Goeken sold Airfone to GTE, which continued to operate the service under its experimental license.

The FCC eventually relented and established a permanent spectrum home for ATG. Verizon later acquired GTE Airfone.

The service is allocated 849–851 MHz (uplink) and 894–896 MHz (downlink). Of the six available licenses, only Airfone and Claircom/AT&T Inflight are in use. (Another licensee, In-Flight Phone Corp., is now closed. In-Flight Network, a satellite-based service of News Corp., is preparing to compete with the ATG incumbents.)

At this writing, Claircom is experimenting with repeater stations that permit customers to make phone calls while the aircraft is at a gate and blocked from ground station reception. The company revealed that it seeks to add packet data to this in-airport service. The experiments use spectrum that has not otherwise been assigned to an ATG licensee.

Novel Air Cellular System

AirCell Inc. established another air-ground telephone system intended for private aircraft and small airlines. It uses existing terrestrial cellular spectrum and infrastructure in cooperation with licensed cellular carriers. AirCell is not itself a Commission licensee.

The company faced opposition from large cellular carriers, but successfully persuaded the FCC to authorize the system, exempting it from a general ban on cellular phone usage aboard an aircraft.

Air-ground telephony for business aircraft has been conducted mainly in the *General Aviation Air-Ground Radiotelephone Service* (see 454–455 MHz).

851-866 MHz

FG: None

Non-FG: NG30 NG31 NG63 FIXED. LAND MOBILE

FCC: Public Mobile (22). Private Land Mobile (90)

The 851–866 MHz band contains the base-to-mobile frequencies for conventional and trunked private land Mobile Radio Services (see 806–821 MHz).

866-869 MHz

FG: None

Non-FG: NG30 NG63 LAND MOBILE

FCC: Private Land Mobile (90)

This band contains base-to-mobile frequencies for public safety licensees under the FCC national public safety radio plan (see 821–824 MHz).

869-894 MHz

FG: None

Non-FG: US116 US268 NG30 NG63 NG151 FIXED. LAND MOBILE

FCC: Public Mobile (22)

Base stations in the *Cellular Radiotelephone Service* transmit to mobile units in this band. Mobile units transmit 45 MHz lower in frequency (see 824–849 MHz).

FCC Part 15 rules permit non-cellular, unlicensed operation in this band at very low-power levels, a provision that to our knowledge has never been used.

Other Part 15 rules forbid the manufacture or importation of scanning receivers that tune 824–849 MHz or 869–894 MHz in order to protect the privacy of cellular users. Cellular signals can be received on older TV sets (channels 70–83) and other devices.

The 869–894 MHz band is one of several in which Lucent Technologies' Bell Laboratories is authorized to experiment with the unusual, high-capacity BLAST technology (see 746–764 MHz).

Footnote US268 provides for Navy ship-borne, long-range search radars in the 890–902 MHz and 928–942 MHz spectrum.

894-896 MHz

FG: US116 US268 G2

Non-FG: US116 US268 AERONAUTICAL MOBILE

FCC: Public Mobile (22)

This is the downlink band for *Commercial Aviation Air-Ground Systems,* also called *Air-to-Ground* (ATG) (see 849–851 MHz).

896-901 MHz

FG: US116 US268 G2

Non-FG: US116 US268 FIXED. LAND MOBILE

FCC: Private Land Mobile (90)

This band contains the mobile-to-base frequencies for the 900 MHz version of *Specialized Mobile Radio* (SMR), which was established in 1986 to alleviate congestion in the 800 MHz SMR service (see 806-821 MHz). Base stations transmit in 935-940 MHz.

Unlike the 806-821 MHz/851-866 MHz SMR and private radio spectrum, the 896-901 MHz 935-940 MHz spectrum was slow to develop.

In 1995, the FCC moved to grant the remaining available licenses in the 900 MHz SMR band by auction. The licenses are based on *Major Trading Areas* (MTAs), geographic zones defined by Rand McNally.

The auction concluded April 15, 1996, raising more than $204 million for 20, ten-channel licenses in each of 51 MTAs. These are *Commercial Mobile Radio Services* (CMRS), in which Nextel is the dominant provider. Nextel terms its digital mobile radio and mobile phone services *Enhanced SMR* (ESMR).

Incumbent licensees in this spectrum will retain the right to operate under their existing authorizations, but will have to obtain the MTA license or the consent of the MTA licensee to be able to expand their systems.

Advanced Personal Service

An unusual service that would have utilized this spectrum was the Citizens Band Land Mobile *Private Radio Communications Service* (PRCS).

The FCC *Personal Use Radio Advisory Committee* (PURAC) of the 1970s was charged, among other things, with exploring the idea of a successor to the amok 27 MHz Citizens Band.

PURAC recommended a new service at 900 MHz, and the FCC issued a proposal based on the PURAC recommendation. The proposal languished, however, until General Electric asked the FCC to expedite it in a petition for rulemaking. The petition engendered a *Notice of Proposed Rulemaking* (NPRM) in Docket GN 83-26.

GE envisioned PRCS as a FM-based consumer service: a radio with advanced features such as mobile telephone, selective calling, and decentralized trunking (see 806–821 MHz).

PRCS was to use 133 channels at 898–902 MHz and 937–941 MHz. GE invested heavily to develop PRCS mobile and base radios. GE demonstrated the system, branded CarFone, in Washington, D.C. to a handful of FCC personnel and reporters. The FCC refused to permit a more extensive market test in Syracuse, New York.

We believe that the FCC rejected the market test because it feared that consumers might eventually bring unwanted pressure on the Commission to formally authorize the service.

PRCS was to employ a base station connecting to the public network at the owner's premises, much like a cordless telephone base unit. It would have been capable not only of phone calls through these stations and via optional, subscription-based repeaters, but also of direct, free mobile-to-mobile and base-to-mobile communication. This feature is impossible to deliver in the public cellular service.

General Mobile Radio Service (GMRS) licensees (see 462.5375–462.7375 MHz) supported PRCS. However, cellular carriers and GE's competitors lobbied bitterly against PRCS. GE itself delivered a serious blow in 1984 by ending PRCS development after financial reviews within the company, but before a formal FCC decision on the matter.

Without a manufacturer to build the radios and with cellular interests determined to halt PRCS, the FCC terminated its PRCS proceeding in a tense public meeting in November 1984.

At the last minute, some Commissioners seemed reluctant to cancel PRCS. This prompted an unprecedented outburst from a former FCC official in the audience. He denounced the proposal, warning that if the public were allowed access to PRCS, the FCC would never be able to control it.

"It appears that PRCS, despite the possibility of its widespread appeal, is in the nature of a luxury or convenience, given the alternatives available," the FCC concluded. "While we do not question the claims made by PRCS proponents that this service could have some safety of life and property utility, in our estimation the predominant use would be for lower priority communications."

In what one official termed an "apologetic palliative," the FCC considered, but later rejected, allowing PRCS to share spectrum allocated to the cellular service in some geographic areas.

901-902 MHz

FG: US116 US268 G2

Non-FG: US116 US268 FIXED. MOBILE

FCC: Personal Communications (24)

The *General Purpose Mobile Radio Service* (GPMRS) was an ephemeral service at 901–902 MHz and 940–941 MHz with no licensees, rules, or operations.

The FCC defended GPMRS against petitions for reconsideration from private mobile radio associations that wanted access to GPMRS spectrum. The Commission never actually implemented GPMRS, however. In 1993, it eliminated the service entirely.

The FCC allocated the 901–902 MHz and 940–941 MHz bands plus 930–931 MHz, a reserve band for paging, to the *Narrowband Personal Communications Services* for a total of three megahertz (N-PCS, Docket GEN 90-314).

The FCC "declined to adopt a restrictive definition of Narrowband PCS, such as limiting this category of PCS to advanced messaging and paging services" in order to promote other potential uses of the service. Nevertheless, most N-PCS systems were designed for two-way messaging and assured-receipt one-way messaging. N-PCS licensee Conxus was devoted to digital voice messaging. It closed in August 1999, discontinuing service to some 80,000 major-market subscribers.

Pioneer Preference

Though 900 MHz allocations for some type of new paging service had been an industry discussion topic for years, the FCC based its decision to create N-PCS primarily on a proposal of *Mobile Telecommunication Technologies Corp.* (Mtel) for a network for portable two-way data devices.

For its development of multicarrier modulation and other N-PCS technologies, the FCC awarded Mtel a Pioneer's Preference (early assurance of a license). The FCC said, "Mtel's development of technology permitting attainment of 24 kb/s data rates for a nationwide simulcast paging and messaging integrated service is a significant communications innovation...

"Mtel has demonstrated the feasibility of its new spectrum-efficient technology and designed a system based upon it that permits using the same infrastructure to provide both substantially improved and more spectrum-efficient paging services and new one-and two-way messaging services."

Mtel was supposed to receive a free license, but the FCC charged the company $33.3 million for the license anyway after receiving congressional authority to auction licenses.

The FCC reasoned that "awarding Mtel's license subject to only nominal [application] fees when its competitors would be required to pay a substantial amount of money to acquire narrowband PCS licenses at auction would have a significant adverse impact on the competitive marketplace."

A federal appeals judge concluded otherwise, saying that the FCC's action was "lawless" and "borders on the outrageous." Congress later withdrew the FCC's authority to provide preferential treatment to applicants who were judged to be pioneers.

Service Structure

The Mtel license, operated by SkyTel, comprised one of the 11 licenses available nationwide.

The FCC established four different types of licenses and originally, four types of service areas for Narrowband PCS. The four license types include two paired 50-kHz channels, a 50-kHz channel paired with a 12.5-kHz channel, 50-kHz unpaired channels, and 12.5-kHz unpaired channels. The four service areas were nationwide: 51 *Major Trading Areas* (MTAs), five regional areas, and 493 smaller *Basic Trading Areas* (BTAs).

At this writing, two N-PCS auctions have been held. In its first-ever license auction in July 1994, the FCC received more than $617 million for the ten remaining nationwide Narrowband PCS licenses. In November 1994, the FCC auctioned 30 regional licenses (six in each of the five regions) for more than $394 million. Some of the licenses were cancelled or returned for nonpayment. The MTA and BTA licenses and unpaired channels designated as response channels had not been auctioned.

In May 2000, the FCC made major changes to N-PCS (Docket GEN 90-314). Convinced that BTAs were too small, it stopped using BTAs in N-PCS and will license the remaining N-PCS spectrum as MTAs. It eliminated the spectrum aggregation limit that prevented a licensee from holding more than three channels in an area.

It also eliminated certain restrictions on eligibility for eight of the N-PCS channels devoted to mobile-to-base response use, permitted flexible geographic partitioning and spectrum disaggregation of N-PCS licenses, and decided to auction licenses in the 930-931 MHz band that had been held in reserve. However, it left open the details of how 930–931 MHz should be channelized and licensed.

902-928 MHz

FG: S5.150 US215 US218 US267 US275 G11 RADIOLOCATION G59

Non-FG: S5.150 US215 US218 US267 US275

FCC: ISM Equipment (18). Private Land Mobile (90). Amateur (97)

Informally known as the "kitchen sink" band, 902–928 MHz contains many applications including cordless phones, listening devices, wireless local area networks, military radars, and commercial Location and Monitoring Service systems.

Government radiolocation is primary in 902–928 MHz. Examples of these military systems include high-power air surveillance radars on aircraft carriers, tracking and telemetry radar systems used in aeronautical flight testing, systems that monitor the positions of missiles, drone and manned aircraft and land units, and perimeter protection devices for intrusion detection at military facilities.

Naval radars use 902–928 MHz because the band's propagation characteristics enable detection of "sea skimmers," fast moving targets over water. Radars in other bands are less effective over water because the sea surface reflects radio energy back to the radar receiver, causing errors in the determination of the speed and distance of the object of interest.

Another radiolocation use in 902–928 MHz is radar wind profiling for weather forecasting, aviation warning, marine observations, and environmental studies. Wind profilers also operate at 449 MHz. NASA uses the band with remotely piloted aircraft that measure atmospheric ozone at high altitudes. This work supports remote sensing for forest fire studies and disaster assessments.

Government fixed and mobile radio systems are in this band on a secondary basis to radiolocation. These include mobile and portable radios, the transmission of images seen by bomb disposal robots, and fixed systems for such purposes as control of power utilities and video links for monitoring entry points at national borders.

Listening and Looking In

Federal law enforcement agents and criminal organizations also use this band, among others, for voice and video eavesdropping and surveillance equipment. "In the United States, the act of electronic eavesdropping—and mere possession of electronic eavesdropping devices—is illegal unless it has been authorized by a court, or has been granted exception status by law," according to counterespionage consultant Kevin Murray of Murray Associates. "Unauthorized frequency occupation by a bug is another legal issue. Legal issues, however, don't concern illegal eavesdroppers, who are concerned with getting the information without getting caught.

"For these reasons wireless bugs have low-power output and use frequencies where the chance of discovery (accidental or otherwise) is slim. They may use unusual modulation techniques such as spread spectrum or subcarrier. They may be remotely controlled, to keep on-air time at a minimum. Predicting their operating frequencies is not practical. Although most transmissions are in the 30 MHz to 2.5 GHz range, low frequency, and microwave transmissions occur too," he continued.

Industrial, Scientific, and Medical (ISM)

ISM devices are highest in priority among non-government uses, with the ISM frequency specified as 915 +/- 13 MHz. The FCC defines ISM equipment as "equipment or appliances designed to generate and use *radio frequency* (RF) energy to perform some work other than telecommunications."

In addition to 902–928 MHz, U.S. ISM bands include 6.765–6.795 MHz, 13.553–13.567 MHz, 26.957–27.283 MHz, 40.66–40.70 MHz, 2400–2500 MHz, 5.725–5.875 GHz, 24–24.25 GHz, 61–61.5 GHz, 122–123 GHz, and 244–246 GHz.

ISM equipment includes industrial heaters that cure glue, inks, and rubber, welding equipment, food equipment such as bacon dryers and donut fryers, and medical instruments for magnetic resonance imaging, diathermy (tissue heating), microwave aided liposuction, and thermotherapy for treatment of prostate disease (see 1240–1300 MHz). Consumer microwave ovens formerly operated at 915 MHz. These products are now in the 2400 MHz ISM band.

Also in 902–928 MHz are the 33-cm Amateur Radio band and unlicensed Part 15 devices. This band has not yet become a major hub of amateur operation as the lower VHF and UHF bands common in amateur mobile and data communications have. Sophisticated 902–928 MHz spread spectrum amateur equipment is under development by the *Tucson Amateur Packet Radio Corp.* (TAPR) at this writing (see 144–146 MHz).

Location Services

The FCC defines the *Location and Monitoring Service* (LMS) as "the use of non-voice signaling methods to locate or monitor mobile radio units. LMS systems may transmit and receive voice and non-voice status and instructional information related to such units." Two types of LMS are available: multilateration and non-multilateration.

A multilateration LMS system is "designed to locate vehicles or other objects by measuring the difference of time of arrival, or difference in phase, of signals transmitted from a unit to a number of fixed points or from a number of fixed points to the unit to be located," the FCC said.

Multilateration LMS for vehicle location (and potentially, the location of people and objects) was supposed to expand in this band after new entrants received licenses at auction. Multilateration LMS is an undeveloped service at this writing that has little competition or consumer awareness nationally.

The FCC made LMS part of the *Intelligent Transportation Systems* (ITS) Radio Services. According to the FCC, LMS "refers to advanced radio technologies designed to support the nation's transportation infrastructure and to facilitate the growth of ITS."

ITS generally refers to technology applications intended to make commercial and consumer transportation safer and more efficient. Enormous amounts of federal funds are devoted to ITS demonstrations and system installations, which is consistent with international trends, yet none has been spent on LMS to our knowledge. Typical ITS services include in-vehicle navigation computers, electronic signage, traveler information and driver alerting systems, and roadway surveillance networks.

LMS licensees may transmit status and instructional messages, but the messages must be related to the location or monitoring functions of the system. The FCC allows LMS real-time telephone interconnection (that is, mobile telephone service) only in emergencies. LMS is allowed to transmit voice mail. These restrictions make it harder for LMS to compete directly with Cellular and Personal Communications Services.

Teletrac is the leading LMS developer and operator, providing fleet tracking and stolen vehicle location services in several U.S. cities. Teletrac acquired licenses under a previous, non-auction regulatory scheme. It was the only such licensee "grandfathered" (allowed to remain on the air) when the FCC switched to auctions for LMS licenses.

Fleet tracking allows dispatchers of taxis, trucks, buses, and ambulances to locate the vehicles and to be notified of emergencies. A vehicle theft alarm can alert Teletrac if not deactivated within a short time after it is triggered. Teletrac then attempts to contact the vehicle owner and police.

A prospective competitor to Teletrac, Pinpoint Communications, held LMS licenses in about 30 major markets. The company was in the beta testing stage when it was liquidated in 1996.

Analysts generally believe that the market for location-oriented wireless services is headed for growth. Peter Rysavy, president of Rysavy Research, told us, "Corporate America has widely endorsed the value of tracking shipments, inventory and vehicles in real time. The next stage is the application of location technology to consumer applications. Any company that can simultaneously address the issues of privacy, ease of use and suitable pricing will have a tremendous opportunity. The applications are limitless, whether assisting parents with knowing where their children are, providing personal safety, or offering select medical services to the elderly or people with unique health conditions."

LMS at 902–928 MHz also should not be confused with the *Stolen Vehicle Recovery Service* (SVRS), operated only by Lo-Jack Corp. SVRS is allocated the single VHF frequency 173.075 MHz (see 162.0125–173.2 MHz).

Some LMS operation is below 512 MHz, but this is mainly for law enforcement. LMS below 512 MHz cannot legally be used for commercial purposes or to offer any service to the public.

Non-Multilateration

The FCC defines non-multilateration LMS as a system that "employs any of a number of non-multilateration technologies to transmit information to and/or from vehicular units." In practice, this usually means short-range data communication with radio tags. Reader devices illuminate the tags affixed to vehicles or other objects.

Passive tags reflect the signal back to the reader's receiver. The tag uses data about the host object in its memory to modulate the reflection (modulated backscatter). Active tags use battery powered transmitters. The reader recognizes tags by the identification of the signals and forwards this data to computers for processing. Tag applications include toll collection, antipilferage, security and access control, shipping, railroad operations, and even identifying lost pets.

For pets, the industry is developing standards for tags implanted in animals. The tags operate under LMS or under Part 15 rules, especially at 2400 MHz.

Non-multilateration systems are prohibited from offering non-vehicular location services. The FCC reasoned that this policy reduces the potential for interference from LMS to amateurs, Part 15 devices, and government operations in the non-multilateration sub-bands. Non-multilateration licenses are assigned on a shared, first-come basis without auctions. Some non-multilateration applications are expected to migrate to *Dedicated Short Range Communications* (DSRC, see 5.85–5.925 GHz).

Bandplan and Licensing

The 902–928 MHz LMS bandplan is as follows:

Sub-band	Frequencies (MHz)	Usage
A	902.000–904.000	Non-multilateration
B	904.000–909.750	Multilateration
C	909.750–919.750	Non-multilateration
D	919.750–921.750	Multilateration and non-multilateration
E	921.750–927.250	Multilateration
F	927.250–927.500	Narrow band associated with sub-band E
G	927.500–927.750	Narrow band associated with sub-band D
H	927.750–928.000	Narrow band associated with sub-band B

A multilateration licensee can aggregate sub-bands *D* and *G*, and *E* and *F* in order to obtain up to 8 MHz in a market, but cannot use more than one of the multilateration sub-bands in a market.

To delineate LMS market areas, the FCC uses *Economic Areas* (EAs) devised by the Department of Commerce. One hundred seventy-two EAs cover the continental U.S. In addition to these, the FCC added four EAs for Alaska and U.S. possessions, plus a Gulf of Mexico territory.

The FCC assigns three multilateration licenses in each EA (for sub-bands *B* and *H*; *D* and *G*; and *E* and *F*), creating a total of 528 EA licenses. The FCC said it will use a case-by-case approach in determining whether a particular LMS offering is a *Commercial Mobile Radio Service* (CMRS) or *Private Mobile Radio Service* (PMRS) (see 30.56–32 MHz). The decision will turn mostly on the extent to which a LMS operation is interconnected to the public telephone network.

The March 1999 LMS license auction raised approximately $3.4 million. Only five bidders participated.

An LMS-like *Personal Location and Monitoring Service* (PLMS) has been proposed for other bands (see 2300–2305 MHz).

Part 15 Devices

Low-power unlicensed devices operating under Part 15 of the FCC Rules in 902–928 MHz have become a sizable business.

Applications include residential and business cordless telephones, fire and security alarms, wireless bar-code readers, data collection terminals for industrial, medical, and financial uses, remote utility meter readers, remote video cameras, and wireless video products for consumers.

Wireless *local area networks* (LANs) for computers are increasingly common in this band and in 2400 MHz (see 2400–2402 MHz). Wireless Internet service providers such as Metricom also use this band. Its service is the first wide-scale attempt to provide a for-profit carrier business using unlicensed spectrum. Its customers use small radios to connect to base stations that are mounted on utility light poles.

Metricom's next-generation service provides faster access with a combination of 902–928 MHz, 2400 MHz, and *Wireless Communications Service* operations (see 2305–2310 MHz).

Figure 9
The robot jockeys of Mid-South Indoor Horse Racing represent one of the earliest commercial applications of spread spectrum communications technology. (C. D. McVean archives)

Betting on Bots

A pioneering commercial use of this band (and easily the most entertaining) was in spread spectrum remote control of robot jockeys.

Developed by Super-Jock, Inc. and Mid-South Indoor Horse Racing of Memphis, the robots made it possible to race breeds too small to support the weight of a human jockey. The smaller animals can race through the tighter turns required by indoor spaces.

Mid-South raced Hackneys, a British breed about 40 percent of the weight of thoroughbreds. The human jockey operated the robot with a handheld device that permits voice command as well as control of reins and crop.

The designers spent considerable effort on preventing jamming or unauthorized control of the robots. They chose spread spectrum for its security and interference immunity. The control signals of all jockeys were recorded and could be synchronized with video replays. "We believe these high-tech systems, taken collectively, have created the least fixable form of pari-mutuel racing possible," the company said.

The FCC authorized experiments, but Tennessee regulators denied Mid-South and others the necessary racing licenses, and the project was discontinued.

Spread Spectrum Techniques

The FCC permits spread spectrum devices to transmit up to one watt of power in the ISM bands, depending on the specific technology used (most devices transmit at lower power levels). Spread spectrum emissions are produced by frequency hopping or direct sequence, which combines data with fast digital codes or a combination of these techniques.

As the name implies, spread spectrum emissions are wider in bandwidth than conventional emissions. This makes the spread spectrum system a lessened interference threat to narrowband devices or services and more immune to interference from them.

Strictly speaking, unlicensed devices are not allocated spectrum, but they are permitted to use the spectrum on a sufferance basis and must accept any interference they may receive from transmitters higher in priority. Nevertheless, manufacturers of 902–928 MHz Part 15 devices became concerned about interference from LMS and the effects of their devices on LMS networks.

Unprecedented Policy

The FCC established a complex regulatory scheme to govern the relationships between licensed and unlicensed emitters in this band. Unlicensed devices have no right to the continued use of any given frequency. An unprecedented "safe harbor" policy, however, deems unlicensed spread spectrum Part 15 devices in 902–928 MHz to not interfere with multilateration LMS if the devices meet antenna, power, and certain usage criteria.

For example, if certain devices employ outdoor antennas and provide "the final link for communications of health care providers that serve rural areas, elementary schools, secondary schools or libraries," then their operations will not be considered to be causing harmful interference to a multilateration LMS system, according to the FCC. (The "safe harbor" policy does not insulate Part 15 and amateur operators from claims of interference from non-multilateration LMS.)

Unlicensed devices may operate outside of the safe harbor provisions, as long as they comply with the ordinary Part 15 rules that apply to such devices, but are not protected from LMS operators' claims of interference. Radio amateurs may also operate under the safe harbor provisions and enjoy the same protection as Part 15 devices.

The policy was not universally praised. FCC Commissioner James Quello described it as "but our first step into a potential bog of interference problems."

Pointing out the difficulty of financial valuation of LMS spectrum (and in acknowledgment of the interference protection provisions of Footnote US218), Teletrac stated, "The locales and times of interference that multilateration LMS systems must accept from devices such as industrial heaters, welding equipment and magnetic resonance scanners cannot be predicted with any degree of certainty, and even informed estimates require time-consuming research into patterns of industrial development...

"LMS systems must also accept interference from federal government Radiolocation, Fixed and Mobile Services. Interference from military radar, data and video links, tracking systems, NASA aircraft and other federal usage may require testing, adjustment and accommodation in affected locales to permit reliable LMS service and will limit the technologies that can be used in the band, thus rendering the spectrum less valuable...

"Amateurs and unlicensed devices operating under Part 15 of the Commission's Rules also use the 902–928 MHz spectrum. Penetration in this band of unlicensed devices such as cordless telephones, alarm systems, wireless Internet and LAN services, remote utility meter readers and remote video cameras and links is thought to be sizable."

LMS auction winners are further encumbered by non-multilateration LMS systems in sub-band D and by the "grandfathered" incumbent LMS operator (Teletrac). The incumbent is supposed to operate on a co-equal shared basis with the geographic auction-based licensee. If this cannot be done, auction winners will have to either geographically isolate their systems or acquire the incumbent systems.

The unlicensed Part 15 device industry is largely shifting to the 2400–2483.5 MHz and 5.725–5.875 GHz ISM bands.

As another alternative for such devices, in 1994 the FCC allocated a dedicated band to *Unlicensed Personal Communications Services* (UPCS) devices at 1910–1930 MHz. This band is being cleared of fixed microwave stations that are operating there. Migration of the fixed stations will ultimately enable the band to be exclusively devoted to low-power unlicensed use by cordless telephone and computer products.

However, the uncertainties, costs, and delays of this band-clearing process prompted computer makers to petition for and receive an additional band for unlicensed wireless data at 2390–2400 MHz. The primary service in that band is Amateur Radio. At this writing, there is no use of that band for UPCS.

The ISM bands are commonly believed to be the only bands available for unlicensed use. Part 15 actually permits very low-power operation in bands through much of the spectrum except for certain restricted bands used by safety, science, or government services. Most manufacturers concentrate instead on higher-power spread spectrum products in the ISM bands.

Radio amateurs and the Part 15 community are concerned about the impact of ISM-band radars on their operations. The *National Telecommunications and Information Administration* (NTIA) has stated, "The expansion of use in the 902–928 MHz band by Federal and non-Federal users, including the operation of wind profiling radars, may make this band untenable for amateur operations in the future."

"The 902–928 MHz band is already an uncomfortable melting pot of uses," according to the American Radio Relay League. "The allocation status of this band was not the result of any apparent plan or design. ...It was, rather, the result of incremental allocation decisions, each one addressed to a particular effort to accommodate individual uses. Each additional incremental addition to the 902–928 MHz allocation, now including wind profiler radars and LMS, brings the band closer to gridlock status."

928–929 MHz

FG: US116 US215 US268 G2

Non-FG: US116 US215 US268 NG120 FIXED FCC: Public Mobile (22). Private Land Mobile (90). Fixed Microwave (101)

This band represents the lowest frequency band available to the Fixed Point-to-Point Services. These services are regulated under several parts of the FCC rules, but principally under Part 101, which governs both private operations in Subpart H and common carrier operations in Subpart I.

They should not be confused with the fixed services that *Commercial Mobile Radio Service* (CMRS) providers may offer, in spectrum as low as 27 MHz (see 30.56–32 MHz).

Since the 1980s, several bands in the 900 MHz spectrum have been used by *Multiple Address Systems* (MAS) for fixed communication between a central point (master station) and surrounding points (remote stations) as needed by private radio and common carrier licensees. Both private internal and for-profit communications operations are licensed in the MAS bands.

Many MAS uses are characterized as *System Control And Data Acquisition* (SCADA). The uses of SCADA and MAS include monitoring of utility and petroleum lines for faults or blockages and remote meter reading of energy usage, sometimes in conjunction with unlicensed Part 15 devices; the control of water supplies, waste water operations and airport runway lights; lottery terminal networks; and credit card verification. Paging carriers use MAS under the Domestic Public Land Mobile (Part 22) rules to control their transmitters. Security companies use MAS to receive alarms from banks and stores.

MAS Revisited

The FCC made sweeping revisions to MAS in Docket WT 97-81, including introduction of auctions, more operational flexibility, and special frequencies for government and private users. MAS systems are permitted in three services within the Fixed Microwave Services. All three may be used for terrestrial point-to-point and point-to-multipoint fixed and mobile operations.

The 928/952/956 MHz Service uses 928–928.85 MHz paired with 952–952.85 MHz, and also makes unpaired frequencies available in 956.25–956.45 MHz. It is licensed on a first-come, first-served, site-by-site basis and is limited to private internal use. Licensees cannot provide service to others on a for-profit or not-for-profit basis. The Commission said that "an urgent need for this spectrum has been demonstrated" and that site-by-site licensing, instead of auctions, will be less disruptive and more expeditious.

The 932/941 MHz Service uses 932.0–932.5 MHz paired with 941.0–941.5 MHz. Five of the 40 channels in this service are set aside for public safety and federal government use. Fifteen are set aside for public safety, federal government, and private internal services. Eligibility for the other 20 channels is not restricted by user type; however, the FCC will assign them by auction. Sixteen of these channels will be awarded as paired blocks of 12.5 kHz each. Four channels are combined and will be auctioned as a single paired 50 kHz license in each Economic Area (EA, see 38.6–39.5 GHz).

The FCC said that this license would facilitate non-traditional MAS Services such as Narrowband PCS (see 901–902 MHz)—a radical concept for MAS, which has been limited to obscure industrial functions. Essentially, the FCC has opened up MAS spectrum for consumer services.

The 928/959 MHz Service uses 928.85–929 MHz paired with 959.85–960 MHz. Licenses in this service will be auctioned by EAs. "EAs appear to mirror the size and development of existing MAS systems and are small enough to provide an opportunity for small businesses to obtain a license," the FCC said. The Department of Commerce and three EA-like areas in U.S. island territories specify 172 EAs, creating a total of 175 service areas.

Existing MAS operators will be "grandfathered:" permitted to continue operating on their current frequencies indefinitely. Incumbents in the 928/959 MHz Service must be protected from interference by EA licensees, but will not be able to expand their systems without permission of the EA licensee for that area.

Incumbents in the 928/952/956 MHz Service will continue operating and can expand their systems. If an incumbent's license is terminated, the spectrum reverts to the EA licensee for that area. It appears that no non-federal MAS operations are in 932–941 MHz.

Flexible Regulations

The MAS rules provide for neither loading (minimum usage) requirements nor a limit on the amount of MAS spectrum an entity may obtain. MAS licenses are also not subject to the 45 MHz spectrum cap for *Commercial Mobile Radio Service* (CMRS) licensees.

As with other new and revised wireless services, EA licensees will be allowed to partition portions of their service area to others and/or disaggregate (sub-authorize) portions of their spectrum to others. EA licensees also will be allowed to combine and subdivide channels without a special showing of need; but licensees in the private internal spectrum will have to offer justification if they want to increase their channel bandwidth beyond 50 kHz.

A longtime freeze, or suspension, has been in effect on the FCC's acceptance of MAS license applications. The wireless and utility industries complained bitterly about the freeze. At this writing, the FCC lifted the freeze partially. It is accepting applications for licenses in the 928/952/956 MHz Service and the channels in the 932/941 MHz Service allocated for public safety and private internal use. The freeze remains in effect for all other MAS spectrum while the FCC prepares to auction licenses.

Footnote US268 provides for Navy ship-borne, long-range search radars in the 890–902 MHz and 928–942 MHz spectrum.

929-930 MHz

FG: US116 US215 US268 G2

Non-FG: US116 US215 US268 FIXED. LAND MOBILE

FCC: Private Land Mobile (90)

This is one of the *Private Carrier Paging* (PCP) bands. The PCP carriers offer paging and messaging like Public Mobile (radio common carrier) paging systems.

Both types of paging come under a common regulatory umbrella: *Commercial Mobile Radio Services* (CMRS) Messaging Services, part of the *Commercial Wireless Radio Services* (CWRS). The FCC still keeps PCP and Public Mobile as separate regulatory categories (see 35–36 MHz).

The FCC converted PCP and Public Mobile paging from station-by-station licensing to geographic-area licensing, as used in other forms of CMRS such as cellular and *Personal Communications Services* (PCS). Geographic-area licensees have considerable flexibility to locate, build, and modify their systems.

"We believe that geographic licensing is particularly suitable for paging," the FCC said, "because the service has evolved away from single-site systems toward multi-site systems that cover large geographic areas."

The 929–930 MHz and 931–932 MHz bands contain nationwide exclusive licenses that the FCC granted to paging carriers, not by auction, but at an earlier stage when the carriers became eligible for these franchises by deploying large numbers of transmitters across the country. The bands also contain other incumbent paging operators licensed under earlier rules.

The rest of the available paging licenses in these bands were assigned by auction. The FCC's March 2000 paging auction offered 12 licenses in 929–930 MHz in each of 51 *Major Economic Areas* (MEAs) for a total of 612 licenses, plus 37 licenses in 931–932 MHz in each MEA for a total of 1,887 licenses. Seventy-eight bidders won 985 licenses for more than $4.1 million in net high bids.

The presence of incumbent licensees in this spectrum undoubtedly reduced participation and bid prices. The FCC pointed out that the licenses may be used for fixed wireless service in addition to the one- and two-way messaging services usually associated with paging.

The 929 MHz and 931 MHz paging bands "are likely to be directly competitive with Narrowband PCS," the FCC said (see 901–902 MHz).

930–931 MHz

FG: US116 US215 US268 G2

Non-FG: US116 US215 US268 FIXED. MOBILE

FCC: Personal Communications (24)

This is one of the three Narrowband Personal Communications Services bands. At this writing, its configuration is still pending in the Second Further Notice of Proposed Rulemaking (FCC 00-159) in Docket GEN 90-314 (Second FNPRM, see 901–902 MHz).

"The demand for spectrum has increased dramatically as a result of explosive growth in wireless communications and there is very little unencumbered spectrum available for new services," the FCC said in the Second FNPRM.

"Thus, consistent with our conclusion...that the Commission must focus on increasing the amount of spectrum available for use, we tentatively conclude that it is in the public interest to proceed with licensing the one megahertz of narrowband PCS spectrum that has been held in reserve. We believe that this spectrum, which is unencumbered, should be made available to those interested in bringing new and innovative services to the public, and that the Commission should not create an artificial shortage of spectrum that might limit service options."

Paging the Planets

The strange Space Shot service once tried to get access to this band. Space Shot billed itself "The Interplanetary Communication System That Links You to The Universe."

Space Shot helped customers "search for extraterrestrial life" by sending love notes or peace messages into space on 903.0125 MHz for $5.00 per "launch." Its purveyors held a temporary, experimental license, but the FCC declined its request for permanent status at 930 MHz in 1985.

Instead of dismissing the ludicrous idea outright, the Commission subjected it to careful engineering analysis. It dryly concluded that the Space Shot transmitter was unlikely to activate receivers at the target locations, which were supposed to be the Sun, the Moon, and "any of the planets in our Solar System."

We considered Space Shot an intriguing test of the FCC's pro-business philosophy. Industry commenters called it "a radio version of the Pet Rock."

Secret Prayer Frequency

A Vermont company resurrected this concept. Its Prayers Heavenbound service transmitted religious messages into space for a fee.

"Our exact communications frequency to God's Heaven is the result of much research and is therefore proprietary," the company told us. "Our prayer uplinking facility is not located on United States soil, or within any of its possessions. As such, its existence or RF emissions are of no jurisdictional concern to the FCC."

Prayers Heavenbound won World Wide Web awards including Worst of the Web and Lycos Top 10 Turkeys.

Space Shot and Prayers Heavenbound have disappeared, but a similar service is available at this writing from the Davis, California-based Bentspace.com. Its patent-pending technology enables customers to send an e-mail to the "Cosmos" for $10.95. An FCC-certified, unlicensed 2400 MHz wireless LAN is the signal source.

"What better way to commemorate a person, event or idea than to broadcast it to the universe for all eternity?" the company said. "Do at least one thing in your life that will live forever."

In a solemn disclaimer, Bentspace warns that it "is only responsible for sending the messages, and not for any consequence as a result of the messages that are sent."

931-932 MHz

FG: US116 US215 US268 G2

Non-FG: US116 US215 US268 FIXED. LAND MOBILE FCC: PUBLIC MOBILE (22)

This band contains frequencies for internal paging within organizations and for *Commercial Mobile Radio Service* (CMRS) paging in the Private and Public Mobile Services.

These include the frequencies assigned to the three original, non-auctioned radio common carrier exclusive nationwide network paging licensees on 931.9375 MHz, 931.8875 MHz, and 931.9125 MHz. There are additional nationwide operations in 929-930 MHz and 931-932 MHz, which in some cases earned nationwide exclusivity (no requirement to share the frequencies with others) through extensive transmitter construction.

Other paging licenses in these bands are incumbent authorizations under earlier FCC rules or were assigned in FCC auctions (see 929-930 MHz).

The 929 MHz and 931 MHz paging bands "are likely to be directly competitive with Narrowband PCS," according to the FCC (see 901–902 MHz).

932–935 MHz

FG: US215 US268 G2 FIXED

Non-FG: US215 US268 NG120 FIXED

FCC: Public Mobile (22). Fixed Microwave (101)

Multiple Address Systems (MAS) operate in this band in the 932/941 MHz Service, providing communications for such uses as credit card verification, alarms, and remote control. MAS may expand beyond industrial uses (see 928–929 MHz).

The *Common Carrier Fixed Point-to-Point Microwave Service* and the *Private Operational Fixed Point-to-Point Microwave Service* are both permitted at 932.5–935 MHz and 941.5–944 MHz. Many of these non-government operations support paging systems.

Government agencies use 932–935 MHz and 941–944 MHz for point-to-point operations. Large users include the Federal Aviation Administration for *Radio Communications Links* (RCL) between air traffic control facilities; the Department of Agriculture, for microwave backbone use; the Department of the Interior for links supporting law enforcement, fire protection, and resource management; and the Department of Energy for control of electric power and weather data transfer.

935–940 MHz

FG: US116 US215 US268 G2

Non-FG: US116 US215 US268 FIXED. LAND MOBILE

FCC: Private Land Mobile (90)

See 896–901 MHz for information about this band.

940-941 MHz

FG: US116 US268 G2

Non-FG: US116 US268 FIXED. MOBILE

FCC: Personal Communications (24)

This is one of three bands allocated to the Narrowband Personal Communications Service (see 901–902 MHz).

941-944 MHz

FG: US268 US301 US302 G2 FIXED

Non-FG: US268 US301 US302 NG120 FIXED

FCC: Public Mobile (22). Fixed Microwave (101)

This band is shared by Fixed Services, including *Multiple Address Systems* (MAS) in the 932/941-MHz Service (see 928–929 MHz and 932–935 MHz).

944-960 MHz

FG: None

Non-FG: NG120 FIXED

FCC: Public Mobile (22). International Fixed (23). Auxiliary Broadcast (74). Fixed Microwave (101)

This band is divided into the 944–952 MHz and 952–960 MHz segments.

Radio and TV stations use 944–952 MHz for such *Broadcast Auxiliary Services* (BAS) uses as *studio-transmitter links* (STL) and *intercity relay links* (ICR). They can lease bandwidth for other purposes (see also 1990–2025 MHz).

NTIA reported in late 1998 that the BAS segment has an approximate 24 percent annual growth rate, due largely to radio stations that are changing from wireline to radio STLs. Nevertheless, it forecast reduced spectrum needs for these types of BAS uses because of digital audio compression on the radio side and the availability of wideband digital network services on the wireline side.

Multiple Address Systems (MAS) in the 928/952/956 MHz Service use this band (see 928–929 MHz), as do private operational fixed microwave systems serving industrial voice and data users.

NTIA promoted the idea of auctions in 944–960 MHz, among other bands, as an alternative to the reallocation of spectrum used by government space programs (see 2025–2110 MHz). It cited Office of Management and Budget estimates of $900 million for license auctions when protection of incumbent users in the band is taken into account.

Yet the presence of existing services in bands such as 944–960 MHz creates potential problems, the FCC has observed. "We are unsure of the feasibility of licensing new services in these bands in a manner that would be attractive to competitive bidders," it said.

960-1215 MHz

FG: S5.328 US224 AERONAUTICAL RADIONAVIGATION

Non-FG: S5.328 US224 AERONAUTICAL RADIONAVIGATION

FCC: Aviation (87)

This band is used worldwide for commercial, space, and military flight operations, and air traffic control, including tracking, navigation, collision avoidance, and landing guidance.

Air Traffic Control Radar Beacon Systems (ATCRBS) use 1030 MHz (ground-to-air) and 1090 MHz (air-to-ground). ATCRBS uses a ground-based transmitter to interrogate airborne transponders and display target information to controllers.

Internationally known as *Secondary Surveillance Radar* (SSR), it includes a transponder employing Mode *A* for aircraft identification; Mode *C* for pressure-altitude information; and Mode *S,* which selectively interrogates aircraft within its field of view and provides data communication (Data Link) between aircraft and ground controllers.

Mode *S* will ultimately provide a nationwide, two-way data capability for air traffic, flight information, and surveillance according to the FAA. A similar technology in this band is the military *Identification Friend or Foe* (IFF).

The *Traffic Alert and Collision Avoidance System* (TCAS) uses these same frequencies. This airborne system obtains data about nearby aircraft (five to 40 miles) from transponders. Various types of TCAS equipment can alert pilots to the presence of other aircraft and can recommend avoidance maneuvers such as climb or dive. TCAS is internationally known as *Airborne Collision Avoidance System* (ACAS).

Distance Measuring Equipment (DME) and the military equivalent *Tactical Air Navigation* (TACAN) operate in this band as the standard systems for determining ranges within radio line of sight. Land-based TACAN will be discontinued when enough aircraft are equipped with GPS capability (phase-down expected to start in 2008); sea-based TACAN will continue until replaced.

A key military use of this band is the *Joint Tactical Information Distribution System* (JTIDS). This multibillion-dollar development is part of a NATO system of secure and jam-resistant frequency-hopping communications, navigation, and identification among airborne and surface-based elements in hostile environments. JTIDS is regarded as a limited presence in this band that was carefully coordinated to avoid interference to aviation.

The *Land Mobile Communications Council* (LMCC), a trade association for the land mobile radio industry, petitioned the FCC to reallocate 85 MHz from this band to the *Private Mobile Radio Services* (PMRS) by the year 2010, with an additional 70 MHz of the band reallocated to public safety uses. "It is recognized that the aeronautical navigation services in this band are of considerable importance," LMCC said.

"On the other hand," the group added, "it is clear that these services will shift to the new *Global Navigation Satellite Services* (GNSS) operations in the not-too-distant future and that this spectrum offers the last chance for PMRS to access spectrum that is both sufficient in scope and low enough in frequency to satisfy foreseeable future needs, including the perceived explosion in demand for advanced, wide bandwidth applications."

The aviation industry, citing the findings of international aviation authorities, disputed those assertions and argued that the existing radar and navigation systems in this band will remain in use well beyond 2015.

A frequency in this band, 1176.45 MHz, will be added to the frequencies transmitted by the *Global Positioning System* (GPS, see 1215–1240 MHz).

The 960–1240-MHz spectrum is one of many restricted bands in which the FCC Part 15 rules permit unlicensed devices to emit only very low level emissions.

The 1000–2000-MHz spectrum is traditionally designated the S-band.

1215-1240 MHz

FG: RADIOLOCATION S5.333 G56. RADIONAVIGATION-SATELLITE (space-to-Earth)

Non-FG: S5.333

The 1215–1400 MHz spectrum is used for defense and civil radar, including battlefield early warning, portable search, and target acquisition, missile detection by ship-borne systems, and scientific observation. Radiolocation has primary status in 1215–1300 MHz and 1350–1390 MHz, and is secondary in 1300–1350 MHz. Aeronautical radionavigation is primary in 1240–1370 MHz.

This spectrum is used by the *Joint Surveillance System* (JSS), a long-range *Air Route Surveillance Radar* (ARSR) network operated by the Air Force and the Federal Aviation Administration for radionavigation, air and fleet defense, and drug interdiction.

Many of these frequency-hopping radars operate at high-peak power levels—tens or hundreds of kilowatts for defense radars and one to five megawatts for FAA radars. Their emissions can extend more than 10 MHz in bandwidth.

ARSR radars can detect small, low-flying aircraft and missiles at long distances and provide height, bearing, and distance data. Through digital signal processing and exotic antenna technology, they can discriminate between targets and background clutter, especially turbulent water.

The Cobra Dane phased-array radar on the far end of the Aleutian Island chain uses this band to observe test launches of ballistic missiles and to provide warning and assessment of missile attacks on the U.S. and Canada. The pyramid-like radar can track up to 300 incoming warheads and up to 200 satellites simultaneously.

Also sharing this spectrum in 1215–1300 MHz are space-borne active microwave sensors, namely *synthetic aperture radars* (SAR) carried on satellites and the Space Shuttle. This technology obtains multi-spectral images of land and water areas. These are categorized as uses of the *Earth Exploration Satellite Service* (EESS) and the *Space Research Service* (SRS).

Space-borne defense radars now in development include *Discoverer II*, used for ground target indication, imaging, and mapping, and the *Tactical Satellite Radar* (TACSRAD) constellation of ten satellites in equatorial orbit.

GPS, A Global Utility

The frequency 1227.6 MHz (Link 2 or L2) is one of three downlink frequencies associated with the *Navstar Global Positioning System* (GPS). The others are L1 (1575.42 MHz) and L3 (1381.05 MHz). L1 and L2 transmit positioning information. L3 broadcasts nuclear alerts (see 1350–1390 MHz).

GPS is the Department of Defense's $10 billion constellation of 24 satellites to provide latitude, longitude, altitude, velocity, and time to land, sea, and air users. Congress mandated that all military land vehicles, vessels, and aircraft incorporate GPS receivers. Even bombs may be fitted with GPS.

GPS receivers are increasingly useful on launch vehicles and in spacecraft. Applications include altimetry, orbit determination, attitude determination for spacecraft and instrument pointing, precise time synchronization, safety, and guidance.

These capabilities are expected to lead to autonomous spacecraft navigation. GPS will provide all navigation functions for the Space Shuttle, the International Space Station, and the Emergency Crew Return Vehicle.

Though created for defense, GPS launched a huge industry in aids to vehicle location, camping and exploration, navigation, surveying and mapping, law enforcement and emergency services, and timekeeping, among its myriad markets. GPS broadcasts free of charge to the entire world.

Controlled by the Air Force Space Command, GPS satellites continuously transmit spread spectrum signals on L1 and L2. The user's GPS receiver measures the time of arrival of signals from the satellites and reports the user's position, velocity, and time.

The receiver displays position information in the form of coordinates on Earth and in some models, maps or the world grid system that is used by radio amateurs.

GPS commanders intentionally degraded the civil GPS signal on L1 to deprive adversaries of high accuracy. Several levels of degradation and techniques (such as encryption and dithering or skewing the timing signals) are available. This degradation is called *Selective Availability* (SA) and reduces location accuracy to 100 meters.

SA Discontinued

After technology was developed to jam GPS on a localized basis, SA was "set to 0" (removed) on May 3, 2000, permitting GPS users location accuracy within 20 meters.

The discontinuance of SA, originally anticipated to occur as late as 2006, is regarded as one of the most important developments in GPS history, lowering costs and increasing value throughout the range of GPS applications.

"Around the world, more and more people and industries rely on GPS for accurate positioning and timing information," President Bill Clinton announced upon terminating SA. "The action we are taking today will dramatically improve civilian use of GPS—by drivers on the road, cellular phone users dialing 911, hikers and boaters, businesses and institutions. ...

"Threat assessments conclude that ending signal degradation at this time will not weaken our national security. We have demonstrated the capability to selectively deny GPS signals on a regional basis when our security is threatened. We will continue to ensure—and to upgrade U.S. defense systems to exploit-—the full military utility of GPS."

Augmenting GPS

In a $400 million initiative, the government plans two new civil GPS signals. One of the signals will be within the L2 spectrum and will become available in 2003; the other frequency, 1176.45 MHz, will begin after launch of a satellite in 2005. The additional signals will enable receivers to more effectively correct for the distorting effects of Earth's atmosphere on GPS signals.

Some types of GPS errors are monitored internally within the GPS satellites and reported via a "health message." Other failures are detectable only by GPS controllers. It can take from 15 minutes to several hours before users are notified of a problem.

Reliance on erroneous GPS signals could be dangerous for some uses such as harbor navigation. Users can overcome this problem and obtain high accuracy by using *differential GPS* (DGPS). This technique is based on the location of a surveyed reference station. Information based on the difference between this reference location and the GPS-computed location is transmitted over non-GPS media to DGPS receivers, which apply the corrections. DGPS permits the use of GPS satellites that are not fully functioning.

The Coast Guard broadcasts differential corrections in U.S. coastal areas, the Great Lakes, Alaska, and Hawaii on 285–325 kHz radiobeacons. Commercial vendors broadcast the corrections on FM radio subcarriers in 88–108 MHz.

Aviation Uses

A gradual transition to GPS-based air navigation incorporates the Federal Aviation Administration's $500 million *Wide Area Augmentation System* (WAAS). WAAS will operate ground reference stations around the U.S.

These stations compute their GPS-derived locations and compare them with the surveyed locations. A master station receives this data and sends corrections to aircraft via satellites over the L1 frequency used by GPS.

The system will permit accuracy to within ten meters. An even more precise *Local Area Augmentation System* (LAAS) will use psuedosats or pseudolites (GPS-like ground transmitters) in areas such as airports where higher accuracy is needed.

GPS also will aid the transition to Free Flight, a future mode that will reduce direction by ground controllers. With this mode, pilots will have more flexibility to choose routes. They will be able to view aircraft separation data in the cockpit from reports of aircraft positions computed with GPS.

Free Flight depends on new communications systems known generically as Data Link (see 136–137 MHz). They incorporate *Automatic Dependent Surveillance* (ADS), which transmits GPS-derived aircraft positions to other aircraft and controllers. ADS also is used in maritime services.

The FAA testified before Congress that the government will save more than $200 million annually when ground-based navigational aids are replaced by GPS.

Galileo

In addition to Russia's GPS-like Russian Federation *Global Orbital Navigation Satellite System* (GLONASS) (see 1610–1610.6 MHz), the GPS may be joined by another system: the European Union's Galileo, a 21-satellite radionavigation system that may work with or integrate GLONASS.

Galileo is expected to begin operation in 2005; at this writing the project was still in the study stage.

Key motivators for Galileo include job creation, promoting the European market for GPS applications, and independence from the GPS and GLONASS military command authorities.

Implants

A company called Digital Angel has patented GPS-aided tracking devices said to be suitable for implantation in young children as well as adults.

"When implanted within the human body, the transceiver is powered electromechanically through the movement of muscles. It can be activated either by the wearer or by a remote monitoring facility. The device also can monitor certain biological functions of the human body—such as heart rate—and send a distress signal to a monitoring facility when it detects a medical emergency," the company said.

The 960–1240 MHz spectrum is one of many restricted bands in which the FCC Part 15 rules permit unlicensed devices to emit only very low level emissions.

1240–1300 MHz

FG: S5.334 RADIOLOCATION S5.333 G56

Non-FG: S5.282 S5.333 S5.334 Amateur

FCC: Amateur (97)

This band is part of the 1215–1400 MHz spectrum used by Air Force and Federal Aviation Administration *Air Route Surveillance Radars* (ARSR) and space-borne active microwave sensors (see 1215–1240 MHz).

As the 23 cm Amateur Radio band, 1240–1300 MHz is used in mobile, video, digital, and experimental communications. These include satellite uplinks in 1260–1270 MHz.

Prostate Treatment

A medical device, the Prostatron, uses 1296 MHz for thermotherapy to treat prostate disease. Thermotherapy delivers microwave energy to destroy prostate cells.

However, 1240–1300 MHz is not allocated for medical uses. The Prostatron's manufacturer sought authorization from the FCC to permit its use in the U.S, asserting that the likelihood of interference to other band users is low. Radio amateurs and radio astronomers opposed introduction of the device in this spectrum.

The inventors said they chose 1296 MHz "specifically because its wavelength is uniquely suited to the treatment process" and that *Industrial, Scientific, and Medical* (ISM) frequencies such as 915 MHz are less effective. A Prostatron competitor uses ISM frequencies (see 902–928 MHz).

A urology journal compared prostate thermotherapy equipment. It concluded that antenna design, not frequency of operation, was significant. "It makes no difference" whether the heating device uses 900 MHz or 1300 MHz, according to the journal. It suggested that 1296 MHz was chosen because Amateur Radio devices that could be repurposed for thermotherapy are available for that frequency.

The *National Telecommunications and Information Administration* (NTIA) consented to the introduction of the Prostatron provided that the manufacturer, customers, and installations comply with a detailed list of conditions intended to protect aviation radars.

NTIA said that it "continues to have reservations about the long-term consequences of permitting access to the spectrum for a relatively high-powered, non-compliant medical application in a frequency band not designated for use by ISM devices, particularly one used for safety-of-flight aeronautical radionavigation services in the United States."

AMSAT Phase 3-D will use this and other amateur bands. The most ambitious amateur space system yet the $4 million satellite, launched at this writing, is funded and constructed by the *Radio Amateur Satellite Corporation* (AMSAT) and its affiliates around the world (see 144–146 MHz).

The 1235–1241 MHz spectrum may be a candidate for wind profiler radars contingent on studies of compatibility with radionavigation satellite usage.

1300-1350 MHz

FG: S5.149 AERONAUTICAL RADIONAVIGATION S5.337 Radiolocation G2

Non-FG: S5.149 AERONAUTICAL RADIONAVIGATION S5.337

FCC: Aviation (87)

This is part of the 1215–1400 MHz aeronautical radar spectrum (see 1215–1240 MHz).

The 1300–1427 MHz spectrum is one of many restricted bands in which the FCC Part 15 rules permit unlicensed devices to emit only very low level emissions.

1350-1390 MHz

FG: S5.149 S5.334 S5.339 US311 G27 G114 RADIOLOCATION G2. Fixed. Mobile

Non-FG: S5.149 S5.334 S5.339

This is part of the 1215–1400 MHz aeronautical radar spectrum (see 1215–1240 MHz).

Revoked Reallocation

Under the 1997 *Balanced Budget Act* (BBA, see 2390–2400 MHz), the 1385–1390 MHz segment was to be reallocated from the government to the private sector in 1999. The Department of Defense, however, believed that these reallocations would have had immense negative impact on military radar and communications investments. The October 1999 Defense authorization bill (P.L. 106-65) cancelled the reallocation of 1385–1390 MHz.

Nuclear Detection

A little-known function of the Global Positioning System (GPS, see 1215–1240 MHz) uses L3, the frequency 1381.05 MHz, to transmit alerts as part of the *Nuclear Detonation Detection System* (NUDET or NDS).

NUDET transmits an alert by ground command or upon detection of a nuclear event. To minimize interference to radio astronomy, satellite transmissions on this frequency are limited to tests, training, and actual atomic explosions.

The global coverage and reliability of GPS make it an excellent platform for the detection and relay of weapons data. GPS satellites carry X-ray sensors, electromagnetic pulse detectors, and the oddly but appropriately named "bhangmeters," sensors that detect bursts of light from detonations.

Other defense satellites also convey data to NUDET. Augmenting NUDET will be a new program of space-based explosion sensors, the Global Verification and Location System, on frequencies in this band.

Atmospheric testing of nuclear devices is banned by international agreement. To monitor clandestine tests as well as aggressive launches, the U.S. employs a vast network of satellites, domestic and overseas ground-based radars, underwater sound detectors on submarines and eavesdropping stations to detect nuclear events.

Air Force data links in 1350–1400 MHz and 1427–1435 MHz broadcast GPS-derived aircraft position during flight testing and training. Unmanned target vehicles are controlled in 1380–1390 MHz. The Army uses 1350–1850 MHz to control unmanned air and ground vehicles and in transportable links between units.

Army, Navy, and Marine Corps uses of this band include headquarters connectivity, communications training, air traffic control, antenna testing, atmospheric research, and tactical radio relay.

Scientific Uses

Remote sensing satellites use 1370–1427 MHz to measure soil moisture and ocean salinity. This band also is important to radio astronomers. Shifts in the 1420.4 MHz hydrogen spectral line caused by motion of celestial bodies places most of the energy from distant galaxies into 1350–1400 MHz.

Spectral line emissions are produced by atoms or molecules that interact with particles and radiation in the environment. Observation of spectral lines can reveal characteristics of the emitting material and the transmission medium through which its energy is received. Radio astronomers have identified about 1,000 spectral lines from molecular clouds in our galaxy.

Docket ET 00-221 would change this band as follows: the Federal Government (FG) allocations would make Fixed and Mobile Services primary, and retain all of the Footnotes. The Non-FG allocations would add Footnote US311 only.

The 1300–1427 MHz spectrum is one of many restricted bands in which the FCC Part 15 rules permit unlicensed devices to emit only very low level emissions.

1390-1395 MHz

FG: S5.149 S5.339 US311 US351 G27 G114 RADIOLOCATION G2. Fixed. Mobile

Non-FG: S5.149 S5.339 US351

Note: 1390–1395 MHz became non-Federal government exclusive spectrum in January 1999.

This military band is used for radar, telemetry, radio relay, and air traffic control. It does not contain non-federal operations.

Under the 1993 *Omnibus Budget Reconciliation Act* (OBRA, see 2390–2400 MHz), 1390–1400 MHz was reallocated from the government to the private sector in 1999 (see LMCS below). Federal radar systems at 17 sites will continue and need interference protection from new uses until 2009.

Only uses that do not interfere with remaining operations or radio astronomy will be allowed. This requirement constrains the commercial development of this spectrum.

According to the FCC, it may be difficult to provide commercial service within hundreds of kilometers of the 17 sites. The protection areas range from desert areas to major urban markets. (From the 1390–1400 MHz band, the FCC carved out 1395–1400 MHz and allocated it to medical telemetry.)

In a November 1999 spectrum policy statement, the FCC announced its intention to allocate 1390–1395 MHz, 1427–1429 MHz, and 1432–1435 MHz to a new *Land Mobile Communications Service* (LMCS). Exclusive licenses in the new service will be auctioned to so-called band managers, which are essentially trade associations or mega-licensees that authorize actual users.

LMCS would answer the private land mobile radio community's longstanding need for additional spectrum (see 30.56–32 MHz), while ensuring that no user could obtain a license on the traditional first-come, first-served basis without the element of mutual exclusivity that drives the auction system.

At this writing, it appears that the FCC has not formally proposed to create LMCS. Instead, in Docket ET 00-221 it proposes only to reallocate this band to general uses for later auction, leaving particulars about service rules and commercial offerings to future proceedings and the decisions of license winners.

As examples of possible uses, the Commission noted certain services (besides LMCS) that could use this and other OBRA bands. These include utility telemetry, the Personal Location and Monitoring Service, and the ArrayComm service (see 2300–2305 MHz), as well as feeder uplinks for Little LEO satellites (see 137–137.025 MHz). The FCC floated several options for bandplans to accommodate these or other future services.

The docket would change this band as follows: the Federal Government (FG) allocation would contain Footnotes S5.149, S5.339, US311, and US351. All other entries would be deleted. The Non-FG allocation would add Footnote US311 and the services FIXED and MOBILE except aeronautical mobile. The FCC Note concerning January 1999 would be deleted.

1395–1400 MHz

FG: S5.149 S5.339 US311 US351 LAND MOBILE US350

Non-FG: S5.149 S5.339 US311 US351 LAND MOBILE US350

FCC: Personal (95)

Under the 1993 *Omnibus Budget Reconciliation Act* (OBRA, see 2390–2400 MHz), 1390–1400 MHz was reallocated from the government to the private sector in 1999.

The FCC later created this separate 1395–1400 MHz band and allocated it to the *Wireless Medical Telemetry Service* (WMTS), one of the Citizens Band Radio Services within the Personal Radio Services category (see 608–614 MHz).

Figure 10
The Allen Telescope Array will use hundreds of antenna dishes in the *Search for Extraterrestrial Intelligence* (SETI). (Ly Ly/SETI Institute illustration)

1400-1427 MHz

FG: S5.341 US246 EARTH EXPLORATION-SATELLITE (Passive). RADIO ASTRONOMY US74. SPACE RESEARCH (Passive)

Non-FG: S5.341 US246 EARTH EXPLORATION-SATELLITE (Passive). RADIO ASTRONOMY US74. SPACE RESEARCH (Passive)

No transmissions are permitted in this receive-only scientific band. In addition to its use for satellite remote sensing and radio astronomy continuum observations, this band contains the hydrogen spectral line at 1420.4 MHz. Astronomers observe this frequency to study the distribution and motion of mass in the universe.

Certain survey programs in the *Search for Extraterrestrial Intelligence* (SETI) concentrate on this area of the spectrum.

A significant SETI development is the Allen Telescope Array, funded by investor and philanthropist Paul Allen and former Microsoft Chief Technology Officer Nathan Myhrvold. The array will be used for both SETI and conventional astronomy. Located at California's Hat Creek Observatory, (a radio quiet zone, protected by FCC allocation footnotes) it will use hundreds of mass produced dish antennas. The $26 million facility will be fully operational in 2005.

"For the first time in our history, we have the ability to pursue a scientifically and technologically sophisticated search for intelligent life beyond Earth at the same time we are doing traditional radio astronomy," according to Allen. The electronic outputs from all of the antennas can be combined to produce high-resolution images of large areas of the sky.

Passive microwave radiometers in this band can measure soil moisture from aircraft. The health of crops is strongly related to soil conditions.

Out-of-band and spurious emissions from commercial uses in 1395–1400 MHz and 1427–1429 MHz will have to be controlled to protect scientific uses in 1400–1427 MHz.

The 1300–1427 MHz spectrum is one of many restricted bands in which the FCC Part 15 rules permit unlicensed devices to emit only very low level emissions.

1427–1429 MHz

FG: S5.341 G30 SPACE OPERATION (Earth-to-space). FIXED. MOBILE except aeronautical mobile

Non-FG: S5.341 SPACE OPERATION (Earth-to-space). Fixed (telemetry). Land Mobile (telemetry and telecommand)

FCC: Satellite Communications (25). Private Land Mobile (90)

Note: 1427–1429 MHz became non-Federal government exclusive spectrum in January 1999.

The main mobile use in this band is military air-to-ground telemetry and ground-to-air telecommand links on test ranges. The main fixed use is by Army and National Guard tactical radio relay systems. The systems have broad tuning ranges including 200–400 MHz, 600–1000 MHz, and 1350–1850 MHz.

The secondary non-government telemetry allocation is used for wireless meter reading by electric, gas, and water utilities. Some of these systems feature two-way communications, for load management and transmission of price and consumption data to customers.

The *Space Operations Service* allocation is not used in order to protect the passive (receive-only) *Earth Exploration Satellite Service* uses in 1400–1427 MHz. At this writing, FCC Docket ET 00-221 inquired into whether any uses exist for this space allocation or whether it should be removed.

Reallocation

Under the *Omnibus Budget Reconciliation Act* of 1993 (OBRA, see 2390–2400 MHz) 1427–1432 MHz was reallocated from the government to the private sector. Military airborne operations in 14 geographic areas will continue to need interference protection in this spectrum until 2004.

These continuing government uses, especially aeronautical uses, as well as the radio astronomy operations in 1400–1427 MHz may make it difficult to accommodate new, commercial uses in 1427–1432 MHz.

Moreover, "the small size of this [1427–1432 MHz] allocation, as well as its remoteness from existing non-federal services, will make it difficult to use this spectrum either as an adjunct to an existing service or to support a new service," a NTIA advisory task force concluded.

Nevertheless, in a November 1999 spectrum policy statement, the FCC announced its intention to allocate 1427–1429 MHz, 1390–1395 MHz, and 1432–1435 MHz to a new *Land Mobile Communications Service* (LMCS).

Docket ET 00-221 proposed bandplans and invited input about the reallocation and configuration of these bands and uses to which they could be applied, including LMCS (see 1390–1395 MHz).

This docket would change the 1427–1429 MHz band as follows: the Federal Government (FG) allocation would contain Footnotes S5.341 and US352 and no other entries. The Non-FG allocation would be SPACE OPERATION (Earth-to-space) (but see FCC inquiry above), and FIXED and MOBILE except aeronautical mobile. Footnotes S5.341 and US352 would govern. The Satellite Communications (25) designator and the FCC Note concerning January 1999 would be deleted.

1429-1432 MHz

FG: S5.341 US352 LAND MOBILE US350

Non-FG: S5.341 US352 LAND MOBILE US350. Fixed (telemetry). Land mobile (telemetry and telecommand)

FCC: Private Land Mobile (90). Personal (95)

This band is allocated to the *Wireless Medical Telemetry Service* (WMTS, see 608-614 MHz).

Defense uses of this spectrum include flight test telecommunications, telecommand of missiles and remotely piloted vehicles, control of ordnance-handling robots, radio relay, and radar cross-section measurement. The aeronautical operations will continue in 1429-1432 MHz at 14 sites until 2004.

Docket ET 00-221 (see 1390-1395 MHz) would remove the Fixed (telemetry) and the secondary Land Mobile (telemetry and telecommand) service from the Non-FG allocation.

At this writing, utility telemetry and other possible uses remain under consideration in ET 00-221 for sharing this band.

1432-1435 MHz

FG: S5.341 G30 FIXED. MOBILE

Non-FG: S5.341 Fixed (telemetry). Land mobile (telemetry and telecommand) FCC: Private Land Mobile (90)

Note: 1432-1435 MHz became mixed-use spectrum in January 1999.

As with other bands in this region of the spectrum, defense aeronautical, radar, and telecommunication uses of 1429-1435 MHz predominate.

Aerostat balloons, used to detect low-flying aircraft suspected of carrying drugs, operate in this spectrum among others. The secondary non-federal telemetry allocation is used for utility meter reading. This band also is used in the *Search for Extraterrestrial Intelligence* (SETI).

Under the Balanced Budget Act of 1997 (see 2390-2400 MHz), 1432-1435 MHz was reallocated for mixed federal and non-federal use. Military airborne operations at 23 sites will continue in the band indefinitely.

At this writing, this is one of the bands under consideration for a new *Land Mobile Communications Service* (LMCS) and other possible applications in Docket ET 00-221 (see 1390–1395 MHz).

This docket proposes to delete Footnotes G30 and the FIXED and MOBILE Services from the Federal Government (FG) allocation, and add Footnote USxxx (the actual number to replace 'xxx' will be determined later). To the Non-FG allocation, it would add Footnote USxxx and the FIXED and MOBILE Services, deleting all other entries. The FCC Note concerning January 1999 would be deleted.

1435–1525 MHz

FG: S5.341 US78 MOBILE (aeronautical telemetry)

Non-FG: S5.341 US78 MOBILE (aeronautical telemetry)

FCC: Aviation (87)

The defense and aerospace industries, represented by the *Aerospace and Flight Test Radio Coordinating Council* (AFTRCC), consider this portion of the so-called L-band paramount for aeronautical test telemetry.

They use it in conjunction with 2360–2385 MHz for flight testing of manned and unmanned aircraft, missiles and space vehicles, and associated communications such as range safety, chase aircraft, and weather data.

The aeronautical and space test uses in 1435–1525 MHz were believed to be so essential to national defense and to the commercial space industry that in 1992 the United States rejected the worldwide allocation to satellite audio broadcasting in 1452–1492 MHz.

"An even more compelling case exists today for preservation of the L-band for flight testing than existed in 1992," according to AFTRCC. "Since the early 90s, flight testing has lost no less than one-third of its spectrum inventory (65 out of 180 MHz). During this period, bandwidth demands for flight test telemetry have grown exponentially, from an average of less than 1 MHz channels, to 3 MHz, toward 5 and even 10 MHz. This trend is a function of the increasing complexity and computerization of newer aircraft. ...

"For example, when the Boeing 707 was flight tested 40 years ago, engineers monitored 300 data points; when the 777 was tested four years ago, no less than 40,000 data points were monitored. The simple fact is that flight testing needs more spectrum, not less."

The decision to forego L-band for broadcasting shocked nations which sought a common market and technical approach with the U.S. for 1400 MHz broadcasting. Canada was especially critical of the U.S. decision and of efforts to develop digital broadcasting within the existing U.S. AM and FM radio bands.

The FCC later allocated 2320–2345 MHz to satellite *Digital Audio Radio Services* (DARS or SDARS) and licensed two firms, Satellite CD Radio (now Sirius Satellite Radio) and American Mobile Radio (now XM Satellite Radio) to provide satellite DARS service in that band.

WorldSpace

In 1991, the FCC authorized the AfriStar 1 satellite on an experimental basis to AfriSpace, Inc. as part of the $700 million WorldSpace system of four satellites to broadcast in the L-band to Africa and the Middle East (AfriStar), Asia (AsiaStar) and Latin America-Caribbean (AmeriStar, sometimes called CaribStar), with a fourth satellite to serve as a spare.

AfriStar began service in October 1999. Although controlled from Washington, D.C. via stations in India and Mauritius, the satellite at 21 degrees East Longitude is not visible from U.S. territory. AsiaStar was launched in March 2000.

WorldSpace broadcasts to special receivers capable of solar power operation and text readout. The receivers connect to other devices such as portable computers for e-mail, fax, and image display.

"The market for the WorldSpace service is enormous," according to the company, "made up of 3.5 billion people that are presently underserved by the poor sound quality of shortwave radio, or coverage and program limitations of AM and FM services."

The WorldSpace system is licensed by four governments: the U.S. FCC for AfriStar 1, Ghana for AfriStar A, Trinidad for CaribStar 1, and Australia for AsiaStar 1.

AfriSpace finally won a regular non-experimental FCC license for 1452–1492 MHz after a 10-year wait, very shortly after federal defense and aviation authorities withdrew their opposition based on interference concerns.

"Using its state-of-the-art studios in Melbourne, Australia, WorldSpace has created its own original programming, from modern rock, contemporary pop hits and global dance, as well as spoken word programming for both children and adults," the company said. The system carries programs from well-known sources as well, including All India Radio, BBC, CNN, and Bloomberg.

At this writing, WorldSpace receiver sales appear to be below projected levels. Some observers attribute slow sales to WorldSpace ads for "receivers," an unfamiliar term in the target countries, instead of radios.

The 1435–1626.5 MHz spectrum is one of many restricted bands in which the FCC Part 15 rules permit unlicensed devices to emit only very low level emissions.

1525-1530 MHz

FG: S5.341 S5.351 US78 MOBILE-SATELLITE (space-to-Earth). Mobile (aeronautical telemetry)

Non-FG: S5.341 S5.351 US78 MOBILE-SATELLITE (space-to-Earth). Mobile (aeronautical telemetry)

FCC: Satellite Communications (25). Aviation (87)

This band is allocated worldwide for mobile satellite downlinks.

Aeronautical telemetry may continue in the band on a secondary basis.

Complex allocations in this and adjacent bands comprise the spectrum available to the *Geostationary Mobile Satellite Service* (GSO MSS) serving satellite phones and data terminals.

These allocations include the so-called lower L-band at 1525–1530 MHz, 1530–1544 MHz, and 1626.5–1645.5 MHz, and the upper L-band at 1545–1559 MHz and 1646.5–1660.5 MHz.

Secret Frequencies

The actual GSO MSS frequency assignments within these bands are considered state secrets. The U.S. and other governments that have licensed GSO MSS providers refuse to reveal their frequency usage. They argue that competitors could use the information to formulate estimates of communications traffic served by these companies.

The confidentiality agreement is between sovereign nations, but the U.S. Department of State's position is that the matter is squarely the FCC's responsibility. The frequency usage varies according to decisions made through a so-called Mexico City process that began with a 1996 meeting of licensees there.

It is an annual negotiation that coordinates, or assigns bandwidth to each carrier: Motient (U.S.), TMI (Canada), INMARSAT, TM Sat (Russia), and SatMex (Mexico). The countries are not always able to reach agreement, and in any case, the result of the meetings, a bandplan for the L-band, is withheld from the public.

Motient

The sole U.S. GSO MSS licensee is Motient Corp., formerly *American Mobile Satellite Corp.* (AMSC), whose $500 million satellite AMSC-1 is one of the most powerful non-government satellites. It offers mobile phone, dispatch, data, position reporting, and facsimile services to land, maritime, and aeronautical mobile and fixed users. The subscriber units are essentially mobile radios and briefcase phones; hand portable units were never developed for this system.

Motient began in 1987 as a consortium of competing license applicants, which the FCC directed to form a single organization. At this writing, its major shareholders were Hughes Communications, Motorola, Singapore Telecom, and AT&T Wireless Services.

Motient is a monopoly in the sense that no other U.S. companies may provide GSO MSS Services from a dedicated U.S.-licensed satellite because of a purported shortage of spectrum. At its founding, most AMSC principals argued that the most pressing market need was for spectrum for voice services. Dissenters argued that data communications were more likely to become the principal need, a position that Motient now appears to support.

Aeronautical Use

The *Aeronautical Mobile Satellite (Route) Service* or AMS(R)S allocation is primary in 1545–1555 MHz and 1646.5–1656.5 MHz.

The (R) for route indicates that the communications concern safety and regularity of flight. The 1555–1559 MHz and 1656.5–1660.5 MHz spectrum is available to MSS with AMS(R)S having pre-emptive priority over the spectrum if needed.

The Motient system is required to accommodate AMS(R)S use should it occur. Similarly, use of the 1530–1545 MHz and 1626.5–1646.5 MHz spectrum requires priority access to distress and safety communications. This spectrum is used by COSPAS/SARSAT, part of the Global Maritime Distress and Safety System (GMDSS) that replaced Morse code use by ships (see 154–156.2475 MHz).

The huge Hughes-built AMSC-1 satellite was launched in April 1995. Motient's Canadian partner TMI Communications launched MSAT-1, a nearly identical satellite, in 1996. In November 1999, the FCC authorized TMI and a U.S. company, SatCom Systems, to serve U.S. customers over MSAT-1, an action Motient vehemently opposed.

In July 1998, Motient filed an updated application for AMSC-2, a satellite to provide telephone and data services via multiple spot beams plus multipoint dispatch services with nationwide coverage using a single beam. AMSC-1 would eventually be used as a spare for AMSC-2. However, AMSC-2 has not been authorized at this writing, nor has AMSC-3, which was applied for in 1993. Motient now refers to AMSC-1 as MSAT-2.

The future of the L-band spectrum is controversial and speculative. Motient does not have access to spectrum in the lower L-band. In the still-pending Docket IB 96-132, the FCC proposed to assign the first internationally coordinated 28 MHz of spectrum in the upper and lower L-band to Motient and to accommodate additional MSS licensees on additional spectrum, if available.

The FCC has authorized Motient, however, to conduct "highly sensitive" experiments in 1631.5–1660 MHz (see 1626.5–1645.5 MHz).

INMARSAT

The London-based *International Maritime Satellite Organization* (INMARSAT) was established in 1979 by treaty as an international cooperative. It was privatized in 1999. INMARSAT's traditional focus is marine telephone service in the L-band, which it offers through local affiliates in each country.

The carrier also offers satellite-based paging, Internet services, aeronautical communications, and asset tracking as well as emergency alerting in the GMDSS (see 1645.5–1646.5 MHz). Several manufacturers make INMARSAT-compatible beacons, phones, and radios of different sizes and capabilities.

INMARSAT is planning its fourth generation of satellites with two main spacecraft and one spare to be launched in 2004. The $1.4 billion *Broadband Global Area Network* (B-GAN) system will use the L-band and will provide multimedia services compatible with IMT-2000 mobile devices (see 1850–1990 MHz). Unlike Motient, INMARSAT has access to both the upper and lower L-band.

Boeing Co. applied to the FCC for a license for a $3 billion, 16-satellite non-geostationary AMS(R)S system for navigation, surveillance, air traffic management, and augmentation of the *Global Positioning System* (GPS, see 1215–1240 MHz). Boeing's traffic information service would permit flight crews to monitor the positions of surrounding aircraft on a global basis as well as weather reports, airport status, navigation aids, and military airspace information.

Boeing applied for a similar system at 2 GHz (see 1990–2025 MHz) and updated its application in November 2000. At this writing, however, the FCC has not resolved all of the issues surrounding AMS(R)S in this band.

The 1435–1626.5 MHz spectrum is one of many restricted bands in which the FCC Part 15 rules permit unlicensed devices to emit only very low level emissions.

1530–1535 MHz

FG: S5.341 S5.351 US78 US315 MOBILE-SATELLITE (space-to-Earth). MARITIME MOBILE-SATELLITE (space-to-Earth). Mobile (aeronautical telemetry)

Non-FG: S5.341 S5.351 US78 US315 MOBILE-SATELLITE (space-to-Earth). MARITIME MOBILE-SATELLITE (space-to-Earth). Mobile (aeronautical telemetry)

This is part of the L-band mobile satellite spectrum (see 1525–1530 MHz). The Air Force uses 1525–1535 MHz for aeronautical telemetry on a secondary basis.

The 1435–1626.5 MHz spectrum is one of many restricted bands in which the FCC Part 15 rules permit unlicensed devices to emit only very low level emissions.

1535-1544 MHz

FG: S5.341 S5.351 US315 MOBILE-SATELLITE (space-to-Earth). MARITIME MOBILE-SATELLITE (space-to-Earth)

Non-FG: S5.341 S5.351 US315 MOBILE-SATELLITE (space-to-Earth). MARITIME MOBILE-SATELLITE (space-to-Earth)

FCC: Satellite Communications (25). Maritime (80)

This is part of the L-band mobile satellite spectrum (see 1525–1530 MHz).

The 1435–1626.5 MHz spectrum is one of many restricted bands in which the FCC Part 15 rules permit unlicensed devices to emit only very low level emissions.

1544-1545 MHz

FG: S5.341 S5.356 MOBILE-SATELLITE (space-to-Earth)

Non-FG: S5.341 S5.356 MOBILE-SATELLITE (space-to-Earth)

FCC: Satellite Communications (25). Aviation (87)

This band is used in the COSPAS/SARSAT system to downlink distress signals (1544.5 MHz). U.S. SARSAT payloads are carried on POES and GOES satellites (see 1670–1675 MHz and 1675–1700 MHz).

The 1435–1626.5 MHz spectrum is one of many restricted bands in which the FCC Part 15 rules permit unlicensed devices to emit only very low level emissions.

Figure 11
The U.S. COSPAS-SARSAT Mission Control Center in Suitland, Maryland. Atop the console are emergency radio beacons. (NOAA/NESDIS photo)

1545-1549.5 MHz

FG: S5.341 S5.351 US308 US309 AERONAUTICAL MOBILE-SATELLITE (R) (space-to-Earth). Mobile-Satellite (space-to-Earth)

Non-FG: S5.341 S5.351 US308 US309 AERONAUTICAL MOBILE-SATELLITE (R) (space-to-Earth). Mobile-Satellite (space-to-Earth)

FCC: Aviation (87)

This is part of the L-band mobile satellite and *Aeronautical Mobile Satellite (R) Service* (AMS(R)S) spectrum (see 1525–1530 MHz).

The 1545–1555 MHz spectrum is internationally allocated to AMS(R)S downlinks, which are devoted to safety and regularity of flight, with uplinks at 1646.5–1656.5 MHz.

Aviation users have pre-emptive access to AMS(R)S frequencies. MSS radios are supposed to identify emergency communications and stop transmitting upon command (thus yielding capacity to AMS(R)S), although they may accomplish this function by remote control by the service provider.

The 1435–1626.5 MHz spectrum is one of many restricted bands in which the FCC Part 15 rules permit unlicensed devices to emit only very low-level emissions.

1549.5–1558.5 MHz

FG: S5.341 S5.351 US308 US309 AERONAUTICAL MOBILE-SATELLITE (R) (space-to-Earth). Mobile-Satellite (space-to-Earth)

Non-FG: S5.341 S5.351 US308 US309 AERONAUTICAL MOBILE-SATELLITE (R) (space-to-Earth). Mobile-Satellite (space-to-Earth)

FCC: Aviation (87)

This is part of the L-band mobile satellite and *Aeronautical Mobile Satellite (R) Service* (AMS(R)S) spectrum (see 1525–1530 MHz).

The 1435–1626.5 MHz spectrum is one of many restricted bands in which the FCC Part 15 rules permit unlicensed devices to emit only very low level emissions.

1558.5–1559 MHz

FG: S5.341 S5.351 US308 US309 AERONAUTICAL MOBILE-SATELLITE (R) (space-to-Earth).

Non-FG: S5.341 S5.351 US308 US309 AERONAUTICAL MOBILE-SATELLITE (R) (space-to-Earth).

FCC: Aviation (87)

This is part of the L-band mobile satellite and *Aeronautical Mobile Satellite (R) Service* spectrum (see 1525–1530 MHz).

The 1435–1626.5 MHz spectrum is one of many restricted bands in which the FCC Part 15 rules permit unlicensed devices to emit only very low-level emissions.

1559-1610 MHz

FG: S5.341 US208 US260 AERONAUTICAL RADIONAVIGATION. RADIONAVIGATION-SATELLITE (space-to-Earth)

Non-FG: S5.341 US208 US260 AERONAUTICAL RADIONAVIGATION. RADIONAVIGATION-SATELLITE (space-to-Earth)

FCC: Aviation (87)

Note: The NTIA Manual (footnote G126) states that differential GPS stations may be authorized in the 1559-1610 MHz band, but the FCC has not yet addressed this footnote.

Link 2 (L2), 1575.42 MHz, is one of the frequencies transmitted by the Global Positioning System (GPS, see 1215-1240 MHz).

The Russian Federation *Global Orbital Navigation Satellite System* (GLONASS) uses 1606-1616 MHz, but is to move to 1598-1605 MHz after 2005 (see 1610-1610.6 MHz).

This band also includes radio altimeters among its uses.

The 1435-1626.5 MHz spectrum is one of many restricted bands in which the FCC Part 15 rules permit unlicensed devices to emit only very low level emissions.

1610-1610.6 MHz

FG: S5.341 S5.364 S5.366 S5.367 S5.368 S5.372 US208 MOBILE-SATELLITE (Earth-to-space) US319. AERONAUTICAL RADIONAVIGATION US260. RADIODETERMINATION-SATELLITE (Earth-to-space)

Non-FG: S5.341 S5.364 S5.366 S5.367 S5.368 S5.372 US208 MOBILE-SATELLITE (Earth-to-space) US319. AERONAUTICAL RADIONAVIGATION US260. RADIODETERMINATION-SATELLITE(Earth-to-space)

FCC: Satellite Communications (25). Aviation (87)

The 1610-1626.5 MHz band is reserved worldwide for air radionavigation, including ground or space-based facilities.

Radionavigation is officially defined as radiodetermination for the purpose of navigation, including obstruction warning. Big LEO systems are authorized to use the same spectrum for user uplinks (see below).

The 1610–1610.6 MHz band is allocated in the U.S. on a co-primary basis to the *Mobile Satellite Service* (MSS) and the *Radiodetermination Satellite Service* (RDSS). Radiodetermination is defined as determination of the position, velocity, and/or other characteristics of an object, or the obtaining of information relating to these parameters by means of the propagation properties of radio waves.

Geostar

RDSS was the brainchild of Geostar Corp., founded by the late futurist and pilot Gerard O'Neill, who was an advocate of space colonization. Another O'Neill company produced some of the first mass-market wireless local area networks.

Geostar enraptured FCC commissioners by promising portable satellite-aided devices for citizens to use to signal police. The devices never materialized. Police did not endorse RDSS. None of the RDSS licensees ever built their own satellites. Geostar declared bankruptcy in 1991, obsoleted by the *Global Positioning System* (GPS, see 1215–1240 MHz).

Geostar did, however, operate a mobile data and position location service using RDSS payloads carried on the GSTAR-3 and Spacenet-3 satellites. This interim service was credited with saving lives by locating remote accident victims.

Other companies took over the interim operations from Geostar. They provided services such as ship-based credit card verification and pipeline monitoring.

Big LEOs

The interim operations had to stop when service began from the *Mobile Satellite Service Above 1 GHz*, also called the *Non-Geostationary Mobile Satellite Service* (NGSO MSS) and the *1.6/2.4 GHz Mobile-Satellite Service*. It is allocated 1610–1626.5 MHz for uplinks from user stations and 2483.5–2500 MHz for downlinks to user stations (Docket CC 92-166).

As if three names were not enough, this also is dubbed the Big LEO service. It uses satellite constellations in low-Earth (about 600 mile) or medium-Earth (about 6000 mile) orbits.

Big LEOs are among the largest and most expensive satellite projects ever attempted. They should not be confused with Little LEOs in the VHF spectrum, which are lower in cost and offer non-voice services (see 137–137.025 MHz).

"The Big LEO service can offer an almost limitless number of services," the FCC said, "including ubiquitous voice and data mobile services, position location services, search and rescue communications, disaster management communications, environmental monitoring, paging services, facsimile transmission services, cargo tracking, and industrial monitoring and control.

"Domestically, this service will help meet the demand for a seamless, nationwide and eventually global communications system that is available to all and that can offer a wide range of voice and data telecommunication services. In addition to enhancing the competitive market for mobile telecommunication services in areas served by terrestrial mobile services, this new mobile satellite service will offer Americans in rural areas that are not otherwise linked to the communications infrastructure immediate access to a feature-rich communications network. ...

"Moreover, Big LEO systems can extend these benefits throughout the world, and can provide those countries that have not been able to develop a nationwide communication service an instant global and national telecommunication infrastructure. ...

"This network can be used to provide both basic and emergency communications to their entire populations. Big LEO systems may prove to be a critical component in the development of the global information highway."

Of benefit to all LEO operators is the FCC policy that eliminated the regulatory distinction between domestic and international satellite service. All U.S. licensed satellite services can provide both domestic and international service. U.S. law currently prohibits auctions of spectrum or orbital positions used by satellites that provide global services.

Satellite service providers must still obtain licenses to operate from the countries in which they intend to provide service. Satellite systems licensed by other nations do not need FCC licenses in order to serve the U.S.; however, they must comply with other FCC filing requirements.

Pioneer: Iridium

The FCC licensed five NGSO MSS operators subject to a sharing plan. Spectrum sharing requirements and changes to international allocations could necessitate changes to the individual systems' authorized band segments listed here.

Iridium, an international consortium founded by Motorola, was the first Big LEO licensee to begin service. Its $5 billion system of 66 satellites and 12 in-orbit spares was called by Reuters "one of the costliest corporate fiascoes of all time."

Iridium is unique among the Big LEOs in that it uses *Time Division Multiple Access* (TDMA) in 1621.35–1626.5 MHz to provide both uplink and downlink service to U.S. subscribers. Iridium does not use the 2400 MHz NGSO MSS allocation. Its satellites can operate across 1616–1626.5 MHz for non-U.S. service. Iridium's inter-satellite links are at 23.18–23.38 GHz, with 19.4–19.6 GHz as the feeder downlink and 29.1–29.25 GHz as feeder uplink.

The idea for Iridium came when a Motorola executive's wife complained that phone service was unavailable in an exotic vacation spot. After gaining some 55,000 customers for its portable phone and paging service, the company filed for bankruptcy in August 1999 and discontinued revenue service March 17, 2000.

Most analysts concluded that Iridium's problems were due to high airtime prices and large, "clunky" phones. Motorola replaced these early phones with sleek, colorful models, but made little apparent attempt to gain attention for them. Iridium's second phone supplier, Kyocera, was beset with manufacturing delays.

Iridium suffered from a flawed marketing campaign and substandard press relations. Its advertising emphasized high-tech imagery and vague sloganeering where it was visible at all. Using identical typography and design to Motorola's, the print ads lumped Iridium with unrelated Motorola products and services.

Only late in the venture did Iridium focus on the remote facilities and offshore industries that market research had consistently shown were most willing to pay for satellite phone service.

These customers are the focus of Iridium Satellite LLC, a new firm that bought the operating assets of Iridium LLC and its subsidiaries for $25 million in late 2000. The new company identified its target markets as government, military, humanitarian, heavy industry, maritime, aviation, and adventure. Motorola agreed to continue to provide subscriber equipment.

Iridium LLC had planned to acquire the Claircom air-ground telephone network from AT&T, but could not complete the acquisition due to its financial problems (see 849–851 MHz).

Other Big LEOs

Loral/Qualcomm established Globalstar, a $1.5 billion, CDMA-based system of 52 satellites using both Big LEO bands. Globalstar experienced a setback when 12 of its satellites were destroyed at once in a failed Russian launch. But more were successfully launched. Service rollout began in 1999 in limited geographic areas. At this writing, more than 100 countries had signed service provider agreements with Globalstar, but the service was well behind its sales targets.

Constellation Communications is authorized for ECCO (formerly Aries), a $1.15 billion system of 46 satellites using both Big LEO bands. Mobile Communications Holdings is authorized for Ellipso, a $564 million system of 16 satellites in both Big LEO bands. Ellipso is backing the Virgo satellite proposal (see 11.7–12.2 GHz).

TRW was authorized for Odyssey, a $1.8 billion system of 12 satellites. But TRW later decided to end its independent pursuit of Odyssey. It became an investor in the ICO system (see 1990–2025 MHz). The partnership ended a long dispute between TRW and ICO over the patented orbit ICO will use.

An attractive prospect is the Big LEO "world phone" that could be used anywhere on Earth. The FCC will only authorize satellite phones capable of global roaming if they are equipped with position determination or other capability to prevent them from transmitting from forbidden locations.

Examples of such locations could be nations that ban Big LEO communications or radio astronomy observatories that are sensitive to interference. The Iridium experience suggests that a more practical wireless world phone, at least in the near term, is a multimode device for use on terrestrial networks.

The Big LEO licensees and other proponents have applied to the FCC for licenses for additional Big LEO-type systems (see 1990–2025 MHz).

The Department of Defense is funding development of the Eagle Eye Tag, a wristwatch-mounted device that allows a person or object to be located on demand via Big LEO satellite anywhere in the world. Applications include tracking patients, children, executives, convicts, military personnel, shipping containers, and vehicles.

Pull Up!

The 1610–1626.5 MHz spectrum also is reserved worldwide for aeronautical navigation and directly associated terrestrial or satellite-based facilities.

Radio altimeters sweep this spectrum or emit pulses that reflect off of terrain to aircraft. The sweeping altimeter determines altitude by the frequency difference between transmitted and received signals. The pulsed altimeter, used especially for high altitudes, measures the delay time of the pulse from the aircraft to the terrain and back. Another altimeter band is 4.2–4.4 GHz.

The radio altimeter is central to the *Ground Proximity Warning System* (GPWS) required on U.S. aircraft. GPWS warns against "controlled flight into terrain," the crashing of a functioning aircraft into the ground, usually on mountainous terrain or land adjacent to the airport, often in low-visibility conditions.

The GPWS sounds a "Pull Up!" alert to the pilot if there is insufficient clearance above terrain, except during situations such as landing. The Enhanced GPWS, now being deployed, compares altimeter and navigation data with a stored terrain database. It depicts terrain under and ahead of the aircraft on a display and can provide much earlier warnings than the conventional GPWS.

Global System

The Russian Federation *Global Orbital Navigation Satellite System* (GLONASS) uses 1606–1616 MHz and is similar in purpose to GPS. GLONASS is to move to 1598–1605 MHz after 2005.

International aviation interests seek complementary operation of GLONASS and GPS. Navigation and position systems that use both satellites could be more accurate than either system could be of itself. Standards are being developed for dual-system GPS/GLONASS receivers.

The combined GPS/GLONASS system is called the *Global Navigation Satellite System* (GNSS). It is being considered as the principal long-term means of aeronautical radionavigation in the U.S. and internationally.

Possible interference issues between GLONASS and NGSO MSS are a concern for Big LEO licensees. Depending on how the issues are resolved, the FCC could conclude that insufficient spectrum exists to accommodate all five licensees, thus calling into question the viability of one or more of the licenses.

The 1435–1626.5 MHz spectrum is one of many restricted bands in which the FCC Part 15 rules permit unlicensed devices to emit only very low level emissions.

1610.6-1613.8 MHz

FG: S5.341 S5.364 S5.366 S5.367 S5.368 S5.372 US208 MOBILE-SATELLITE (Earth-to-space) US319. RADIO ASTRONOMY. AERONAUTICAL RADIONAVIGATION US260. RADIODETERMINATION-SATELLITE (Earth-to-space)

Non-FG: S5.341 S5.364 S5.366 S5.367 S5.368 S5.372 US208 MOBILE-SATELLITE (Earth-to-space) US319. RADIO ASTRONOMY. AERONAUTICAL RADIONAVIGATION US260. RADIODETERMINATION-SATELLITE (Earth-to-space)

FCC: Satellite Communications (25). Aviation (87)

This is part of the Big LEO spectrum (see 1610-1610.6 MHz).

Radio astronomers use this band and other nearby bands in observations of hydrogen and hydroxyl radical spectral lines.

An agreement between Iridium and the National Astronomy and Ionosphere Center, operators of Arecibo Observatory (see 2360-2385 MHz), provides for the coordination of scheduling between the two operations in order to minimize interference from the Iridium downlinks to astronomy observations.

The 1435-1626.5 MHz spectrum is one of many restricted bands in which the FCC Part 15 rules permit unlicensed devices to emit only very low-level emissions.

1613.8-1626.5 MHz

FG: S5.341 S5.364 S5.365 S5.366 S5.367 S5.368 S5.372 US208 MOBILE-SATELLITE (Earth-to-space) US319. AERONAUTICAL RADIONAVIGATION US260. RADIODETERMINATION-SATELLITE (Earth-to-space). Mobile-satellite (space-to-Earth)

Non-FG: S5.341 S5.364 S5.365 S5.366 S5.367 S5.368 S5.372 US208 MOBILE-SATELLITE (Earth-to-space) US319. AERONAUTICAL RADIONAVIGATION US260. RADIODETERMINATION-SATELLITE (Earth-to-space). Mobile-satellite (space-to-Earth)

FCC: Satellite Communications (25). Aviation (87)

This is part of the Big LEO spectrum (see 1610–1610.6 MHz).

The 1435–1626.5 MHz spectrum is one of many restricted bands in which the FCC Part 15 rules permit unlicensed devices to emit only very low level emissions.

1626.5-1645.5 MHz

FG: S5.341 S5.351 US315 MOBILE SATELLITE (Earth-to-space). MARITIME MOBILE-SATELLITE (Earth-to-space)

Non-FG: S5.341 S5.351 US315 MOBILE SATELLITE (Earth-to-space). MARITIME MOBILE-SATELLITE (Earth-to-space)

FCC: Satellite Communications (25). Maritime (80)

This is part of the L-band mobile satellite spectrum (see 1525–1530 MHz).

Motient Corp., the sole U.S. *Geostationary Mobile Satellite Service* (GSO MSS) licensee, is authorized to use 1631.5–1660 MHz to experiment with earth stations using high-gain antennas. The company called these "highly sensitive experimental operations, the goal of which is to find new ways to use currently licensed spectrum."

The FCC did not reveal the details. "If Motient's competitors were to obtain knowledge of these experiments or the scientific methodology underlying these experiments, Motient could lose any competitive advantage and intellectual property rights in the new technology," the company said.

The Space Shuttle, International Space Station, and other spacecraft use this band to connect experimenters on the ground with their spaceborne packages and instruments, independently of NASA's mission communications network.

The *Spacehab Universal Communications Systems*, known as SHUCS, are TCP/IP based links between the spacecraft, the INMARSAT satellites, and the earth.

1645.5-1646.5 MHz

FG: S5.341 S5.375 MOBILE SATELLITE (Earth-to-space)

Non-FG: S5.341 S5.375 MOBILE SATELLITE (Earth-to-space)

FCC: Satellite Communications (25). Maritime (80)

This band is used for maritime distress signaling via the INMARSAT E service using geostationary satellites in the *Global Maritime Distress and Safety System* (GMDSS), which mandates carriage of emergency beacons on board various types of vessels.

The *International Maritime Satellite Organization* (INMARSAT) is a worldwide mobile satellite voice and data carrier (see 1525–1530 MHz).

The INMARSAT E device, a type of *Emergency Position Indicating Radio Beacon* (EPIRB, see 406–406.1 MHz), contains a *Global Positioning System* (GPS) receiver that permits transmission of the vessel's location in an emergency.

The "float-free" function allows the EPIRB to begin transmitting when it contacts water. Some models include a flashing light and 121.5 MHz transmitter for local homing, or close-in direction-finding by rescuers.

This is a restricted band in which the FCC Part 15 rules permit unlicensed devices to emit only very low level emissions.

1646.5-1651 MHz

FG: S5.341 S5.351 US308 US309 AERONAUTICAL MOBILE-SATELLITE (R) (Earth-to-space). Mobile-Satellite (Earth-to-space)

Non-FG: S5.341 S5.351 US308 US309 AERONAUTICAL MOBILE-SATELLITE (R) (Earth-to-space). Mobile-Satellite (Earth-to-space)

FCC: Aviation (87)

This is part of the Mobile Satellite/Aeronautical Mobile Satellite (R) spectrum (see 1525–1530 MHz).

1651-1660 MHz

FG: S5.341 S5.351 US308 US309 MOBILE-SATELLITE (Earth-to-space). AERONAUTICAL MOBILE-SATELLITE (R) (Earth-to-space)

Non-FG: S5.341 S5.351 US308 US309 MOBILE-SATELLITE (Earth-to-space). AERONAUTICAL MOBILE-SATELLITE (R) (Earth-to-space)

FCC: AVIATION (87)

This is part of the L-band Mobile Satellite/Aeronautical Mobile Satellite (R) spectrum (see 1525–1530 MHz).

1660-1660.5 MHz

FG: S5.149 S5.341 S5.351 US308 US309 AERONAUTICAL MOBILE-SATELLITE (R) (Earth-to-space). RADIO ASTRONOMY

Non-FG: S5.149 S5.341 S5.351 US308 US309 AERONAUTICAL MOBILE-SATELLITE (R) (Earth-to-space). RADIO ASTRONOMY

FCC: Aviation (87)

This is part of the L-band Mobile Satellite/Aeronautical Mobile Satellite (R) spectrum (see 1525–1530 MHz). Bands in 1660–1670 MHz also are allocated for radio astronomy.

The 1660–1710 MHz spectrum is one of many restricted bands in which the FCC Part 15 rules permit unlicensed devices to emit only very low-level emissions.

1660.5-1668.4 MHz

FG: S5.341 US246 RADIO ASTRONOMY US74. SPACE RESEARCH (passive)

Non-FG: S5.341 US246 RADIO ASTRONOMY US74. SPACE RESEARCH (passive)

No transmissions are permitted in this receive-only scientific band.

Radio astronomers use the band to observe the spectral lines of *OH*: the hydroxyl radical, the first molecule to be detected at radio frequencies (1665.402 MHz and 1667.359 MHz).

Observation of these spectral lines is important for understanding the interstellar medium and star formation in galaxies.

The 1660–1710 MHz spectrum is one of many restricted bands in which the FCC Part 15 rules permit unlicensed devices to emit only very low level emissions.

1668.4–1670 MHz

FG: S5.149 S5.341 US99 METEOROLOGICAL AIDS (radiosonde). RADIO ASTRONOMY

Non-FG: S5.149 S5.341 US99 METEOROLOGICAL AIDS (radiosonde). RADIO ASTRONOMY

This band is mostly used for radio astronomy in the U.S. Radiosonde operations in the band would have to be engineered to avoid conflicting with astronomy uses.

The 1660–1710 MHz spectrum is one of many restricted bands in which the FCC Part 15 rules permit unlicensed devices to emit only very low-level emissions.

1670–1675 MHz

FG: S5.341 US211 METEOROLOGICAL AIDS (radiosonde). METEOROLOGICAL-SATELLITE (space-to-Earth)

Non-FG: S5.341 US211 METEOROLOGICAL AIDS (radiosonde). METEOROLOGICAL-SATELLITE (space-to-Earth)

Note: 1670–1675 MHz became mixed-use spectrum in January 1999.

Weather satellites use this band for forecasting and research. The private sector will obtain access to it as well through the 1993 reallocation of the band in the Omnibus Budget Reconciliation Act (OBRA-93, see

2390–2400 MHz). The 1670–1675 MHz band may be paired or associated with 2385–2390 MHz in future auctions.

NTIA pointed to auction of licenses in 1670–1675 MHz, along with 1390–1400 MHz and 1427–1432 MHz, as a way to avoid reallocation of spectrum used by government space programs (see 2025–2110 MHz). Those three bands would fetch $1.57 billion in auction receipts according to the Office of Management and Budget.

A Notice of Proposed Rulemaking in Docket ET 00-221 offered several options for the increased use of 1670–1675 MHz. "In allocating this band, we must consider the need to protect extremely sensitive radio astronomy receivers in the subjacent 1660–1670 MHz band, as well as the need to protect meteorological-satellite earth stations at the Wallops Island and Fairbanks sites. For these reasons, this band will not be allocated to any airborne or space-to-Earth applications," the FCC said (see "Meteorological Uses," below).

The *Land Mobile Communications Council* (LMCC) has identified the band as a candidate for *Private Mobile Radio Services* (see 1390–1395 MHz). MicroTrax identified the band as desirable for its *Personal Location and Monitoring Service.* The FCC also cited the AeroAstro SENS satellite system as a possible application for this band (see 2300–2305 MHz).

Internationally, 1670–1675 MHz was designated in 1992 for aeronautical telephone service. The U.S. has declared no intention of using it for that purpose. U.S. air-ground telephone services use other bands (see 454–455 MHz and 849–851 MHz). Mobile satellites will provide aeronautical telephone service also.

Meteorological Uses

The 1670–1675 MHz band contains downlinks for the *Geostationary Operational Environmental Satellite System* (GOES), which is operated by the *National Environmental Satellite, Data, and Information Service* (NESDIS) of the *National Oceanic and Atmospheric Administration* (NOAA).

GOES, in operation since 1974, is the main source of the satellite weather imagery displayed on TV newscasts. GOES uplinks are in the 2025–2035 MHz segment of the 2025–2110 MHz U.S. space control band.

NOAA normally operates two GOES satellites to monitor the Atlantic and Pacific coasts plus a third in standby mode. NESDIS earth stations that are responsible for GOES are at Wallops Island, Virginia; Greenbelt, Maryland; and Fairbanks, Alaska. The Wallops and Fairbanks stations will remain in the band and require interference protection indefinitely.

GOES views Earth at all times to produce temperature and moisture profiles of the atmosphere above the U.S.. The satellite imager can detect land features, which helps forecasters track weather systems in relation to ground position. A small-scale scan feature enables the satellite to take pictures of local weather trouble spots. Hydrologic data gathered by GOES and Doppler radars (see 2700–2900 MHz) is used in water resource allocation.

GOES also monitors the Sun and its effects on the electromagnetic environment. Changes in the Sun can dramatically affect radio communications, electric power transmission, and the safety of space crews.

On 1544.5 MHz, GOES relays the 406.025 MHz distress signals received from people, aircraft, or vessels in the COSPAS/SARSAT system. Ground stations analyze these signals and alert rescue centers (see 406–406.1 MHz).

Docket ET 00-221 would change this band as follows. To the Federal Government (FG) allocation, it would add Footnote USyyy (the actual number to replace yyy would be decided later), retain Footnotes S5.341 and US211, and delete all other entries. In the Non-FG allocation, it would add Footnote USyyy, retain Footnotes S5.341 and US211, and convert the band to FIXED and MOBILE except aeronautical mobile services. The FCC Note regarding January 1999 would be deleted.

The 1660–1710 MHz spectrum is one of many restricted bands in which the FCC Part 15 rules permit unlicensed devices to emit only very low level emissions.

1675–1700 MHz

FG: S5.289 S5.341 US211 METEOROLOGICAL AIDS (radiosonde). METEOROLOGICAL-SATELLITE (space-to-Earth)

Non-FG: S5.289 S5.341 US211 METEOROLOGICAL AIDS (radiosonde). METEOROLOGICAL-SATELLITE (space-to-Earth)

Meteorological satellites use this band in forecasting and research.

This band was connected to 1670–1675 MHz. The FCC separated the two bands when it identified 1670–1675 MHz as subject to new commercial uses.

This band is used for downlinks by the *Polar-orbiting Operational Environmental Satellite* (POES, sometimes also called the *Television Infrared Observation Satellite*, TIROS). Its command uplink is in the 148-MHz band. The POES infrared and visible-spectrum radiometry instruments create the images transmitted to forecasters, scientists, educators, and hobbyists around the world.

The *National Oceanic and Atmospheric Administration* (NOAA) normally operates at least two POES satellites in polar orbit. From their 450-mile altitude, the satellites observe temperature, cloud cover, water, ice and snow, vegetation, and other environmental parameters.

They also monitor platform sources such as radio-equipped buoys, remote weather stations, and radiosondes in the ARGOS data collection system (see 401–402 MHz). They convey the data to stations in Virginia and Alaska for processing and dissemination.

Radiosondes are weather instruments lofted by balloons. They measure air pressure, temperature, and relative humidity. The data received from radiosondes, called *soundings*, are distributed globally and used in climate studies. About half of the world's radiosondes use this band; the others are lower in the spectrum (see 401–402 MHz).

The radiosonde ground station uses a tracking antenna that follows the balloon during all phases of flight and a receiver that is capable of following the radiosonde's drift in frequency. If the receiver loses the radiosonde signal, it scans the entire band looking for it.

The *National Weather Service* (NWS) launches about 80,000 radiosondes a year. Each carries a message asking the finder to return it, postage prepaid, but few are returned. The devices are available on the electronic surplus and swapmeet market. The government spends about $5 million annually for replacement radiosondes.

Defense Applications

The *Defense Meteorological Satellite Program* (DMSP) performs a similar mission for military users, offering data through the Air Force Global Weather Central system.

DMSP satellites measure wind speed over the ocean, monitor precipitation to determine mobility of forces, aid in artillery and missile targeting, forecast weather over unfriendly territory, and provide data on the space environment that affects radar and ground communications. The DMSP imager can make observations at extremely low light levels and can monitor fires.

Congress required the Department of Defense and NOAA to combine their polar orbiting meteorological satellite programs. Eventually the DMSP and POES functions will be carried on the same spacecraft. The combined *National Polar-Orbiting Operational Environmental Satellite System* (NPOESS) is expected to launch its first satellites in 2006. NPOESS will cost $5.3 billion through 2018.

POES satellites function in the international COSPAS/SARSAT search and rescue system. POES relays on 1544.5 MHz emergency beacon signals received on 121.5 MHz, 243 MHz and 406.025 MHz. Ground stations analyze these signals and alert rescue centers (see 406–406.1 MHz).

The *Geostationary Operational Environmental Satellite* has some downlinks in this band (GOES, see 1670–1675 MHz). On 1690.6 MHz, GOES distributes the *Emergency Managers Weather Information Network* (EMWIN), a source of continuous weather data including graphics and files. EMWIN data transmissions are rebroadcast to the public in other bands (see 150.05–150.8 MHz).

This band and 1700–1710 MHz are also used for downlinks in the Earth Exploration Satellite Service (EESS, see 8.025–8.175 GHz).

The 1660–1710 MHz spectrum is one of many restricted bands in which the FCC Part 15 rules permit unlicensed devices to emit only very low level emissions.

1700–1710 MHz

FG: S5.289 S5.341 FIXED G118. METEOROLOGICAL-SATELLITE (space-to-Earth)

Non-FG: S5.289 S5.341 METEOROLOGICAL-SATELLITE (space-to-Earth). Fixed

See 1670–1675 MHz for this band's meteorological uses.

Fixed government and secondary, non-government industrial microwave stations use this band.

From 1985 to 1988, Motorola campaigned to permit wireless *local area networks* (LANs) in this band, enlisting the FCC's support. The FCC eventually was forced to reject the proposal, however, after strong opposition from the Department of Commerce, representing meteorological satellite users.

Motorola later successfully petitioned to permit these products in *Digital Termination System* (DTS) spectrum and launched a business there that it eventually exited (see 18.8–19.3 GHz).

The 1660–1710 MHz spectrum is one of many restricted bands in which the FCC Part 15 rules permit unlicensed devices to emit only very low level emissions.

1710–1755 MHz

FG: S5.341 US256 FIXED. MOBILE

Non-FG: S5.341 US256

Note: Proceeds from the auction of the 1710–1755 MHz mixed-use band are to be deposited no later than September 30, 2002.

The federal 1700 MHz spectrum extended from 1710 to 1850 MHz and is traditionally used for fixed microwave links and defense systems.

The Balanced Budget Act of 1997 (BBA, see 2390–2400 MHz) requires the auction of licenses in 1710–1755 MHz, as indicated by the FCC Note above. This band will be cleared of non-exempt federal users by January 2004; they are entitled to compensation for their relocation to other bands.

In a November 1999 spectrum policy statement, the Commission announced its intention to apply this band and 2160–2165 MHz to an *Advanced Mobile and Fixed Communications Service* (AMFCS) that could include IMT-2000, a global personal communications service via terrestrial and satellite links (see 1850–1990 MHz). The FCC also said it will consider adding 2110–2150 MHz to this allocation.

Later, the FCC's *Third Generation* (3G) Notice of Proposed Rulemaking (Docket ET 00-258) proposed that 1710–1755 MHz be allocated for mobile and fixed services on a co-primary basis, which "would allow this band to be used for the introduction of new advanced mobile and fixed communications services, including 3G systems."

Transition Issues

As a "mixed-use" band, new users will have to coordinate with exempt federal systems that will remain in the band indefinitely (see 1755–1850

MHz). Zones that need coordination, centered on defense facilities, encompass much of the U.S. East and West coasts and considerable inland areas. Also exempted for continuation in the band will be fixed microwave systems operated by federal electric power agencies.

An advisory task force to the NTIA found that existing federal systems in 1710–1755 MHz could be relocated or retuned to 1755–1850 MHz at a cost of approximately $350 million. The task force recommended that the government share this band with compatible sharing partners, which it identified as non-subscriber based, *Private Mobile Radio Services* (PMRS) instead of *Commercial Mobile Radio Services* (CMRS). However, we believe that the powerful momentum of IMT-2000—a public service— could dim the prospects for PMRS in this spectrum.

Lucent Technologies' Bell Laboratories is authorized to experiment in 1710–1910 MHz with the unusual, high-capacity *Bell Labs Layered Space-Time* (BLAST) technology (see 746–764 MHz).

Astronomy Uses

The 1718.8–1722.2 MHz segment is allocated internationally on a secondary basis for radio astronomy and is used for observation of the 1720.53 MHz hydroxyl radical spectral line. In the United States, radio astronomy in this segment is unprotected, but other users are encouraged to minimize potential interference to radio astronomy.

Accordingly, 1718.8–1722.2 MHz is one of many restricted bands in which the FCC Part 15 rules permit unlicensed devices to emit only very low level emissions.

1755–1850 MHz

FG: G42 FIXED. MOBILE

Non-FG: None

International Mobile Telecommunications-2000 (IMT-2000) is a global personal communications service via terrestrial and satellite links (see 1850–1990 MHz). Among the spectrum identified for IMT-2000 by the *World Radio Conference-2000* (WRC-2000) was 1710–1885 MHz.

This is a critical band in the defense infrastructure. Federal frequency assignments in 1755–1850 MHz are extensive and nationwide. Certain key sites will need special interference protection from prospective IMT-2000 deployments. In the FCC's words, "potentially serious sharing problems" arise between IMT-2000 and federal systems in this band.

The *Air Force Satellite Control Network* (AFSCN) *Space-Ground Link Subsystem* (SGLS) in 1761–1842 MHz (uplink) and 2200–2290 MHz (downlink) controls numerous defense and research satellites, including British and NATO systems and the *Global Positioning System* (GPS) as well as Space Shuttle functions.

U.S.-based SGLS primary stations are in Colorado and California, with tracking stations nationwide and at Diego Garcia, Greenland and the United Kingdom. SGLS also uses transportable stations, normally operated from military bases, to provide coverage for launch and orbit operations.

Mobile Subscriber Equipment (MSE, Army) and *Digital Wideband Transmission System* (DWTS, Navy) are microwave links for use in several bands including 1755–1850 MHz. Video teleconferencing, e-mail, telephone and Internet connectivity are among the services they provide.

Air Combat Training Systems (ACTS) and *Joint Tactical Combat Training Systems* (JTCTS) use this spectrum to compute and transmit altitude, velocity, and weapons status in simulations. The JTCTS provides aircrew training using only links between aircraft, without the need for instrumented ground facilities.

Other defense uses in this part of the spectrum include tactical radio relay; television from aircraft-mounted cameras for remote piloting and monitoring of civil disturbances; wireless local area networks for combat troops; intrusion detection; dismounted soldier identification; and telecommand of robots.

Another interesting use of this spectrum is in control of precision weapons used against "high-value" and hardened enemy targets. Equipped with cameras or infrared sensors, GPS receivers, and wireless control links, the weapons are guided to the target by an officer watching on video.

Aerostat balloons used to detect low-flying aircraft in drug interdiction use voice and data links in this band. The Navy operates high-speed interceptor boats that transmit video in this band to nearby warships.

This is key government spectrum for non-defense, fixed microwave links. Among the fixed service users are the Department of Energy for control and sensing in electric utility operations; the Department of Justice and the Department of Treasury for linking law enforcement land mobile systems; and the Department of the Interior and the Department of

Agriculture for forest and park management, and emergency communications in remote areas. The Army Corps of Engineers uses this band to control hydropower stations and provide flood control and maintenance communications at inland waterway facilities. The fixed-link portions of Coast Guard safety systems (see 154–156.2475 MHz) use this band. Many of these systems are being moved to higher bands.

IMT-2000 Issues

The incumbency and wide coverage areas of these federal operations, especially SGLS, MSE/DWTS, ACTS and fixed systems, greatly impacts the introduction of IMT-2000 into the 1700 MHz spectrum, particularly the 1805–1850 MHz segment.

A variety of techniques will be needed to prevent interaction between IMT-2000 and defense systems, including band segmenting, exploitation of antenna capabilities, siting of base stations, changes to power levels, possible "keep away" beacons, and cooperative scheduling of operations. Receivers already on satellites could face degradation from IMT-2000 transmitters as market penetration for the new service increases.

The need for these precautions will likely impact the revenues that IMT-2000 license auctions can generate. Government users who must relocate their stations will be entitled to compensation by new entrants.

1850–1990 MHz

FG: None

Non-FG: FIXED. MOBILE

FCC: RF Devices (15). Personal Communications (24). Fixed Microwave (101)

This is one of several bands traditionally used by fixed microwave systems. The FCC reallocated it to the broadband and unlicensed parts of the *Personal Communications Service* (PCS). The FCC established a separate PCS service, Narrowband PCS, for data and paging including digital voice messaging (see 901–902 MHz).

Typical licensees in the *Private Operational Fixed Point-to-Point Microwave Service* (POFS) include state and local governments, public safety agencies, petroleum, railroad, and utility companies. They use the service to transmit telemetry and control signals and to provide links in mobile radio systems. The FCC's Part 101 rules define POFS as "a private radio service rendered by fixed and temporary fixed stations on microwave frequencies for the exclusive use or availability for use of the licensee or other eligible entities for communication between two or more designated points."

This definition distinguishes POFS from the *Common Carrier Fixed Point-to-Point Microwave Service*, where the licensee leases communications capacity to customers. The FCC allows common carrier licensees to provide private services, but private licensees cannot provide common carrier services or lease reserve capacity to common carriers for their traffic (this rule is under reexamination). Otherwise, the two types of services share a number of regulations, allocations, and technical standards.

Historically, fixed microwave licenses were issued upon application, provided that frequency coordination showed that adequate spectrum was available. The 1997 Balanced Budget Act (see 2390–2400 MHz) directed the FCC to auction all mutually exclusive applications for licenses except for a short list of exemptions such as digital TV licenses.

In a Notice of Proposed Rulemaking in Docket WT 00-19, the FCC is considering how to implement auctions in the fixed microwave spectrum. "While spectrum above 2 GHz is becoming scarcer," the FCC said, "demand for it is growing. Microwave is used as the backbone infrastructure for cellular, PCS, and other CMRS [Commercial Mobile Radio Service] providers, which are expanding rapidly. Microwave spectrum may also be used for fixed point-to-multipoint service backbone support, such as for Local Multipoint Distribution Service [LMDS, see 27.5–29.5 GHz].

"Finally, the spectrum above 2 GHz is fertile ground for advanced telecommunications applications. These competing forces must be addressed in our effort to comply with the Congressional intent to ensure that spectrum is used for the purposes the public interest requires."

The FCC proposed several options for auctioning fixed microwave licenses, including licensing spectrum on a geographic-area basis, permitting incumbent licensees to retain their status but not to expand their systems, moving incumbents to other spectrum in order to clear bands for auctioned licenses, and various schemes for sharing spectrum between existing and new users.

Reallocation to Mobile Services

Finding that fixed microwave licensees could migrate to other bands such as 3.7–4.2 GHz, 5.925–6.425 GHz, 6.525–6.875 GHz, 10.55–10.6 GHz, 10.6–10.68 GHz, and 10.7–11.7 GHz, the FCC in 1994 reallocated 1850–1990 MHz to a so-called Emerging Technologies category for advanced services. Other Emerging Technologies bands encompass 2110–2150 MHz and 2160–2200 MHz.

Emerging Technology providers must share the bands with fixed service incumbents, and pay them to move out of the bands where necessary. Detailed regulations govern the rights of incumbents and new entrants. Industry clearinghouses administer relocation and compensation, which also applies to unlicensed systems in this band.

Defining PCS

The 1850–1990 MHz band hosts the first Emerging Technology service: *Personal Communications Services* (PCS), best known as the digital wireless telephone service that competes with cellular telephony (see 824–849 MHz).

The FCC originally defined PCS as "radio communications that encompass mobile and ancillary fixed communication services that provide services to individuals and businesses and can be integrated with a variety of competing networks."

It later removed the requirement that fixed PCS be offered only as an "ancillary" service. As a form of *Commercial Mobile Radio Service* (CMRS), PCS may now offer fixed wireless services to the public on a co-primary basis with or without offering mobile service.

This means, for example, that PCS may offer wireless local loop: basic telephone service to fixed subscribers in competition with local telephone companies. In 1997, AT&T announced plans to use some of its PCS license holdings to provide residential fixed wireless telephone service and Internet access. This service, formerly Project Angel and now dubbed AT&T Digital Broadband, began in Fort Worth, Texas in March 2000. By the end of 2001, the company plans to have some 10 million homes within its coverage.

PCS Origins

PCS started at the FCC in 1989 when wireless entrepreneur Matthew Edwards (see 30.56–32 MHz) petitioned the agency to allocate spectrum for telepoint services. Telepoint was already underway in Europe as a part of modernized residential and business cordless phone services. Telepoint phones worked with base units in the home and through telepoints or public base stations in places such as train platforms and gas stations.

Telepoint was based on 800 MHz digital technology and inexpensive infrastructure. It fizzled in the United Kingdom, one of its first markets.

A telepoint user could make calls but not receive them because the system was not designed to locate users and ring their phones. This one-way calling was a marketing disaster. It was widely misinterpreted to mean that the user could talk into a phone, but would hear nothing.

Telepoint still attracted attention as a low-cost competitor to cellular. A telepoint-like, but more sophisticated service, the *Personal Handyphone System* (PHS), became immensely popular in Japan. Attempts to promote PHS in the United States failed (see 2305–2310 MHz).

Looking beyond telepoint, British regulators established the *Personal Communications Network* (PCN), a 2-GHz portable phone service. PCN fully competes with cellular and wireline local telephone services.

The FCC adopted the term PCS to encompass PCN and telepoint (if anyone cared to offer it). Excited by telepoint and PCN, more than 200 firms applied to the FCC for experimental PCS licenses. This prompted the Commission to formalize the new service.

Finalizing PCS

The FCC's bandplan for PCS incorporates allocations for licensed PCS carriers and for unlicensed PCS devices. The band plan provides for three 30-MHz licenses (blocks *A, B,* and *C*) and three 10-MHz licenses (blocks *D, E,* and *F*).

The *A* and *B* blocks (1850–1865/1930–1945 MHz and 1870–1885/1950–1965 MHz, respectively) are licensed in 51 *Major Trading Areas* (MTAs). These are metropolitan regions identified by Rand McNally based on geography, population, and economic measures.

The *C* (1895–1910/1975 –1990 MHz), *D* (1865–1870/1945 –1950 MHz), *E* (1885–1890/1965 –1970 MHz), and *F* (1890–1895/1970 –1975 MHz) blocks are licensed within 493 smaller *Basic Trading Areas* (BTAs).

Based on their PCS research and development, three companies received early assurance of MTA PCS license award and discounts on auction payments under the FCC's Pioneer Preference program. Awardees were Omnipoint Corp. for New York, Cox Enterprises for Los Angeles and San Diego, and American Personal Communications for the Washington

D.C./Baltimore area. Each of these licensees later merged with other carriers.

Congress later withdrew the FCC's authority to provide preferential treatment to applicants judged to be pioneers in their respective fields. But Qualcomm Inc. secured a Pioneer's Preference in 2000 after years of litigation. An appeals court ordered the FCC to grant Qualcomm a preference that ultimately became an auction discount voucher of more than $125 million for use in the license auction of its choice.

Auction Results

The FCC auction for the 99 regional Broadband PCS licenses concluded March 13, 1995. Including payments by the three Pioneer Preference awardees of more than $701 million, the total revenue from the A and B blocks was more than $7.7 billion, the largest auction in history at the time.

The initial C-block auctions, intended for entrepreneurs and small businesses, were held in May and July 1996. The FCC had to reauction some C-block licenses after some bidders failed to make necessary payments. Continuing financial problems of C-block participants led to extensive litigation and prompted the FCC to discontinue its practice of offering installment payments for auctioned licenses. The FCC permitted some of the C-block licensees to reduce the size of their licenses to 15 MHz.

The initial D, E and, F-block auction concluded in January 1997, and an April 1999 auction offered C, E, and F-block licenses that had returned to the FCC. At this writing, more than 200 C and F-block licenses have been involved in bankruptcy proceedings and license payment defaults.

The FCC made several changes that affect subsequent PCS license auctions (Docket WT 97-82). From an allocations perspective, the most important change was the reconfiguration of remaining 30 MHz C-block licenses into three 10-MHz block licenses. These include 1895–1900/1975–1980 MHz, 1900–1905/1980–1985 MHz, and 1905–1910/1985–1990 MHz. The licenses can be aggregated to form larger blocks.

The decision also divided BTAs into two population-based tiers, and removed eligibility restrictions that had prevented large entrants from bidding on some of the licenses.

The first auctioned PCS system began revenue operation in Honolulu in February 1996, and was operated as VoiceStream by Western Wireless Corp.

Technical Standards

Unlike first generation analog cellular telephony, where U.S. carriers adhere to a uniform national standard, Broadband PCS is available in several technological flavors depending on the digital second generation (2G) standard chosen by carriers in each market. These standards include the *Time Division Multiple Access* (TDMA) approaches *Global Standard for Mobile Communications* (GSM) and TDMA-EDGE (Enhanced Data Rates for Global Evolution), and the *Code-Division Multiple Access* (CDMA) standard called cdmaOne.

3G and IMT-2000

Third Generation (3G) PCS is under development by manufacturers and standards groups. This advanced PCS is intended to support Web browsing, electronic mail, and voice services, and a seamless infrastructure across diverse wireless media as a last mile connection that competes with wireline telephone and cable data delivery.

The center of 3G development is *International Mobile Telecommunications-2000* (IMT-2000), the *International Telecommunication Union* (ITU) program to standardize multimedia wireless services capable of global roaming.

IMT-2000 became the new, more attractive name for what was called the *Future Public Land Mobile Telecommunications System*, abbreviated FPLMTS and usually pronounced "Flumpits" unofficially.

IMT-2000 will offer high quality digital voice transmission and short message service similar to that available from current wireless providers. Its data capability is intended to deliver speeds of two megabits per second or higher for indoor traffic; 384 kb/s or higher for pedestrian traffic and 144 kb/s or higher in vehicular use. Several different radio interfaces and modulation techniques, employing code-, time-, and frequency-division multiplexing in various combinations, are under intense investigation for the service.

High levels of interoperability and ease of roaming, highly standardized user billing and carrier exchange of user data, and terminal position location also are planned for IMT-2000. "IMT-2000 users will not, in most circumstances, notice that a radio link is used to connect their terminal to the world's telecommunication networks," the FCC has stated. However, "We do not believe it is necessary or desirable to define specifically what is or is not a '3G' or 'advanced' wireless service," the Commission said in opening its IMT-2000 Notice of Proposed Rulemaking (NPRM, Docket ET 00-258).

Numerous Bands Identified

The *World Administrative Radio Conference-1992* (WARC-92) identified the bands 1885–2025 MHz and 2110–2200 MHz for IMT-2000 worldwide. The *World Radiocommunication Conference-2000* (WRC-2000) made additional allocations to IMT-2000: 806–960 MHz, 1710–1885 MHz, and 2500–2690 MHz for terrestrial use, and 1525–1559 MHz, 1610–1660.5 MHz, 2483.5–2500 MHz, 2500–2520 MHz, and 2670–2690 MHz for satellite IMT-2000. WRC-2000 took no action regarding the 1800 and 2100 MHz bands identified for IMT-2000 at WARC-92; these bands may still be used for the service.

In a November 1999 spectrum policy statement, the Commission announced its intention to propose to allocate 50 MHz at 1710–1755 MHz and 2160–2165 MHz to an *Advanced Mobile and Fixed Communications Service* (AMFCS) that could include IMT-2000, and to consider adding 2110–2150 MHz to this allocation.

"There currently is no global consensus as to how the frequency bands identified at WARC-92 and WRC-2000 will be used to implement 3G, or whether common global bands for use by 3G systems are achievable," the FCC said. And in its IMT-2000 NPRM, the FCC said, "it is not Commission policy to set aside a certain amount of spectrum restricted to a given technology, such as 3G. Instead, we intend to identify a flexible allocation for the provision of advanced wireless services."

Presidential Action

In October 2000, former President Bill Clinton signed an Executive Memorandum directing interagency cooperation in finding suitable spectrum for 3G services. "The action I am taking today will help U.S. high-tech entrepreneurs compete and win in the global marketplace," he said. "It also will allow consumers to enjoy a wide range of new wireless tools and technologies, such as handheld devices that combine services like a phone, a computer, a pager, a radio, a customized newspaper, a GPS locator, and a credit card.

"I am confident that federal agencies, working with the private sector, can develop a plan for identifying the spectrum that will meet the needs of the wireless industry and is fully consistent with national security and public safety concerns. ...

"Time is of the essence. If the United States does not move quickly to allocate this spectrum, there is a danger that the U.S. could lose market share in the industries of the 21st century. If we do this right, it will help ensure continued economic growth, the creation of new high-tech jobs, and the creation of exciting new Internet and telecommunications services."

The plan is intended to lead to an FCC Report and Order making a formal allocation of 3G spectrum by the end of July 2001, to be followed by a *Further Notice of Proposed Rulemaking* (FNPRM) to establish rules for the new service. The FCC plans to finish the FNPRM by December 15, 2001, to auction 3G licenses June 15, 2002, and to finish assigning the licenses by September 30, 2002.

Licensees in the *Multipoint Distribution Service* and *Instructional TV Fixed Service* in the 2500 MHz spectrum are not keen on the prospect of forced participation in IMT-2000. The Wireless Communications Association stated, "We are confident that the U.S. Government will continue to recognize the irreplaceable value of MMDS/ITFS spectrum for broadband Internet access to unserved and under-served Americans, and for state-of-the-art interactive educational programming. This is consistent with the U.S. stand at the WRC-2000, and its results. We believe that there is more than enough spectrum to accommodate 3G needs without displacing existing ITFS and MMDS operators. These operators are well-positioned to provide the most advanced broadband services to large populations, and are in harmony with the recent fixed wireless MMDS allocations by such important communications partners of the U.S. as Mexico, Canada and Brazil."

Location Services

The FCC requires CMRS carriers such as PCS and cellular to permit public safety authorities to locate users who dial 911 in emergencies. These regulations are called Enhanced 911 or E911. Phase I of E911, which began in 1998, sends the emergency caller's callback phone number and cell site or base station location to the *public service answering point* (PSAP).

Phase II begins October 1, 2001. It requires carriers to provide the location of a 911 call by geographical coordinates. Developers promote various technologies to locate users, including GPS-derived information and processing of signal arrival geometries.

Beyond emergency calling, mobile commerce should flourish as carriers develop specialized information and navigation services based on user location.

Unlicensed PCS

To provide for indoor, low-power and short-range voice and data communications, the FCC designated 1910–1930 MHz for *unlicensed PCS* (UPCS) devices under its Part 15 rules, hence the listing of *Radio Frequency* (RF) Devices (15) in the FCC use designator for this band.

This spectrum designation resulted from the efforts of Apple Computer, Inc., which envisioned unlicensed Data-PCS as an inexpensive wireless local area network, and by the *Wireless Information Networks Forum* (WINForum), an association of manufacturers.

These petitioners estimated that at least 40 MHz should be allocated to UPCS. The FCC eventually divided 1910–1930 MHz into two 10 MHz portions (20 MHz total, half of the estimate) corresponding to two general categories of unlicensed devices.

The 1910–1920 MHz segment is devoted to asynchronous applications, principally wireless data, that transmit data at irregular time intervals and that are relatively insensitive to time delays. The 1920–1930 MHz segment is for isochronous uses that transmit regular, periodic signals such as in voice telephony.

The FCC requires all UPCS devices to obey a WINForum-developed coexistence scheme or "spectrum etiquette" containing provisions for both types of devices. This approach permits devices of different manufacturers to share the unlicensed PCS bands by means of signal detection and automated channel selection without a need for communication between different devices. A compliance test procedure for the etiquette was issued as a formal standard.

The UPCS 1910–1930 MHz segment contains the fewest incumbent fixed users of the entire 1850–1990 MHz spectrum. Nevertheless, the industry must collect money from manufacturers of unlicensed devices and use it to compensate the incumbents for the relocation of the fixed stations in 1910–1930 MHz.

Relocation is particularly important for unlicensed data at 1910–1920 MHz. Unlike desktop wireless networks, Data-PCS is supposed to be portable or nomadic in use. Users could operate Data-PCS units in any location, including places where operation could threaten incumbents with interference.

The expense, delay, and difficulty of relocating fixed systems led Apple Computer and Microsoft (also a Data-PCS supporter) to promote another band for nomadic data: 2390–2400 MHz. This band is allocated to the Amateur Radio Service. The FCC eventually made this band available to Data-PCS as well. Apple's initial wireless product, AirPort, does not use

Data-PCS, however. It uses standard, spread-spectrum wireless LAN technology instead.

Relocation is somewhat less pressing for other UPCS products that operate under the control of a fixed base station. Cordless telephone systems are an example. Such non-nomadic products can be sold and used, but they must be coordinated in frequency and location so as not to interfere with incumbent stations.

Hewlett-Packard Co. first proposed frequency coordination as a means toward the early (pre-relocation) deployment of unlicensed devices. It was implemented by UTAM Inc., a UPCS industry services organization.

Frequency coordination imposes financial, technical, and legal obligations on UPCS products and suppliers. These responsibilities constitute a de facto type of licensing. UTAM is solely authorized by the FCC to ensure that UPCS products are only used in places where they will not interfere with incumbents.

UPCS products must only operate at UTAM-approved customer locations until incumbents have moved out of 1910–1930 MHz. If a product moves away from its coordinated location, it may not be reactivated unless UTAM approves operation at the new location.

UTAM assesses a fee ($20 at this writing) on each radiating UPCS product and directs the money toward the relocation effort.

Unlicensed voice PCS is available in cordless phones in 1920–1930 MHz. These products, marketed for commercial and institutional users, are more sophisticated and expensive than the cordless phones sold in stores to consumers.

Obtaining Additional Spectrum

"The UPCS allocation has been tremendously successful," according to WINForum. "Since allocation of the UPCS band, a number of isochronous systems have been developed by at least ten manufacturers, certified by the Commission, and marketed to the public. Based on market forecasts, over 1.5 million UPCS handsets will have been sold by 2002, at which time the UPCS handsets will account for more than half of all in-building wireless handsets sold."

On the other hand, WINForum told the FCC, no asynchronous UPCS devices [using 1910–1920 MHz or 2390–2400 MHz] have been fielded. This may be due, it said, to the larger bandwidth and/or transmit power limits available in other bands—namely the 902–928 MHz, 2400–2483.5 MHz, and 5.725–5.85 GHz bands used under the FCC spread spectrum rule 15.247, as well as the constraints imposed by the asynchronous part of the etiquette and the requirement that UPCS devices in 1910–1920 MHz be coordinated with the incumbent fixed microwave systems through UTAM.

(More than 80 percent of the nation's area had been cleared of microwave systems in 1910–1920 MHz at this writing, according to UTAM.)

WINForum petitioned the FCC in 1999 (RM-9498) to allow access to the 1910–1920 MHz segment by isochronous devices. The group reasoned that such devices, especially high-density business cordless telephone systems, need the extra spectrum more than asynchronous devices, which have other bands and have avoided this segment entirely.

"Clearly, there is no hard number on the spectrum required for isochronous UPCS systems," according to WINForum president Jay E. Padgett. "For some situations (low densities, poor propagation), the current 10 MHz is adequate, at least for a single air interface that is not attempting to coexist with a system that uses a different UPCS air interface. At the other extreme, there will undoubtedly be situations in which even 20 MHz will not be enough to meet the nominal wireless communications needs (high densities, large numbers of users, good propagation, other UPCS systems nearby). However, in many cases, the additional 10 MHz may allow more efficient use of system resources, improve performance, and facilitate coexistence with other systems using different air interfaces."

In late 2000, Ascom Wireless Solutions asked the FCC to waive its rules and permit isochronous devices in 1910–1920 MHz (DA 00-2833). The company wants to use this segment to expand the capacity of its UPCS phones used on the trading floors of the American Stock Exchange, the Chicago Board Options Exchange and Chicago Board of Trade, the Pacific Stock Exchange (San Francisco), and the Philadelphia Stock Exchange. The FCC invited comments on the request.

WINForum also proposed alternative etiquette rules for the 2390–2400 MHz UPCS band. The alternative rules would not require devices to listen before transmitting, but would impose a duty cycle limit to help prevent any unit from dominating the band. Authorization of any asynchronous devices in 1910–1920 MHz would be foreclosed after a certain date.

Experimental Use Halted

The FCC no longer grants experimental licenses for use of 1910–1930 MHz except for experiments that directly involve unlicensed PCS. Some licensed PCS operators wanted to perform propagation tests in the band because of the lower concentration of incumbent fixed microwave systems there.

However, according to UTAM, "particularly during the early period, public perception of the utility and reliability of UPCS products is critical. Random shut-downs or microwave incumbent complaints (either of which could result from the testing of equipment other than UPCS devices) could cause irreparable harm and prevent this industry from bringing its full benefits to the public."

The FCC agreed, and ordered non-UPCS experiments in 1910–1930 MHz off the air in 1996.

1990–2025 MHz

FG: None

Non-FG: NG156 MOBILE-SATELLITE (Earth-to-space)

FCC: Satellite Communications (25)

The 1990–2130 MHz spectrum was allocated to the *Broadcast Auxiliary Services* (BAS), especially for on-scene *Electronic News Gathering* (ENG). These mobile operations include news vans and sky traffic aircraft that feed receive sites, which in turn relay the signal to TV stations.

Other BAS Services include *studio-transmitter links* (STL); the *Cable Television Relay Service* (CARS), used to interconnect cable TV facilities; *Point of View* (POV) cameras attached to racecars, skiers, and bobsleds; and the *Local Television Transmission Service* (LTTS), where common carriers provide linking service to TV stations.

New Mobile Satellites

To make more spectrum available for *Mobile Satellite Services* (MSS) and consistent with international and regional allocations, the FCC allocated 1990–2025 MHz (uplink) and 2165–2200 MHz (downlink) to a service that it calls *2 GHz MSS* (Docket IB 99-81).

The Commission said that 2 GHz MSS services are "essentially the same as provided by the Big LEOs (i.e., voice, data, and fax via MSS) and…the proposed 2 GHz MSS system designs are similar to those used by the Big LEOs."

Big LEOs, such as Iridium and Globalstar, are the satellite telephone services known as Non-Geostationary or NGSO MSS, or *1.6/2.4 GHz Mobile Satellite Service* (see 1610–1610.6 MHz).

To release spectrum for 2 GHz MSS, the FCC established Phase I, in which MSS licensees clear(by retuning radios, for example) BAS operations out of 1990–2008 MHz, moving them to 2008–2025 MHz. In Phase II, which satellite licensees will start if they need more spectrum, BAS users would move from 2008–2025 MHz to 2025–2110 MHz. Phase II will require new BAS equipment.

MSS operators must bear the costs of moving incumbent users out of MSS spectrum into new spectrum "regardless of the nationality, service, or technology of the new [MSS] entrant," the Commission said, rejecting protests against this policy by MSS interests. Subsequent MSS entrants will have to compensate the first entrant for expenses of clearing. MSS may be able to operate without interfering with some fixed service licensees in the downlink spectrum, so no relocation is needed in those instances.

"The system proponents in this proceeding propose both NGSO and GSO mobile satellite systems," the FCC said. "Each type of system has technical advantages and disadvantages. For example, because NGSO satellites orbit close to the Earth's surface, time delays during radio transmissions from the Earth to the satellite and back are shorter than for GSO systems.

"Conversely, because GSO satellites are at high altitudes, approximately 22,300 miles, a single GSO satellite has a very large potential coverage area, compared to a single NGSO satellite. Our goal is to provide an opportunity for both types of systems to compete in the marketplace to provide users with the best combination of services and prices. ...We will accommodate both types of systems in our band plan."

The term NGSO encompasses *low-Earth orbit* (LEO), *medium-Earth orbit* (MEO), and highly elliptical orbits, the FCC noted.

Auctions Avoided

The FCC does not anticipate holding auctions for 2-GHz MSS licenses as it forged a bandplan that should accommodate all of the 2-GHz license applicants. The plan divides 1990–2025 MHz and 2165–2200 MHz into segments called, "selected assignments," of equal bandwidth to be allocated among the licensees. The exact bandwidth will depend on the number of applicants at the time it authorizes the systems and on the bandwidth selections of the first system.

A MSS licensee could operate only in its selected assignment, only outside its selected assignment (on a secondary basis), or both. The plan provides for expansion spectrum to be used when a licensee exceeds its original assignment as well as incentives for licensees to serve rural areas.

The FCC received nine, 2-GHz MSS license applications, representing both LEO and geostationary satellites. The applications include three *letters of intent* (LOI) from non-U.S. proponents.

New ICO

One LOI came from ICO Services Ltd. ICO planned a $4 billion constellation of 12 satellites to provide international service to handheld, vehicular, and fixed wireless telephones. ICO emerged from bankruptcy in May 2000 after a $1.2 billion investment led by telecom magnate Craig McCaw.

McCaw rechristened the firm New ICO and is integrating it with the Teledesic "Internet-in-the-Sky" project (see 18.8–19.3 GHz). New ICO plans to deliver broadband Internet and *Third Generation* (3G) services (see 1850–1990 MHz).

The *International Mobile Satellite Organization* (INMARSAT), which spawned ICO, itself filed a LOI for a four-satellite MSS system called Horizons. It later withdrew the LOI, stating that instead of Horizons, it will build the INMARSAT 4 system, a *Broadband Global Area Network* (B-GAN) in already-authorized L-band spectrum (see 1525–1530 MHz). The first of these satellites will be launched in 2004. A Canadian satellite operator, TMI, was the third LOI filer.

GMPCS

These systems are classified as Global *Mobile Personal Communications by Satellite* (GMPCS). The *International Telecommunication Union* (ITU) sponsors agreements that are supposed to permit global circulation and transborder use of GMPCS terminals without burdensome regulation. Compliance with this GMPCS *Memorandum of Understanding* (MoU) will be indicated by a logo on authorized GMPCS devices.

Celsat America proposed a single geostationary satellite, with two more that will be added later, to provide a "mobile hybrid personal communications service, consisting of mobile voice, data, paging, messaging, facsimile, imaging, video, and other digital services." Personal Communications Satellite Corporation proposed a two-satellite system; the company is a subsidiary of GSO MSS licensee Motient (formerly American Mobile Satellite Corp., see 1525–1530 MHz).

Other 2 GHz MSS applicants include Boeing, for 16 satellites providing *Aeronautical Mobile Satellite (Route) Service* (AMS(R)S, see 1525–1530 MHz); Iridium, for a Macrocell system of 96 satellites; Globalstar, which proposed GS-2, a system of 68 satellites including four geostationary satellites; Constellation Communications, which proposed Constellation-II with 46 satellites; and Mobile Communications Holdings, for Ellipso 2G, with 26 satellites.

These systems require feeder links for gateway communications and links for satellite monitoring and control. The exact frequencies to be made available for such links had not been resolved at this writing.

2025–2110 MHz

FG: S5.391 S5.392 US90 US222 US346 US347 SPACE OPERATION (Earth-to-space) (space-to-space). EARTH EXPLORATION-SATELLITE (Earth-to-space) (space-to-space). SPACE RESEARCH (Earth-to-space) (space-to-space)

Non-FG: S5.392 US90 US222 US346 US347 FIXED NG23 NG118. MOBILE S5.391

FCC: TV Auxiliary Broadcasting (74F). Cable TV Relay (78). Local TV Transmission (101J)

The 2-GHz *Broadcast Auxiliary Services* (BAS) spectrum extended from 1990 to 2130 MHz (see 1990–2025 MHz).

To comply with the 1997 *Balanced Budget Act* (BBA), the FCC in Docket ET 95-18 moved BAS out of 2110–2130 MHz, leaving it with seven channels in the 2025–2110 MHz band. The BBA (see 2390–2400 MHz) requires the FCC to reallocate spectrum and auction licenses in 2110–2150 MHz by September 30, 2002.

Relocation of BAS will be funded by 2-GHz *Mobile Satellite Service* (MSS) licensees, who will be using 1990–2025 MHz for uplinks and therefore must move BAS out of that spectrum as well. A key BAS service is *Electronic News Gathering* (ENG).

Space Uses

The 2025–2110 MHz band is also the primary command and control band for U.S. civil space programs.

Control of spacecraft is generally termed TT&C: Telemetry, Tracking, and Command. "TT&C operations control and monitor a satellite's electrical function and orbit from a fixed earth station and allow information to be transmitted to the satellite that can alter the electrical function of the spacecraft and, if necessary, adjust its orbit," the FCC has explained. "TT&C functions may be conducted in bands allocated to the Space Operation Service or in the band on which the underlying service is being provided."

High-profile systems that use this band include the *Tracking and Data Relay Satellite System* (TDRSS), the Space Shuttle, the *Hubble Space Telescope* (HST), the International Space Station, the GOES and POES weather satellites, and the Landsat remote sensing system. These systems generally downlink and return link in 2200–2300 MHz.

The TDRSS relays communications between ground stations and low-orbiting spacecraft, including manned spacecraft, when the spacecraft are over the horizon and not capable of directly using ground stations. TDRSS represents the most complex non-military communications satellites ever built.

The HST, an orbiting, unmanned observatory, was launched in 1990 for an expected lifetime of 15 years. It receives in 2025–2110 MHz and outputs data in 2200–2290 MHz. All HST communications go through TDRSS.

Future of 2025–2110 MHz

The FCC had said it intended to identify 15 MHz of spectrum for competitive bidding within 2025–2110 MHz. However, the *National Telecommunications and Information Administration* (NTIA), on behalf of the President, cited a "profoundly negative impact" of such a decision on $87 billion in existing and planned U.S. space program investment, as well as on foreign space systems in which the U.S. participates.

NTIA cited an Office of Management and Budget conclusion that auction of licenses in 15 MHz at 2025–2110 MHz would obtain "negligible" receipts for the government.

"Levels of radio interference to federal space services would be unacceptable if the FCC were to assign any 15 MHz portion of the 2025–2110 MHz band by competitive bidding for a use different in character than the ENG service," NTIA said.

High-density mobile services, such as *Personal Communications Services* (PCS), were ruled out as too dissimilar to the low-density mobile and fixed ENG operations with which space users have successfully shared the band.

NTIA also pointed out that NASA may, in the future, provide TT&C to commercial low-Earth orbit remote sensing systems that will operate in this band.

2110-2150 MHz

FG: US252 (in 2110–2120 MHz; None in 2120–2200 MHz)

Non-FG: US252 NG153 FIXED NG23. MOBILE

FCC: Public Mobile (22). Fixed Microwave (101)

Note: 2110–2150 MHz must be auctioned by September 30, 2002.

The 2-GHz *Broadcast Auxiliary Services* (BAS) spectrum extended from 2025 to 2130 MHz. A principal BAS use is *Electronic News Gathering* (ENG). In Docket ET 95-18, the FCC reduced the BAS spectrum to 2025–2110 MHz (see 1990–2025 MHz and 2025–2110 MHz).

The 2110–2130 MHz segment of this band is widely used by common carrier fixed microwave links; systems in the *Local Television Transmission Service* (LTTS) used for broadcast networks; and control and repeater stations in paging systems. Some of these systems use paired links at 2160–2165 MHz.

The 2130–2150 MHz segment, which is paired with 2180–2200 MHz, is devoted to private, fixed service licensees such as state and local governments, railroads, and utilities. Many of these links will migrate to other bands or media.

The World Radiocommunication Conference-1992 identified 2110–2150 MHz as one of several bands for *Third Generation* (3G) services, also called *International Mobile Telecommunications-2000* (IMT-2000, see 1850–1990 MHz).

The FCC 3G Notice of Proposed Rulemaking (Docket ET 00-258) proposes to "designate" 2110–2150 MHz and 2160–2165 MHz for *Advanced Mobile and Fixed Communications Services* that could include IMT-2000. Existing users would have to relocate to other bands if necessary, except for the *Space Research Service* (SRS) (see below). The 1997 *Balanced Budget Act* (BBA) requires the FCC to reallocate spectrum and auction licenses in 2110–2150 MHz by September 30, 2002 (see 2390–2400 MHz). "Due to similarities in allocation, usage, and current licensing, we propose to auction the 2160–2165 MHz band in this same timeframe," the FCC said.

Footnote US252 allocates the 2110–2120 MHz segment of this band to interplanetary uplinks from NASA's Deep Space Network station at Goldstone, California.

This SRS station transmits up to 400 megawatts of power and could disrupt IMT-2000 users in the area around the station or those using equipment in adjacent spectrum.

2150-2160 MHz

FG: None

Non-FG: FIXED NG23

FCC: Domestic Public Fixed (21). Fixed Microwave (101)

The *Multipoint Distribution Service* (MDS) was intended for data transmission. The band became used for video when the FCC extended the high end of the band to 2162 MHz. This created room for two video channels, MDS 1 at 2150–2156 MHz and MDS 2 at 2156–2162 MHz, for use in the top metropolitan areas.

These actions established the original microwave pay-TV service that evolved into "wireless cable." The two-channel MDS was hard-hit when cable TV arrived in many communities with a greater number of channels.

Outside major markets, channel 2A (2156–2160 MHz) can be used for voice and data, as the channel is too small for a 6 MHz video channel. At this writing, Internet access was available in several cities via 2150–2162 MHz.

The 2150–2160 MHz band also is available to the *Private Operational Fixed Point-to-Point Microwave Service* (POFS or OFS).

In 1983, the FCC launched the modern era of wireless competition to cable TV when it established the multichannel MDS, which evolved from a video entertainment service into broadband fixed access (see 2500–2655 MHz).

2160-2165 MHz

FG: None

Non-FG: NG153 FIXED NG23. MOBILE

FCC: Domestic Public Fixed (21). Public Mobile (22). Fixed Microwave (101)

This band is devoted to the same types of common carrier usage as in 2110–2130 MHz (see 2110–2150 MHz).

The 2160–2162 MHz segment is part of the 2156–2162 MHz *Multipoint Distribution Service* channel 2 (see 2500–2655 MHz). The FCC reallocated this segment, however, to the Emerging Technologies category for future unspecified uses. MDS use of that segment after 1992 is secondary.

The FCC 3G Notice of Proposed Rulemaking (Docket ET 00-258) proposes to "designate" 2110–2150 MHz and 2160–2165 MHz for *Advanced Mobile and Fixed Communications Services* that could include *International Mobile Telecommunications-2000* (IMT-2000, see 1850–1990 MHz). Existing users would have to relocate to other bands if necessary.

The 1997 Balanced Budget Act requires the FCC to reallocate spectrum and auction licenses in 2110–2150 MHz by September 30, 2002 (see 2390–2400 MHz). "Due to similarities in allocation, usage, and current licensing, we propose to auction the 2160–2165 MHz band in this same timeframe," the FCC said.

2165-2200 MHz

FG: None

Non-FG: NG23 NG168 MOBILE-SATELLITE (space-to-Earth)

FCC: Satellite Communications (25)

This band is used for common carrier fixed operations, such as microwave links in telephone networks (2165–2180 MHz), and private fixed links used by utilities, railroads, and local governments (2180–2200 MHz).

This band is now part of the 2 GHz *Mobile Satellite Services* (MSS) downlink spectrum.

MSS licensees, to the extent they are unable to avoid interfering with incumbents in this band, will have to relocate themselves to other bands or media (see 1990–2025 MHz). The fixed service transmitters also could cause interference to MSS user handsets.

2200–2290 MHz

FG: S5.392 US303 SPACE OPERATION (space-to-Earth) (space-to-space). EARTH EXPLORATION-SATELLITE (space-to-Earth) (space-to-space). FIXED (line-of-sight only). MOBILE (line-of-sight only including aeronautical telemetry, but excluding flight testing of manned aircraft). SPACE RESEARCH (space-to-Earth) (space-to-space)

Non-FG: US303

As its detailed allocation implies, this government band supports tracking, telemetry and control for space programs, conventional fixed microwave systems and military aviation, and tactical and training operations.

The *Air Force Satellite Control Network* (AFSCN) uses *Space-Ground Link Subsystem* (SGLS) facilities in 1761–1842 MHz (for uplinks, see 1710–1755 MHz) and 2200–2290 MHz (downlinks) to control communication, navigation, missile warning and meteorological satellites, as well as Space Shuttle functions for U.S., British, and NATO defense and research users in Guam, Hawaii, New Hampshire, Colorado, California, and from transportable stations that provide coverage for launch and orbit operations.

The principal AFSCN operations center is at Falcon Air Force Base in Colorado, and is connected with many other domestic and international installations.

Key missions controlled by AFSCN include the *Defense Meteorological Satellite Program* (DMSP, see 1675–1700 MHz), the *Global Positioning System* (GPS, see 1215–1240 MHz), the *Defense Support Program* (DSP, see 7.9–8.025 GHz), and the *Military, Strategic, and Tactical Relay satellite* (Milstar, see 20.2–21.2 GHz).

Figure 12
This winter paradise is an *Air Force Satellite Control Network* (AFSCN) remote tracking station at Thule, Greenland, one of several such stations throughout the world. (Department of Defense photo)

Intelligence Gathering

Another user believed to operate in this band is the little-known *Naval Ocean Surveillance System* (NOSS, also called White Cloud, Classic Wizard, or Parcae after the mythological daughters of Zeus). It tracks vessel locations by intercepting emissions from their communications, navigation, and weapons control systems.

Ground observers sometimes mistake the triangular Parcae objects for UFOs. However, according to James Oberg, considered the dean of aerospace journalists, NOSS/Parcae sightings probably do not account for very many of the triangle UFO sightings.

Some Navy and Air Force surveillance programs will be consolidated into a future *Space Based Wide Area Surveillance System* (SBWAS) for signals intelligence.

For its "customer," the *National Security Agency* (NSA), the *National Reconnaissance Office* (NRO) operates classified systems that are believed to use this band, including KH-11 film-return satellites, the so-called KH-12 real-time video satellite, Satellite Data System relay satellites, immense Lacrosse radar imaging satellites, and other intelligence-gathering spacecraft.

From the Blossom Point Tracking Facility in Maryland, the Naval Research Laboratory operates its own satellite control network for classified and scientific missions.

The largest civil satellite control network is the NASA Space Network, operated from White Sands, New Mexico. NASA uses 2200–2290 MHz intensively for space operations and research. It is a key downlink band for scientific and engineering data from spacecraft, including the Space Shuttle, via the *Tracking and Data Relay Satellite System* (TDRSS, see 2025–2110 MHz).

Another civil program is Landsat. Since 1972, Landsat satellites have photographed most of Earth's surface.

The *Earth Observation Satellite Corporation* (EOSAT) operates the current Landsats for the government. New Landsats provide fast access to low-cost color and black-and-white imagery and should stimulate numerous commercial uses of this data in agriculture, mining, and oil exploration, environmental and land use monitoring, mapping, and water management.

This band also provides telemetry in support of airborne and seaborne target tracking and scoring, flight testing, inter-ship data relay, missile range operations, and for radar-equipped balloons (aerostats) tethered along the southern border and Caribbean areas for use in drug interdiction.

This is one of the main bands for air-to-ground video transmission on military test ranges. The video can display missiles and other aircraft being tested.

The 2200–2300 MHz spectrum is one of many restricted bands in which the FCC Part 15 rules permit unlicensed devices to emit only very low-level emissions.

2290–2300 MHz

FG: FIXED. MOBILE except aeronautical mobile. SPACE RESEARCH (deep space) (space-to-Earth)

Non-FG: SPACE RESEARCH (deep space) (space-to-Earth)

This is a primary NASA *Deep Space Network* (DSN) band for telemetry data from probes outside Earth's orbit. The DSN is controlled from the Jet Propulsion Laboratory in Pasadena, California.

The DSN stations in Goldstone, California; Madrid, Spain; and Canberra, Australia are located at points selected to provide continuous observation and overlap for the transfer of communications from one location to the next.

DSN uplinks are at 2110–2130 MHz. The spacecraft also use 8.4–8.5 GHz (see 8.4–8.45 GHz) and 31.8–32.3 GHz (see 31.8–32 GHz).

The FCC defines "deep space" as space at distances from Earth equal to, or greater than, two million kilometers (1,242,742 miles): more than five times the distance from Earth to the Moon.

Using bands protected by the deep space provision enables planetary spacecraft to communicate with Earth without their signals being overwhelmed by the stronger signals of spacecraft closer to Earth.

Long Shot

DSN receivers pick up minute signal levels from planetary spacecraft such as *Cassini*, *Pioneer*, *Ulysses*, and *Voyager*. Signal transit time is measured in hours. Communication requires very large antennas and ultra-low-noise, cryogenically-cooled amplifiers.

(At this writing, *Voyager 1* is the most remote man-made object in modern history, still functioning at more than 7 billion miles from Earth. See also 2360–2385 MHz.)

This band is used for *Very Long Baseline Interferometry* (VLBI) observations using multiple radio telescopes. The band also is important for measuring the polarization of space radio emissions. These measurements provide clues to radio sources and the fields through which their emissions pass.

The 2200–2300 MHz spectrum is one of many restricted bands in which the FCC Part 15 rules permit unlicensed devices to emit only very low level emissions.

2300–2305 MHz

FG: G123 [Docket ET 00-221 would remove Footnote G123.]

Non-FG: Amateur

FCC: Amateur (97)

Note: 2300–2305 MHz became non-Federal government exclusive spectrum in August 1995.

The 2300–2305 MHz band is not allocated to any service on a primary basis. The Air Force uses it for electronic warfare training, target scoring, and telemetry, but federal use of 2300–2310 MHz is considered to be light compared to other government bands.

The current non-federal user of this spectrum is the Amateur Radio Service on a secondary basis only. Uses of this 13-cm band include weak-signal work (such as moonbounce, or *Earth-Moon-Earth* [EME]) around 2304 MHz as well as fixed and mobile operations. Other nearby amateur bands provide the interference protection accorded by primary status at 2390–2400 MHz and 2402–2417 MHz.

A November 1999 FCC policy statement affirmed the reserve status of this band along with two others: "The five megahertz in the 2300–2305 MHz band has generally been transferred from Government use, but any operations in this band would be subject to significant constraints in order to protect the reception of signals from the Government's Deep Space Network *[in 2290–2300 MHz]* that will remain in the band.

"The 35 megahertz in the 2400–2402 MHz and 2417–2450 MHz bands are currently used by Industrial, Scientific and Medical equipment and very low-power radio devices. This existing use restricts the availability of the bands for new services given current sharing techniques. In view of these considerations relating to existing uses, we believe it is reasonable to reserve the 2300–2305 MHz, 2400–2402 MHz, and 2417–2450 MHz bands until a future time, when new technology or other changes may increase the opportunities for new operations in these bands. Nevertheless, we will be receptive to petitions for reallocation of the reserve spectrum bands." A 1996 petition to upgrade the 2300–2305 MHz Amateur Radio allocation to primary status was still pending at this writing.

Satellite Messaging and Wireless Internet

At this writing, two experimental operations were in this band and one petition of interest. AeroAstro received an FCC experimental license to develop ground-based technology for the *Satellite Enabled Notification System* (SENS). The company described SENS as "fundamentally a one-way data messaging service that enables users to transmit short data messages from any location on the globe, for receipt via the Internet in near real-time. ...

"Using inexpensive GPS technology, SENS can report basic position as well as status data on millions of deployed, highly miniaturized, autonomous terminals. ...SENS terminals may ultimately become a standard feature in children's shoes, military ID bracelets, and even clothing of people who travel or recreate in wilderness areas." A demonstration SENS satellite will be launched for free on an Arianespace flight in 2001-2002.

ArrayComm, the developer of spatial-diversity "smart antenna" systems, was authorized in early 2000 to conduct an experimental market test (that is, with paying customers) of its i-Burst high-speed data communications network for Internet service providers. The company positions i-Burst as an alternative to *Digital Subscriber Line* (DSL) or cable modems for Internet access.

"An important objective of this experiment is to demonstrate the i-Burst capabilities in the 2300–2305 MHz band," the company said. It selected 2300–2305 MHz because the band has not been allocated for a primary use and is not heavily encumbered with existing users; the propagation characteristics, including building penetration losses, are acceptable; components for devices in this band are available or can be manufactured; and the band is near bands being considered for *third generation* (3G) wireless services (see 1850–1990 MHz).

The 2300–2305 MHz band is adjacent to the *Wireless Communications Service* (WCS, see 2305–2310 MHz). ArrayComm noted that few, if any WCS licensees have deployed services and suggested that i-Burst could be viable for that service.

ArrayComm told the FCC that the projected costs of i-Burst are "immense" and far beyond its ability to finance; thus, it must experiment by selling service to several thousand paying customers in order to satisfy investors that the technology is worthwhile.

The FCC authorized ArrayComm to conduct a commercial trial of its system in San Diego, beginning in mid-2001. "Full-scale commercial deployment blanketing the nation's top 100 metropolitan areas will follow," the company said.

Personal Location Service

MicroTrax proposed in petition RM-9797 a *Personal Location and Monitoring Service* (PLMS) for 2300–2305 MHz among several bands. The PLMS technology "will provide location, tracking, and monitoring so effectively and at such reasonable cost that it can be deployed for applications affecting everyday situations confronted by individual consumers and small businesses," the company said.

"Among these applications are enhanced 911 location information for wireless telephone users; asset tracking of items as small and portable as a woman's purse or a youngster's bicycle; improved offender monitoring; child monitoring and tracking; and even pet tracking."

(Tracking of children and small assets was also the original objective of the largely forgotten *Location and Monitoring Service* (LMS) at 902–928 MHz.)

"Other bands, such as the LMS bands, have been allocated to services primarily oriented to vehicular use," MicroTrax said. "These other bands allow relatively high transmit power from the mobile (e.g., 30 W for LMS) which makes them quite useful in the vehicular context, but virtually unavailable to personal, portable services. PLMS devices must be sufficiently low in power usage that they may be safely body worn yet store sufficient power for long-term usefulness."

At this writing, the FCC is considering spectrum allocations that may accommodate the MicroTrax and ArrayComm proposals, among others, in Docket ET 00-221 (see 1390–1395 MHz).

2305-2310 MHz

FG: US338 G123 [Docket ET 00-221 would remove Footnote G123.]

Non-FG: US338 FIXED. MOBILE except aeronautical mobile. RADIOLOCATION. Amateur

FCC: Wireless Communications (27). Amateur (97)

With a fund-raising mandate from Congress that identified specific bands for auction, the FCC allocated 2305–2320 MHz and 2345–2360 MHz to a flexible new service.

It received the generic placeholder name *Wireless Communications Service* (WCS) after an internal FCC naming contest failed to attract any non-frivolous entries. The FCC later added 700 MHz spectrum to WCS and renamed it the Miscellaneous Wireless Communications Services (see 746–764 MHz), although the service is still generally called WCS.

The FCC also proposed to add frequencies in the 4.9 GHz band to WCS: spectrum formerly intended for the *General Wireless Communications Service* (GWCS, see 4.94–4.99 GHz).

Opinions Mean Nothing

FCC auctions chief Jerry Vaughan, presenting WCS to prospective bidders, said, "Here is some spectrum. Do whatever you want with it, with a few very minor exceptions. Our opinions mean nothing 15 seconds after you buy the spectrum."

The April 1997 auction of 128 WCS licenses to 17 winning bidders raised more than $13 million for the Treasury, an amount far less than predicted by congressional budget experts. Some licenses went for as little as $1.00. At this writing, it appears that one of the earliest WCS license winners to place WCS systems in operation is Metricom (see 902–928 MHz).

Various theories about the low receipts emerged. Some charged that the FCC simply had already auctioned more spectrum than the wireless industry could comfortably absorb. Moreover, legislation required the FCC to begin auctioning licenses in these bands no later than April 15, 1997, giving the industry little time to develop equipment or business plans after the WCS allocation order of February 19, 1997. It required the FCC to deposit all funds raised no later than September 30, 1997, ruling out long-term payment plans.

A reason for the relative lack of interest in WCS licenses probably was the FCC's insistence on extremely tight out-of-band emissions limits for WCS transmitters, in order to prevent interference from WCS to reception of the satellite *Digital Audio Radio Service* (SDARS or DARS) in 2320–2345 MHz. The filtering and shielding necessary to achieve the limits could make WCS devices too unwieldy and expensive to compete with tiny wireless phones. Wireless fixed data, telephone, and entertainment transmission, however, remain a possibility.

"WCS licensees themselves will determine the specific services they will provide within their assigned spectrum and geographic areas," the FCC said. "The services that can be provided, however, will be subject to specific technical rules...to prevent interference to other services. We emphasize that with the current state of technology there is a substantial risk that these rules will severely limit, if not preclude, most mobile and mobile radiolocation uses. Fixed uses will be less severely affected, but still will require equipment that will meet technical standards higher than those used for similar purposes on comparable bands, and therefore may be more costly."

The FCC later relented somewhat and relaxed the out-of-band emissions limits in order to accommodate prospective WCS licensees who wanted to deploy services based on the *Personal Access Communications System* (PACS), a low-mobility handheld phone standard. No PACS-based WCS systems are operating.

According to Bell Atlantic NYNEX Mobile, which won major eastern WCS markets for a total of $1.6 million, "some of the potential uses for the new spectrum include data communications, Internet access, and low mobility wireless services for customers who need 'walkaround' wireless access in a limited geographic area, such as a local neighborhood or business campus."

Licensing

The FCC licensed WCS as two 10-MHz channel blocks plus two 5-MHz blocks. "The record suggests that the 10-MHz channel blocks represent the minimum amount of spectrum needed to support certain data and wireless local loop services, including wireless Internet access," the agency said.

"In addition, we believe that providing for 10 MHz of spectrum on a paired basis would allow for the introduction of both one-way and two-way services and would facilitate the implementation of a variety of technologies. In the spectrum adjacent to the satellite DARS band, however, we believe that WCS mobile operations may be prohibitively expensive and technologically infeasible for a substantial period of time."

The WCS channel blocks include *A*, 2305–2310 MHz and 2350–2355 MHz; *B*, 2310–2315 MHz and 2355–2360 MHz; *C*, 2315–2320 MHz; and *D*, 2345–2350 MHz.

The *A* and *B* blocks were licensed in 53 *Major Economic Areas* (MEAs) and the *C* and *D* blocks were licensed in six continental *Regional Economic Area Groupings* (REAGs), plus six other REAGs for U.S. possessions plus Alaska, Hawaii, and the Gulf of Mexico.

Like other holders of FCC licenses for exclusive territories, WCS licensees can partition, or assign parts of their geographic service areas, and disaggregate portions of their licensed spectrum to other entities, including those that did not participate in the auction. These geographic partitioning and spectrum disaggregation rules, "while not a substitute for licensing directly by the Commission, nevertheless will help to eliminate market entry barriers…by providing smaller, less capital-intensive areas and spectrum blocks which are more accessible by small business entities," the FCC said.

WCS licensees were granted the most liberal construction, or build-out requirement ever adopted by the FCC. Licensees are simply required to provide substantial service to their service area within 10 years. "Given the undeveloped nature of equipment for use in this band and the technical requirements we are adopting to prevent interference, we are concerned that strict construction requirements might have the effect of discouraging participation in the provision of services over the WCS spectrum," the Commission said.

It suggested that substantial service could include four links per million population for fixed WCS services, or 20 percent coverage of the population in the service area for mobile services. Other possible substantial service factors could include specialized or technologically sophisticated service or service to niche markets or unserved areas.

WCS Satellites

In late 1998, WCS Radio Inc. of Menlo Park, California, a venture associated with WCS licensees, applied for FCC permission to use their licenses to provide SDARS service from two geostationary spacecraft in 2310–2320 MHz and 2345–2360 MHz.

The FCC's WCS rules do in fact contemplate SDARS in some of the WCS spectrum, as one of a number of possible WCS services. In theory, licensees participating in WCS Radio were in a desirable position because they required neither new spectrum allocations, new service rules, nor auctions in order to establish rights to provide SDARS services.

WCS Radio would have provided "up to 100 channels of high quality music and talk radio and innovative data services throughout the contiguous United States," the company told the FCC. "Users in motor vehicles, on boats, in the air, on foot, or in buildings will be able to receive the programming with a low-cost receiver and a small low-gain, omnidirectional antenna. The digital technology combined with a unique scheme of path diversity and time diversity will provide music quality superior to current and proposed terrestrial radio systems and will rival the quality of stereo compact disks."

The company intended to use its WCS frequencies for terrestrial repeaters that would rebroadcast its satellite signal to improve reception. Service was supposed to begin in 2002. But in May 1999, revealing significant, if brief, information about its future intentions, WCS Radio withdrew its proposal from FCC consideration.

"WCSR discovered both that there were more obstacles to launching an SDARS system than anticipated, and that the market for SDARS services was changing," the company said. "WCSR has now determined that terrestrial-based WCS networks could create a more robust and cost-effective system for delivering audio, Internet and streaming services to handheld and mobile devices than could a satellite system operating under the restraints imposed by WCS licensees."

The 2305–2310 MHz portion of the WCS allocation is shared on a secondary basis with Amateur Radio, one of the few places in the spectrum where ham operators are supposed to share with auction winners.

2310-2320 MHz

FG: S5.396 US327 US328 Fixed. Mobile US339 Radiolocation G2 G120

Non-FG: S5.396 US338 FIXED. MOBILE US339. RADIOLOCATION. BROADCASTING-SATELLITE US327

FCC: Wireless Communications (27)

This band is allocated to the *Wireless Communications Services* (see 2305–2310 MHz).

The 2310–2390 MHz spectrum is one of many restricted bands in which the FCC Part 15 rules permit unlicensed devices to emit only very low level emissions.

2320-2345 MHz

FG: S5.396 US327 US328 Fixed. Mobile US339. Radiolocation G2 G120

Non-FG: S5.396 BROADCASTING-SATELLITE US327. Mobile US276 US328

The principal worldwide band for audio broadcasting direct to the public from satellites (*Broadcasting-Satellite Service* (Sound) or BSS) is the so-called L-band, 1452–1492 MHz.

That band falls within aeronautical test telemetry spectrum in the U.S. (see 1435–1525 MHz).

As an alternative, the *International Telecommunication Union* (ITU) allocated the S-band, 2310–2360 MHz, for domestic satellite audio broadcasting in the U.S. Other nations, especially Canada, criticized this action as detrimental to the realization of a uniform worldwide service in the L-band.

The ITU also allocated 2520–2670 MHz for BSS national and regional systems for community reception. (India and Mexico also are authorized to use the S-band for BSS.)

Against ferocious opposition from conventional broadcasters, the FCC eventually allocated 2320–2345 MHz to satellite *Digital Audio Radio Services* (SDARS or DARS). "Satellite DARS will provide continuous radio service of compact disc quality for all listeners and will offer an increased choice of over-the-air audio programming," the FCC said.

No other significant terrestrial U.S. users are in the S-band, but adjacent countries operate terrestrial fixed point-to-point, fixed point-to-multipoint, and aeronautical mobile telemetry systems in the band. Satellite DARS operators must take precautions to avoid interference with the systems of other nations.

Sirius and XM

There are two SDARS licensees : Sirius Satellite Radio (2320–2332.5 MHz), formerly Satellite CD Radio and XM Satellite Radio (2332.5–2345 MHz), formerly American Mobile Radio.

These companies won their licenses by bidding $83 million and $89 million, respectively, at an April 1997 auction. The FCC concluded that only enough spectrum existed in the S-band for two SDARS licensees of 12.5 MHz each. The FCC requires that SDARS receivers be capable of picking up broadcasts from both of the licensees.

The licensees must deploy hundreds of terrestrial repeaters, ground transmitters that relay broadcasts from satellites. Most of the gap-fillers will be in urban areas where obstructions inhibit satellite reception.

Sirius will use three satellites in inclined elliptical orbits will offer a service directed mainly to vehicle radios. Sirius-1 and Sirius-2 had been launched at this writing. The major investors in Sirius include Ford and Loral.

XM will use two geostationary satellites (officially designated "Rock" and "Roll") that are directed both to vehicle and portable radios. XM's major investors include GM and its DirecTV business; Clear Channel Communications; and Liberty Media, in addition to its founder, Motient Corp., formerly American Mobile Satellite Corp.

Uplink stations in 7.025–7.075 GHz feed these satellites. They use frequencies in the S-band in 3.7–4.2 GHz, and 5.925–6.425 GHz for telemetry and control, operation during transfer to final orbit, and for contingency purposes.

Other SDARS-Related Issues

Sirius and XM might have faced competition from SDARS service in the *Wireless Communications Service* (WCS) band. A group of WCS licensees applied for permission to use their licenses to provide SDARS (see 2305–2310 MHz). They later abandoned the idea.

The FCC licensed, originally on an experimental basis only, the WorldSpace SDARS system to broadcast in the L-band (see 1435–1525 MHz). It later granted full authorization to the Washington, D.C.-based WorldSpace, but does not permit the company to serve U.S. audiences.

NASA's Goldstone Solar System Radar operates at 2320 MHz and 8.56 GHz in the Mojave Desert northeast of Los Angeles. Scientists used it to observe Comet Hyakutake when it passed within 9.3 million miles of Earth in 1996.

SDARS interference is expected to make radar astronomy operations at 2320 MHz "nearly impossible," according to NASA's Jet Propulsion Laboratory.

The 2310–2390 MHz spectrum is one of many restricted bands in which the FCC Part 15 rules permit unlicensed devices to emit only very low level emissions.

2345-2360 MHz

FG: S5.396 US327 US328 Fixed. Mobile US339 Radiolocation G2 G120

Non-FG: S5.396 FIXED. MOBILE US339 RADIOLOCATION. BROADCASTING-SATELLITE US327

FCC: Wireless Communications (27)

This band is allocated to the *Wireless Communications Services* (WCS, see 2305–2310 MHz).

The 2310–2390 MHz spectrum is one of many restricted bands in which the FCC Part 15 rules permit unlicensed devices to emit only very low-level emissions.

2360-2385 MHz

FG: G120 MOBILE US276 RADIOLOCATION G2 Fixed

Non-FG: MOBILE US276

The most powerful radio emitter on Earth is probably the 2380-MHz planetary research radar of the Arecibo Observatory at the National Astronomy and Ionosphere Center.

The 1000-foot diameter dish reflector permits a one-megawatt transmitter to achieve an effective radiated power of 20 terawatts. The dish sits in a sinkhole in the hills of northwest Puerto Rico.

Radar echoes from objects such as comets, planets, and the Moon convey information about surface properties, orbit, and object size. This technique enables spaceflight controllers, for example, to accurately guide spacecraft to specific positions on other planets. The radar also can detect asteroids that may potentially threaten civilization. It could detect a golf ball at the distance of the moon.

Arecibo's achievements include radar maps of Venus, the discovery of ice caps on Mercury, images of the satellites of Jupiter, and observations in galaxies far beyond the Milky Way. The observatory also was instrumental in the discovery of the first planets outside the solar system.

ET Communication

In 1975, this radar sent the famed Arecibo Message aimed at the M13 globular cluster. The 1,679-bit message described our species and DNA composition, our planetary system, and the system transmitting the message.

Other official attempts to convey information to aliens have included the gold-plated pictorial plaque affixed to the *Pioneer 10* and *11* spacecraft, launched in the early 1970s; copper recordings of images, human and animal sounds, with instructions, attached to the *Voyager 1* and *2* probes, launched in 1977 (see 2290–2300 MHz); and SpaceArc, a disc inside a *Direct Broadcast Satellite* (DBS, see 12.2–12.7 GHz).

The 2360–2390 MHz band is otherwise used for defense and commercial aerospace purposes in conjunction with 1435–1525 MHz for telemetry in the flight testing and operation of aircraft, spacecraft, missiles, and scientific balloons at U.S. military and NASA centers. An example system that uses this band is *Theater High-Altitude Area Defense* (THAAD, see 10–10.45 GHz).

2385-2390 MHz

FG: G120 MOBILE US276. RADIOLOCATION G2. Fixed

Non-FG: MOBILE US276

Note: 2385–2390 MHz will become non-Federal government exclusive spectrum in January 2005.

As described by the official FCC Note above, this band will become available to the private sector in 2005. The band will be auctioned before 2002 as required by the Balanced Budget Act of 1997 (see 2390–2400 MHz). Federal users, mainly flight test telemetry operations, will require interference protection at 17 military sites until 2007.

Yet in Docket ET 00-221, the FCC stated, "We are uncertain of how much of this band is used for aeronautical telemetry, and of how many licensees use this service. ...Commenters are invited to address the possibility of moving aeronautical telemetry to another spectrum band, reducing its status to secondary, or providing protection for telemetry in limited areas of the United States."

This docket proposes to reallocate 2385–2390 MHz to the Fixed and Mobile Services generally for flexible and unspecified uses (see 1390–1395 MHz). Accordingly, the band would be changed as follows: USzzz would be the sole Footnote in the FG and Non-FG allocations (the actual number to replace 'zzz' will be decided later). The FG allocation would be changed to None, and the Non-FG allocation would be made to the FIXED and MOBILE Services. The FCC Note regarding the January 2005 date would be deleted.

The FCC has also suggested that 2385–2390 MHz may be paired with or associated with 1670–1675 MHz, another segment slated for auction.

The 2310–2390 MHz spectrum is one of many restricted bands in which the FCC Part 15 rules permit unlicensed devices to emit only very low level emissions.

2390-2400 MHz

FG: G122

Non-FG: AMATEUR FCC: RF Devices (15). Amateur (97)

This is part of the 2390-2450 MHz Amateur Radio Service band, used for repeater links, satellite, amateur TV, and data applications. Most amateur microwave links in 2390-2400 MHz are paired with frequencies at 2300-2310 MHz.

The Air Force and Navy use this band for target identification, telemetry, and certification of navigation systems at high speeds.

Data-PCS

The 2390-2400 MHz band is shared by the Amateur Service and Data-PCS, a class of *Unlicensed Personal Communications Service* (UPCS) data transceivers that observe a coexistence behavior (the spectrum etiquette for asynchronous devices) developed by the Wireless Information Networks Forum (WINForum) and prescribed in FCC Part 15 rules.

The etiquette, among other aspects, requires devices to listen before transmitting. WINForum later concluded that this requirement may not be necessary in data communications (see below).

Apple and Microsoft Involvement

The 2390-2400 MHz band is among the first spectrum reallocated from government to private sector use as established by the 1993 *Omnibus Budget Reconciliation Act* (OBRA, see below). The FCC said that "[A]llocating the band for unlicensed data-PCS, and providing for use of 2402-2417 MHz by Part 15 devices, will provide for the continued development and implementation of a new generation of advanced communications devices and services, such as wireless local area networks, digital cordless telephones, electronic article surveillance equipment, utility metering devices, fire and security alarm devices, and wireless bar code readers."

Apple Computer, Inc. began the Data-PCS allocation process with a rulemaking petition in 1991. "Before [this allocation] action," the company said in a news release, "the few radio bands available for wireless computing were overloaded with industrial transmitters or occupied by microwave stations; and, thus were unusable without massive relocation costs (estimated to be as much as $500 million) and ten to twelve years of band-clearing. ...The new frequency allocation for Data-PCS—2390 to 2400 MHz—is shared only with ham radio operations, which are compatible with Data-PCS devices.

"As a result of the FCC's allocating the new band without further procedural or administrative delay, starting immediately, manufacturers can produce radio modems so educators and other users can set up their own wireless networks."

Microsoft told the FCC that Data-PCS "will operate in an entirely different manner than the technology we view as mobile or portable at present. ...Users of notebook and handheld computers might choose to transmit data between themselves. Ad-hoc work groups might be established as the need arises. ...Low-power communicating devices might operate as digital door locks, providing authentication and access to nomadic environments such as automobiles."

The FCC forbids airborne operation of Data-PCS devices in order to protect space research operations at the National Astronomy and Ionosphere Center Arecibo Observatory in Puerto Rico (see 2360–2385 MHz).

Apple's first wireless LAN product, AirPort, uses conventional Part 15 spread spectrum technology and frequencies, not Data-PCS.

Stalled Service

The other band available for Data-PCS is 1910–1920 MHz. That band contains fixed microwave stations (see 1850–1990 MHz). The fixed stations are moving to other bands under a frequency coordination and microwave relocation program administered by an industry trade group, UTAM Inc. This program, however, does not apply to 2390–2400 MHz.

The objective of introducing Data-PCS products into 1910–1920 MHz has not succeeded, even though more than 80 percent of the nation's area is available for unencumbered deployment in that band.

Despite UTAM's reports to the FCC that "the market for products that operate in this band continues to illustrate all the signs of success," no manufacturers are selling products in 1910–1920 MHz or 2390–2400 MHz at this writing.

WINForum petitioned the FCC in early 1999 to offer alternative UPCS coexistence rules in 2390–2400 MHz to make it more attractive for manufacturers.

The alternative rules would not require devices to listen before transmitting, but would impose a duty cycle limit to help prevent any unit from dominating the band. WINForum also proposed to foreclose authorization of any asynchronous devices in 1910–1920 MHz after a certain date.

OBRA and BBA

The 1993 *Omnibus Budget Reconciliation Act* (OBRA), prompted by findings that the federal government underutilizes some of its spectrum, required the government to release at least 200 MHz of spectrum to the FCC for allocation to the private sector.

Of course, availability to the FCC and reallocation do not necessarily make a given band available to the public for licensing or use, nor are they guarantees of commercial success or even commercial interest. The same political and economic forces influence reallocated bands as any others. In our experience, manufacturers and service providers often do not know that the OBRA bands are supposed to be available, and do not press the government for expeditious access to them. License auction winners often fight the appearance of additional spectrum on the market that competitors might obtain.

Some bands were cancelled or changed after OBRA took effect, due to Department of Defense objections. In one example, the President withdrew 4.635–4.66 GHz and 4.66–4.685 GHz from OBRA reallocation and substituted 4.94–4.99 GHz, a band which the FCC said has a "paucity of interest."

The *Balanced Budget Act* of 1997 (BBA) slated an additional 20 MHz for transfer to the private sector. The National Defense Authorization Act of 1999 provided that new entrants must compensate federal agencies for costs they incur in vacating spectrum. This legislation applies to BBA bands, to 1710–1755 MHz, and to future reallocations, but does not apply to bands transferred to private-sector use under the 1993 OBRA.

In a November 1999 spectrum policy statement, the Commission announced its intention to propose to allocate 50 MHz at 1710–1755 MHz and 2160–2165 MHz to an *Advanced Mobile and Fixed Communications Service* (AMFCS, see 1850–1990 MHz and 2110–2130 MHz). The FCC also said it will consider adding the 2110–2150 MHz spectrum to this allocation.

Spectrum Reallocations of Interest Including OBRA and BBA Bands

138–144 MHz	2130–2150 MHz
216–220 MHz	2160–2165 MHz
698–746 MHz	2300–2305 MHz
1350–1390 MHz	2385–2390 MHz
1390–1400 MHz	2390–2400 MHz
1427–1429 MHz	2400–2402 MHz
1432–1435 MHz	2402–2417 MHz
1670–1675 MHz	2417–2450 MHz
1710–1755 MHz	3.65–3.7 GHz
2110–2130 MHz	4.94–4.99 GHz

2400–2402 MHz

FG: S5.150 G123 [Docket ET 00-221 would remove Footnote G123.]

Non-FG: S5.150 S5.282 Amateur

FCC: ISM Equipment (18). Amateur (97)

This band, which was used for military radar development, was reallocated exclusively to the private sector for non-government uses. The band has "limited usefulness" for military purposes, according to NTIA, due to its small size and location between two non-Federal bands.

Amateur Radio is the only service that holds a domestic allocation for 2400–2402 MHz. The allocation is secondary, but amateurs have asked the FCC to elevate it to primary.

A principal amateur use of the band is for satellite engineering beacons that transmit spacecraft status to ground controllers. This is a use of the Amateur Satellite Service, which is incorporated into the Amateur Radio Service and is not a distinct regulatory entity.

Consumer and commercial wireless devices use the 2400–2483.5 MHz spectrum. These include microwave ovens, *Industrial, Scientific, and Medical* (ISM) equipment (FCC Part 18), and unlicensed narrowband and spread spectrum communications products operating in Part 15 (see 902–928 MHz). Many wireless Internet service providers use unlicensed spread spectrum wireless LAN technology at 2400 MHz.

Reserve Band

A November 1999 FCC spectrum policy statement declared the reserve status of this band along with two others. "The 35 megahertz in the 2400–2402 MHz and 2417–2450 MHz bands are currently used by Industrial, Scientific and Medical equipment and very low-power radio devices," the Commission said.

"This existing use restricts the availability of the bands for new services given current sharing techniques. In view of these considerations relating to existing uses, we believe it is reasonable to reserve the 2300–2305 MHz, 2400–2402 MHz, and 2417–2450 MHz bands until a future time, when new technology or other changes may increase the opportunities for new operations in these bands. Nevertheless, we will be receptive to petitions for reallocation of the reserve spectrum bands."

Petition RM-9949 from the *American Radio Relay League* (ARRL) asks the Commission to elevate the domestic status of the Amateur Radio Service and Amateur Satellite Service from secondary to primary in 2400–2402 MHz.

The Amateur Service currently is primary in 2390–2400 MHz (one of the two Data-PCS bands) and in 2402–2417 MHz. The requested status change would give amateurs a contiguous 27 MHz band and help to forestall commercial development in this spectrum.

"It is urgent to protect the 2400–2402 MHz band due to the extensive reliance by the Amateur Satellite Service on the future development of satellite uplinks and downlinks in that segment in particular," ARRL told the FCC.

"The Amateur Satellite Service can continue to accommodate Part 15 and Part 18 devices at 2400–2402 MHz, since that segment is located at the lower edge of the segment in which Parts 15 and 18 devices are typically deployed, and because of geographic separation typically encountered between Amateur Satellite stations and higher-powered Parts 15 and 18 devices. It is appropriate to accommodate present and future Amateur Satellite Service operation at 2400–2402 MHz through creation of a primary allocation for the Amateur and Amateur Satellite Services domestically at 2400–2402 MHz."

Commercial Devices

In addition to the U.S., spectrum in 2400–2483.5 MHz is available for spread spectrum radios in Canada, some of Europe, Pacific Rim countries, and South America, with differing technical and authorization requirements in each market. Some of these countries accept products tested to FCC specifications.

Although some of the applications in the two bands are similar, the rules governing Part 15 devices in 2400–2483.5 MHz are different from those in the adjacent 2390–2400 MHz band. In 2390–2400 MHz, the FCC requires monitoring and band sharing (spectrum etiquette) by Data-PCS. The rules preclude combining 2390–2400 MHz and 2400–2483.5 MHz into a single, large Part 15 band.

Networking Standards

The adoption of wireless local area network standards by the Institute of Electrical and Electronics Engineers (IEEE) stimulated the market for spread spectrum radios and chipsets in 2400–2483.5 MHz. Interoperable radios, including PC card units and access point base stations, are widely available. The standards include 802.11, a frequency-hopping standard at 1 Mbit/s and 802.11b, a direct-sequence 11 Mbit/s standard. (The 5 GHz, 802.11a standard uses *orthogonal frequency division multiplex* (OFDM) for a data rate of 54 Mbit/s. See 5–5.225 GHz.)

For 802.11b, interoperability certification, "Wi-Fi", is available from the *Wireless Ethernet Compatibility Alliance* (WECA). Wi-Fi products are increasing in market acceptance as Internet connectivity enablers and replacements for wired or licensed wireless links between buildings.

Wireless LAN technology is not limited to short-range use. One watt of power can achieve metropolitan-area links up to about 30 miles, depending on station design and installation. In practice, most of these devices operate at levels only in the few tens of milliwatts, in order to conserve power and increase frequency reuse. For certain longer-range spread spectrum systems, the FCC imposes directional antenna gain restrictions in 2400–2483.5 MHz. The systems are unlicensed, but the FCC holds the operator or installer responsible for ensuring that they are used only for fixed point-to-point operation.

Figure 13
A field researcher accesses the High Performance Wireless Research and Education Network with an antenna mounted on a tripod and pointed towards a relay site, while connected to a networking card in a computer. This San Diego County, California project uses unlicensed *Institute of Electrical and Electronics Engineers* (IEEE) 802.11 radios to provide high-speed Internet access to researchers from several disciplines (geophysics, astronomy, ecology) and educational opportunities for users such as rural Indian reservations and schools. (National Science Foundation/HPWREN photo)

Bluetooth and HomeRF

The IEEE 802 working group also hosts a wireless *personal-area network* (PAN) standards effort, 802.15, focused on 2400 MHz for very short-range 1 Mbit/s systems. This work essentially will transform an industry-developed specification, Bluetooth, into an official IEEE standard.

Bluetooth was intended to replace cabling between portable computers, peripheral devices such as printers and cameras, and wireless modems and phones, but more advanced applications are likely. The specification incorporates "profiles," specific protocols for applications such as cordless telephones, direct peer-to-peer voice (intercom), headsets for phones and music players, fax, and object exchange for such uses as wireless publishing.

Bluetooth enjoys support by numerous companies including the major computer, semiconductor, and personal communications manufacturers, as well as members as diverse as British Airways and Lego Systems. Bluetooth uses frequency hopping spread spectrum in small, inexpensive transceivers with 1 mW power for a typical 10-meter range. Versions with greater range and capacity are under development. Researchers have predicted enormous markets for this technology.

Interaction between Bluetooth and IEEE 802.11b networks is possible as the two systems operate in the same band. Research suggests that some degradation of performance can occur depending on the proximity of network nodes, base stations, and interferers.

Interference in Unlicensed Devices

"The relatively uncontrolled—but, note, not unregulated—nature of the unlicensed bands is often a source of concern to wireless users, as it should be," according to Craig Mathias, a consultant with Farpoint Group. "After all, there is great potential for mutual interference between devices operating in these bands. Users have no recourse if destructive interference does occur, which can and should give pause to anyone contemplating the deployment of systems using these frequencies, mission-critical or not.

"But it should be noted that two conditions must hold true for interference to take place. First, the two radios must be within relatively close range of one another, as unlicensed transmissions usually have a fairly limited range. Second, they must be transmitting at the same moment in time, which is less common in most applications than one might think. Thus, it's possible for many radios to exist in close proximity with only minimal interference resulting.

"Spread-spectrum techniques, which are commonly used in the unlicensed bands, can be quite effective in combating interference. Moreover, the higher-level protocols used over the radio link can detect and even in some cases correct errors, using retransmission when necessary. But just a little planning can eliminate concerns in most cases—for example, one might use a 2.4-GHz wireless LAN in the same location as a 900-MHz or 1.8-GHz unlicensed PCS cordless phone, thus avoiding the opportunity for interference altogether."

HomeRF and SWAP

The Home RF Working Group developed another 2400-MHz frequency hopping specification. HomeRF describes its *Shared Wireless Access Protocol* (SWAP) as "an open industry specification that allows PCs, peripherals, cordless telephones, and other consumer electronic devices to share and communicate voice and data in and around the home without the complication and expense associated with running new wires."

SWAP received a boost from the FCC ruling in Docket ET 99-231, which permitted channels up to 5 MHz wide for frequency hopping systems in 2400–2483.5 MHz. These systems must have at least 15 channels and are limited to 125 mW transmitter output power. Frequency hopping systems with at least 75 channels can use up to 1 W transmitter output power, as can direct sequence systems.

The decision effectively permitted SWAP to increase its data rate from 1–2 Mbit/s to 10 Mbit/s, making it a more effective competitor to IEEE 802.11b systems. However, neither standard may be sufficient for multi-user broadband home networking, which may require higher data rates more suited to 802.11a and *ultrawideband* (UWB) technology.

"The Commission action to allow frequency hopping spread spectrum systems in the 2.4 GHz band to use wider hopping channels will facilitate development of new high-speed data devices for business and consumer applications such as transmission of CD-quality audio and video streams from home PCs to portable devices," according to FCC Chairman William Kennard. "The wider bandwidths will permit these systems to provide higher data speeds, thereby enabling the development of new and improved consumer products such as wireless computer local area networks and wireless cable modems."

Other ISM Uses

Communication in 2400–2483.5 MHz must contend with interference from ISM devices centered on 2450 MHz, especially microwave ovens. NTIA believes that the 2400–2402 MHz band is least impacted by microwave oven emissions.

Radio astronomers searching for space signals have reportedly monitored microwave oven emissions for hours, eventually tracing their source to the observatory kitchen.

Medical ISM uses include heating body tissues, magnetic resonance imaging, radiometry for cancer detection, and even thawing of frozen organs for transplant. Industrial RF heaters in this band treat food, chemicals, packaging, textiles, and construction materials. Microwave lamps are used to manufacture fiber optic cable, automobile glass and headlamps, no-wax floor tiles, and to dry printing on containers. The National Air and Space Museum in Washington, D.C. is illuminated by microwave lighting.

Proponents of ISM uses question the FCC's policy of permitting telecommunications in ISM spectrum. They seem to fear that wireless networks will be too susceptible to interference from proliferating ISM equipment, resulting in complaints and demands to shut down the ISM equipment. The FCC repeatedly proclaims the non-status of Part 15 communications devices, which must accept any interference they receive from ISM sources.

Another non-communication application in this spectrum is local positioning. These radio systems precisely locate personnel, inventory, and equipment in facilities such as hospitals or construction sites. Workers and assets are equipped with small identification tags that respond when queried by nearby base stations. Computers measure signal propagation from the tags to establish their locations, which can be displayed on a screen or used to trigger alarms.

2402-2417 MHz

FG: S5.150 G122

Non-FG: S5.150 S5.282 AMATEUR

FCC: RF Devices (15). ISM Equipment (18). Amateur (97)

The FCC allocated this band to Amateur Radio on a primary basis and permitted unlicensed Part 15 devices to continue to share it on a non-interference basis. *Industrial, Scientific and Medical* (ISM) products have access to the center ISM frequency 2450 MHz (see 2400–2402 MHz).

The FCC includes 2402–2483.5 MHz among the bands that are available for experimental use by students and schools.

Active microwave road condition sensors use this band to measure ice, snow, salinity, and traction.

2417-2450 MHz

FG: S5.150 G124 Radiolocation G2

Non-FG: S5.150 S5.282 Amateur

FCC: ISM Equipment (18). Amateur (97)

This military radar band was reallocated for shared use with the private sector for future commercial operations. Military radar is secondary in 2417–2450 MHz.

A November 1999 FCC spectrum policy statement affirmed the reserve status of this band along with two others; "The 35 megahertz in the 2400–2402 MHz and 2417–2450 MHz bands are currently used by Industrial, Scientific and Medical equipment and very low-power radio devices. This existing use restricts the availability of the bands for new services given current sharing techniques.

"In view of these considerations relating to existing uses, we believe it is reasonable to reserve the 2300–2305 MHz, 2400–2402 MHz, and 2417–2450 MHz bands until a future time, when new technology or other changes may increase the opportunities for new operations in these bands. Nevertheless, we will be receptive to petitions for reallocation of the reserve spectrum bands."

This is part of the 2400–2483.5 MHz spectrum used by unlicensed devices. Most of the radio noise from ovens and *Industrial, Scientific, and Medical* (ISM) uses peaks at 2450 MHz, making this spectrum difficult for future commercial services covering large areas (see 2400–2402 MHz).

The FCC includes 2402–2483.5 MHz among the bands that are available for experimental use by students and schools.

2450–2483.5 MHz

FG: S5.150 US41

Non-FG: S5.150 US41 FIXED. MOBILE. Radiolocation

FCC: ISM Equipment (18). Private Land Mobile (90). Fixed Microwave (101)

Unlicensed communications devices are allowed in 2400–2483.5 MHz, whereas *Industrial, Scientific, and Medical* (ISM) products have access to the center ISM frequency 2450 MHz (see 2400–2402 MHz).

According to the FCC, this part of the 2400–2483.5 MHz band "is the only portion of the radio spectrum allocated to the Part 90 Private Land Mobile Radio Services that permits wide bandwidth operation. This wide bandwidth is critical for the operation of police surveillance systems and other functions, such as video links used with bomb disposal systems."

The band also is favorable for fixed links in the Gulf of Mexico, away from urban oven interference.

Government radiolocation is permitted in 2450–2500 MHz on a non-interference basis to non-Government systems.

The FCC includes 2402–2483.5 MHz among the bands that are available for experimental use by students and schools.

2483.5–2500 MHz

FG: S5.150 753F US41 MOBILE SATELLITE (space-to-Earth) US319 RADIODETERMINATION SATELLITE (space-to-Earth). S5.398

Non-FG: S5.150 753F US41 NG147 MOBILE SATELLITE (space-to-Earth) US319 RADIODETERMINATION SATELLITE (space-to-Earth). S5.398

FCC: ISM Equipment (18). Satellite Communications (25). Private Land Mobile (90). Fixed Microwave (101)

This is now the *Non-Geostationary Mobile Satellite Service* (NGSO MSS or Big LEO) downlink band (see 1610–1610.6 MHz). Non-satellite licensing has not been allowed in the band since 1985.

In 1998, the FCC deleted 2483.5–2500 MHz from the bands designated for experimental use by students and schools.

This band is no longer normally assigned for experimental use of any kind because of the need to protect satellite systems from interference. The Commission replaced it with 2402–2450 MHz and 10–10.5 GHz. The 2450–2483.5 MHz band remains available for educational purposes.

The FCC is concerned about potential interference to Big LEO downlinks from *Instructional Television Fixed Service* (ITFS) stations operating at 2500 MHz and *Industrial, Scientific, and Medical* (ISM) devices. ISM products have access to the center ISM frequency 2450 MHz (see 2400–2402 MHz).

This is one of many restricted bands in which the FCC Part 15 rules permit unlicensed devices to emit only very low level emissions.

2500-2655 MHz

FG: S5.339 US205 US269

Non-FG: S5.339 US269 FIXED S5.409 S5.411 US205. FIXED-SATELLITE (space-to-Earth) NG102. BROADCASTING-SATELLITE NG101

FCC: Domestic Public Fixed (21). Auxiliary Broadcasting (74)

The 2500–2655 MHz and 2655–2690 MHz bands are devoted to the *Multipoint Distribution Service* (MDS) and *Instructional Television Fixed Service* (ITFS), providing broadband access and educational video. There is very limited use of this spectrum by non-MDS/ITFS rural radiotelephone and video stations, and no satellite use. There is some radio astronomy use of this spectrum.

The name MDS refers to single channel and multichannel stations in the Multipoint Distribution Service. Stations on the E and F channel groups (see below) were formerly called *Multichannel Multipoint Distribution Service* (MMDS) stations. MMDS is obsolete as a distinct service identity, but the industry still generally uses the name MMDS.

These stations historically were devoted to wireless cable video programming. Internet access services, however, largely supplanted the commercial video services.

Subscribers use directional, rooftop antennas and set-top boxes to access the signal. This spectrum is suitable for fixed broadband service because it propagates for longer effective distances and is more immune to rain effects than higher frequency bands. The wavelengths of those bands approximate the dimensions of raindrops more closely, which can limit link performance.

Large Carriers

Worldcom, Sprint, and Nucentrix Broadband Networks obtained substantial MDS coverage nationwide, including major markets, through license purchases. These companies are parties to an agreement that permits their systems to exceed FCC-specified interference levels.

"A major advantage in using this band is that the physical need to cellularize is much less than in other major bands," according to Sprint. "With appropriate terrain characteristics, a single MMDS cell can cover a 35-mile radius, or 3,850 square miles. CPE [customer premises equipment] costs and coverage areas make this band appropriate for full coverage of large geographical areas, and provisioning of service to small businesses and homes not in dense clusters, as well as higher clustered businesses."

Allocations

The FCC has described MDS and ITFS spectrum use as "an amalgam of different channels and geographic boundaries that vary from location to location."

Licensees have access to channel 1 at 2150–2156 MHz; channel 2 at 2156–2162 MHz; the 20 ITFS channels A1–A4 (2500–2524 MHz), B1–B4 (2524–2548 MHz), C1–C4 (2548–2572 MHz), D1–D4 (2572–2596 MHz), G1 (2644–2650 MHz), G2 (2656–2662 MHz), G3 (2668–2674 MHz), and G4 (2680–2686 MHz); the MDS channels E1 (2596–2602 MHz), E2 (2608–2614 MHz), E3 (2620–2626 MHz), and E4 (2632–2638 MHz); F1 (2602–2608 MHz), F2 (2614–2620 MHz), F3 (2626–2632 MHz) and F4 (2638–2644 MHz); and former *Operational Fixed Service* (OFS) channels H1 (2650–2656 MHz), H2 (2662–2668 MHz) and H3 (2674–2680 MHz). The 2686–2690 MHz segment of the 2655–2690 MHz band contains 125 kHz response channels, one for each of the 32 MDS and ITFS broadcasting channels.

Outside major markets, channel 2A (2156–2160 MHz) can be used for voice and data as it is too small for a 6 MHz video channel.

In 1992, the 2160–2162 MHz segment was reallocated to Emerging Technologies for future applications. Any subsequent MDS use of those 2 MHz will be secondary. At this writing, the FCC proposed to auction licenses in 2160–2165 MHz (see also 2110–2150 MHz).

Not all MDS and ITFS channels are available to all operators in all locations. Complex regulations control the filing of license applications and channel sharing among MDS and ITFS.

MDS licenses are now awarded by competitive bidding. A March 1996 auction raised more than $200 million for MDS licenses in 493 *Basic Trading Areas* (487 principal BTAs as established by Rand McNally, plus six U.S. island areas added by the FCC).

Interference avoidance is critical in the many cases where MDS auction winners must avoid interfering with the 35-mile *Protected Service Areas* of licensees already serving the area won at auction. Accordingly, the FCC has described MDS as "a service heavily encumbered by previously authorized and proposed MDS facilities."

Instructional TV

ITFS is defined as "a nonbroadcast service intended to be used primarily for the transmission of instructional and cultural material to specified receive sites for the formal education of students enrolled in accredited schools." Educational institutions licensed in the ITFS use it to transmit instruction to remote locations. They are required to transmit a minimum number of educational program hours as a condition of license.

Most ITFS licensees lease their unused capacity to commercial MDS operators. The ITFS operator uses the channels during the school day whereas the MDS operator broadcasts in the evening TV hours. The MDS operator often pays for construction of the ITFS facilities.

Mutually exclusive applications for ITFS stations are resolved in a hearing by means of a point accumulation system. The FCC awards points based on factors that include whether the applicants are local, or from outside the area; educational accreditation; use of the minimum number of channels; amount of educational programming proposed; and expansion of existing service.

FCC rules permit MDS operators to be licensed on channels allocated to ITFS where MDS channel shortages exist. ITFS entities have the right to demand access to such ITFS channels licensed to MDS operators. MDS and ITFS operators may also swap channels, which could help establish two-way service.

MDS Evolution

The MDS industry competes with other Internet access services, notably cable TV modems, *Digital Subscriber Line* (DSL), and satellite-based services. Early MDS Internet experiments used conventional telephone lines or unlicensed wireless devices to return data from the subscriber to the operator, who would convey this data to the Internet.

MDS licensees also may operate response channels in 18.58–18.82 GHz and 18.92–19.16 GHz, but these channels have been called "impractical for use in a consumer environment" because of the need to conduct frequency coordination and to apply for licenses at each subscriber location.

The industry later requested, and the FCC granted (in Mass Media Docket MM 97-217), sweeping updates to the MDS and ITFS regulations to better enable licensees to provide new services.

The FCC expanded the definition of MDS to represent what it called "the reorientation of the regulatory treatment of MDS, no longer regarding it as a one-way service with two-way service permitted on a limited basis, but instead as a fully flexible service in which licensees can provide either one-way or two-way service in response to the demands of the competitive marketplace."

The decision cleared the way for licensees to turn around their formerly one-way channels and use them for transmission to and from subscribers.

To provide an upstream data path from subscribers, the FCC authorized "response station hubs," special stations that receive transmissions from subscriber response stations. It designated a group of 125 kHz I channels in 2686–2690 MHz for use by response stations.

The FCC also gave licensees freedom to combine and subdivide channels within a system, and between systems licensed to different parties. They can aggregate channels for greater bandwidth, or increase the number of channels to help overcome transmission impairments.

ITFS licensees, which must serve an educational purpose, will be able to meet educational obligations with a variety of video, voice, and data services—not just instructional TV. "Both individual and business customers will be able to use the high-speed and high-capacity data transmission and Internet service that will be available through the new systems," the FCC said. "Also, consumers will be able to take advantage of new videoconferencing, distance learning and continuing education opportunities. Commenters have also suggested cutting-edge applications like telemedicine for the new two-way systems."

The Commission dismissed claims by licensees in the 218–219 MHz Service (see 216–220 MHz) that they "obtained at auction a monopoly to provide certain two-way services that other providers, that is, MDS and ITFS licensees, could not." The FCC pointed out that two-way operation by MDS and ITFS was allowed before the 1994 IVDS auction, and said that it has never promised a 218–219 MHz monopoly.

Representing MDS/ITFS issues at the FCC and Congress is the *Wireless Communications Association International* (WCA). "Among WCA's major issues in recent years have been fostering a regulatory and financial environment whereby carriers could freely adapt their mix of services to the changing demands of the market," according to association president Andrew Kreig. "One illustration was playing a strong role in the evolution of the MMDS industry from one-way multichannel video to two-way broadband services now permitted by newer and more flexible regulations in the United States and elsewhere.

"Another longtime effort has been the struggle on behalf of emerging carriers to win non-discriminatory access rights to customers and potential customers who work or live in multi-tenant environments. In these situations, access to rooftops and interior facilities on a basis that does not discriminate against new entrants (especially compared to incumbent former monopolies) is critical to wireless and other competitive carriers," he said.

IMT-2000

Developers of *International Mobile Telecommunications-2000* (IMT-2000) identified 2500–2690 MHz as one of the candidate bands for *Third Generation* (3G) wireless services (see 1850–1990 MHz).

The U.S. position articulated at the *World Radiocommunication Conference-2000* (WRC-2000) is that "The United States uses the 2500–2690 MHz band for important fixed point-to-point and point-to-multipoint operations that provide video and telecommunications services to homes, schools, colleges, universities and businesses. These important existing uses present significant challenges to the United States as it examines their potential use by advanced mobile communications including IMT-2000."

"We are confident that the U.S. government will continue to recognize the irreplaceable value of MMDS/ITFS spectrum for broadband Internet access to unserved and under-served Americans, and for state-of-the-art interactive educational programming," WCA said. "This is consistent with the U.S. stand at the World Radiocommunication Conference-2000, and its results. We believe that there is more than enough spectrum to accommodate 3G needs without displacing existing ITFS and MMDS operators."

At this writing, possible sharing between IMT-2000 and MDS/ITFS spectrum users is under intense study in Docket ET 00-258. The objectives of IMT-2000, with its high-speed data and roaming features, appear to collide with the commercial and technological evolution now taking place in MDS and ITFS toward interactive broadband services. FCC engineers stressed that each geographic area and each MDS and ITFS system configuration, including lease agreements and channel swaps, must be individually analyzed to determine the prospects for MDS/ITFS and 3G interaction.

An interim report by the FCC concluded that "sharing between 3G systems and ITFS/MDS operations is extremely problematic. At this point, there does not appear to be enough spectrum in the 2500–2690 MHz band in the populated areas to support a viable 3G service. Voluntary partitioning between incumbent users and 3G operators, however, could offer some promise of sharing as an interim measure."

Under some possible options, the report found that already auctioned MDS spectrum would have to be "retrieved" to provide spectrum for 3G operators.

2655-2690 MHz

FG: US205 US269 Earth Exploration-Satellite (passive). Radio Astronomy. Space Research (passive)

Non-FG: US269 FIXED US205 NG47. FIXED-SATELLITE (Earth-to-space) NG102. BROADCASTING-SATELLITE NG101. Earth Exploration-Satellite (passive). Radio Astronomy. Space Research (passive)

See 2500–2655 MHz for information about this band.

Instructional Television Fixed Service (ITFS) and *Multipoint Distribution Service* (MDS) channels H1-H3 and G2-G4 occupy 2650–2686 MHz. The 2686–2690 MHz segment contains response station channels: one 125 kHz channel for each of the 32 ITFS and MDS channels.

This band and 2500–2655 MHz are among those under investigation for *International Mobile Telecommunications-2000* (IMT-2000).

Passive radiometers on satellites use this band to observe Earth's surface and atmosphere. They detect the electromagnetic radiation emitted by matter, corresponding to temperature and surface characteristics. Dry and wet soil, for example, emit energy differently, as do ice and water. Radio astronomers perform continuum observations in 2655–2700 MHz.

The 2655–2900 MHz spectrum is one of many restricted bands in which the FCC Part 15 rules permit unlicensed devices to emit only very low level emissions.

2690-2700 MHz

FG: US246 EARTH EXPLORATION-SATELLITE (passive). RADIO ASTRONOMY US74. SPACE RESEARCH (passive)

Non-FG: US246 EARTH EXPLORATION-SATELLITE (passive). RADIO ASTRONOMY US74. SPACE RESEARCH (passive)

No transmissions are permitted in this receive-only scientific band.

Its uses include observations of galactic and extragalactic radio sources, solar observations by the Air Force Radio Solar Telescope Network, and observations of soil moisture, ocean salinity, and temperature.

The U.S. Naval Observatory Interferometer at Green Bank, West Virginia uses the band to produce accurate position determinations.

The 2655–2900 MHz spectrum is one of many restricted bands in which the FCC Part 15 rules permit unlicensed devices to emit only very low level emissions.

2700-2900 MHz

FG: S5.423 US18 G15 AERONAUTICAL RADIONAVIGATION S5.337 METEOROLOGICAL AIDS. Radiolocation G2

Non-FG: S5.423 US18

Airport Surveillance Radars (ASR, also called terminal radars) and military ground approach radars use this band in the control of aircraft departure and landing.

The *next-generation Doppler radar* (NEXRAD) transmits in this band in locations where it does not conflict with aviation radars. It is a project of the Department of Defense, the Federal Aviation Administration, and the *National Weather Service* (NWS).

NEXRAD detects the presence and intensity of rain, sleet, and snow; determines the speed and direction of wind in storms; and identifies conditions hazardous to aviation. At this writing, there are 164 NEXRAD installations around the country.

The Doppler radar can detect a tornado's development high above the earth before it touches down. NEXRAD aided NWS to greatly increase its lead time for tornado warnings.

Local Opposition

The advent of the NEXRAD program angered homeowners worried about the radar's emissions and aesthetic impact. Movie and TV celebrities became some of the first protesters after the NWS installed a NEXRAD station overlooking their Ojai, California homes. The stars appeared on the TV talk show circuit to confront NWS officials and raised thousands of dollars to fight the radar.

Figure 14
The NEXRAD radar represents a breakthrough in severe weather warning. To residents who have protested the radar in some localities, it probably looks like something from another planet. (National Weather Service photo)

Long-range plans in this spectrum include the development of multipurpose radars that would combine the functions of ASR and weather radars.

NASA uses range safety surveillance radar in this band to detect aircraft and ships during airborne tests and space launches.

The daily *Solar Flux Index*, an indicator of solar activity that affects ionospheric propagation of radio signals, is measured daily at 2800 MHz in Ottawa, Canada and reported around the world to scientists and spectrum users.

The 2655–2900 MHz spectrum is one of many restricted bands in which the FCC Part 15 rules permit unlicensed devices to emit only very low level emissions.

Frequency Allocations

SHF

2900-3100 MHz

FG: S5.427 US44 US316 MARITIME RADIONAVIGATION. Radiolocation G56

Non-FG: S5.427 US44 US316 MARITIME RADIONAVIGATION. Radiolocation

FCC: Maritime (80)

The Navy and Coast Guard operate navigation radars and positioning aids in this band.

Radar transponder beacons (RACONs) use this band as required by international treaty. RACONs work with shipborne navigation radars to electronically identify maritime obstructions and navigation points.

NEXRAD weather radars use this band where using 2700–2900 MHz is not practical.

Military air traffic control, artillery and rocket tracking, long-range air search and surveillance, and range safety radars use this band. The Navy's Cobra Judy shipborne phased-array system uses this band to provide missile launch data in support of arms control agreements.

3.1-3.3 GHz

FG: S5.149 RADIOLOCATION S5.333 US110 G59

Non-FG: S5.149 Radiolocation S5.333 US110

The 3.1–3.6 GHz spectrum contains more defense radar investment than any other radiolocation band.

Although radars in this part of the spectrum are on fixed, land mobile, and ship platforms, most are airborne. The best known is AWACS, the *Airborne Warning and Control System* radar that tracks ships, aircraft, and other emitters.

We believe it operates in or near this band. The actual AWACS operating band is classified.

The radar portion of the Navy's premier fleet air defense system, Aegis, operates in 3.1–3.5 GHz from guided missile cruisers and destroyers.

Other defense uses of this spectrum include aircrew bombscoring, battlefield weapon locating, carrier precision approach control, and range safety.

The 3.1–3.3 GHz band is one of those used for multi-spectral imaging of Earth by spaceborne active microwave sensors, especially *synthetic aperture radars* (SAR).

The 3.26–3.267 GHz segment is one of many restricted bands in which the FCC Part 15 rules permit unlicensed devices to emit only very low level emissions.

3.3–3.5 GHz

FG: S5.149 RADIOLOCATION US108 G31

Non-FG: S5.149 S5.282 Amateur. Radiolocation US108

FCC: Amateur (97)

The 3.3–3.65 GHz spectrum contains radar systems that the Department of Defense has stated are some of its "most important assets." These include multifunction systems on Navy cruisers and destroyers, and highly mobile Air Force radars which the FCC has described as "extremely high powered."

Uses also include avionic stationkeeping equipment that permits aircraft to fly in formation under low visibility conditions, countermeasures (jamming), radar experimentation, ground-based radio markers used in precision air-land operations, and data communication with hypersonic aircraft. Upgrades to some of the radars are being considered for part of the National Missile Defense System.

This spectrum is being targeted for broadband *Fixed Wireless Access* (FWA) equipment in other countries. FWA manufacturers have investigated using this spectrum in the U.S., a proposition that defense users firmly rebuffed.

"FWA encroachment on radar spectrum would reduce or eliminate any flexibility we may currently have to incorporate new capabilities into these vital radars and would make the testing and training needed to meet new missions virtually impossible," the Defense Department said.

The FCC did, however, allocate substantial, FWA-suitable spectrum in a nearby band (see 3.65–3.7 GHz).

The Amateur Radio Service 9-cm band shares this portion of the 3.1–3.6 GHz defense radar spectrum on a secondary basis.

The 3.332–3.339 GHz and 3.3458–3.358 GHz segments are restricted bands in which the FCC Part 15 rules permit unlicensed devices to emit only low level emissions.

3.5–3.6 GHz

FG: US245 RADIOLOCATION US110 G59. AERONAUTICAL RADIONAVIGATION (Ground-based) G110

Non-FG: Radiolocation US110

This part of the spectrum is an expansion band for the NEXRAD radar (see 2700–2900 MHz). See 3.3–3.5 GHz for other information about this military radar band.

3.6–3.65 GHz

FG: US245 RADIOLOCATION US110 G59. AERONAUTICAL RADIONAVIGATION (Ground-based) G110

Non-FG: FIXED-SATELLITE (space-to-Earth) US245. Radiolocation US110

See 3.3–3.5 GHz for information about this band.

The 3.625–3.7 GHz extended C-band was traditionally a government radar band. Intelsat and Inmarsat use it for international *Fixed Satellite Service* (FSS) downlinks that were added to the band in the mid-1980s (companion uplinks are at 5.85–5.925 GHz).

No one ever used the non-government radiolocation allocation.

This 3.6–3.65 GHz band was separated from 3.65–3.7 GHz in November 1999, in recognition of the reallocation of the latter segment to mixed federal and non-federal use, but mainly commercial use.

This also is an expansion band for civil air traffic control radar that could not be accommodated in 2700–2900 MHz.

The 3.6–4.4 GHz spectrum is one of many restricted bands in which the FCC Part 15 rules permit unlicensed devices to emit only very low level emissions.

3.65-3.7 GHz

FG: US245 US348 US349

Non-FG: US245 US348 US349 FIXED. FIXED-SATELLITE (space-to-Earth) NG169. MOBILE except aeronautical mobile NG170

This is part of the satellite "extended C-band" (see 3.6–3.65 GHz and 3.7–4.2 GHz). Satellite systems that downlink signals in this band typically uplink in 5.85–5.925 GHz and 6.425–6.525 GHz.

This band was separated from 3.6–3.65 GHz in November 1999 in recognition of the transfer of 3.65–3.7 GHz from government to commercial use under the 1993 Omnibus Reconciliation Budget Act (OBRA, see 2390–2400 MHz).

The 3.65–3.7 GHz band became "mixed-use" spectrum; government coastal radars in Maryland, Mississippi, and Florida will remain in the band and must be protected from interference from any new users (see Footnote US348).

In October 2000, the FCC reallocated this band to the Wireless Communications Service (WCS, Docket ET 98-237, see 2305–2310 MHz) and launched a further proceeding to examine licensing and service rules (Docket WT 00-32). The FCC also has referred to this incipient service as the 3650–3700 MHz Service.

"The explosive growth in wireless communications has created tremendous demand for spectrum," FCC Chairman William Kennard said when the Commission reallocated the band. "The steps we take today will result in 50 MHz of additional spectrum being inserted into the marketplace for companies to provide new and exciting services to consumers."

"This spectrum was reallocated from federal government use and will enable a wide variety of commercial fixed and mobile services, including fixed wireless access to the Internet, which will facilitate the provision of advanced broadband services across the country, particularly in rural areas."

Fixed Wireless Access (FWA) in this area of the spectrum already is offered in some other countries, including Canada and the United Kingdom.

Service and Technical Rules

The FCC did not specify exactly which services could be provided in 3.65–3.7 GHz, nor will licensees have to describe their services in detail. Fixed, mobile, and satellite services will be possible, with new *Fixed Satellite Service* (FSS) operations permitted only on a secondary basis.

At this writing, the FCC was accepting public comment on the number and sizes of geographic area licenses to be auctioned and the amount of spectrum conveyed by each license. It proposed, for example, that a single license could cover as much as 50 MHz.

Any new service in this band will have to co-exist with the powerful military radars throughout 3.3–3.65 GHz (see 3.3–3.5 GHz), as well as the remaining in-band installations. NTIA recommended receiver selectivity standards to minimize interference interactions with the adjacent-band and in-band radars.

The FCC, as is its typical practice, declined to mandate such standards. "We continue to believe that this matter is best left to market forces," the Commission said. "Specifically, we believe that, by making the appropriate technical information available to manufacturers, they will, as a matter of course, take into account the electromagnetic environment when designing and building equipment for the 3650–3700 MHz band."

Transition from Satellite

The FCC frequently "freezes" (suspends acceptance of) new applications and requests for major modifications to stations operating in a band that is undergoing substantial regulatory changes. It instituted a freeze affecting FSS earth stations in the extended C-band, angering satellite operators such as AT&T that depend on it for service and control links that alleviate capacity shortages.

In the end, the FCC decided to "grandfather" (retain existing) FSS stations in the band indefinitely. Also to accommodate FSS, it temporarily accepted new applications for FSS stations within 10 miles of the grandfathered stations, to operate on a co-primary basis in the band; and it permitted additional FSS earth stations on a secondary basis.

Spectrum use in this band for FSS isn't subject to auctions. The FCC reasoned that FSS is, or may be, used for international services. It applied the 2000 Open-Market Reorganization for the Betterment of

International Telecommunications Act (Orbit Act) which withholds FCC authority to auction any spectrum or orbit slots "used for the provision of international or global satellite communication services."

The FCC also elected not to permit mobile stations to transmit in this band. Because of their transient nature, mobile stations could threaten FSS earth stations with interference. The locations of base stations are known, controllable and limited, and therefore represent less of an interference threat.

This limitation does not preclude the band from use for mobile communications. Mobile service providers will have to employ the band in transmission from base stations to mobile units and not in the other direction.

As a companion band, the FCC proposed that 4.94–4.99 GHz be made available at the same time as 3.65–3.7 GHz. Both bands are 50 MHz wide and could be paired as is commonly done in radiotelephony.

The Office of Management and Budget predicted auction receipts of $1.2 billion for 3.65–3.7 GHz, after deductions for interference, continued band sharing with federal operations, and commercial implications of the high frequency of the segment.

3.7–4.2 GHz

FG: None

Non-FG: FIXED NG41. FIXED-SATELLITE (space-to-Earth)

FCC: International Fixed (23). Satellite Communications (25). Fixed Microwave (101)

The "C-band" is the workhorse of the *Geostationary Fixed Satellite Service* (GSO FSS), with 3.7–4.2 GHz the downlink band and its associated uplink 5.925–6.425 GHz. More generally, the 4–8 GHz spectrum is designated as C-band.

Cable TV networks use the C-band to provide programming to cable systems. Program package providers directly serve viewers with home satellite dish receivers (Direct-to-Home FSS or DTH-FSS, a service also available in the Ku-band, see 11.7–12.2 GHz).

Many radio stations receive programming by C-band satellite, and international telephone and data traffic travel over the satellites of the

International Telecommunications Satellite Organization (Intelsat) and its competitors in the 4/6 and 11/14 GHz bands.

The C-band also is used for *tracking, telemetry, and control* (TTC) of satellites while transferring them to final orbit position. Satellite operators want to expand TTC to adjacent spectrum (see 3.65–3.7 GHz).

Fixed Operations

This band hosted the first long-haul network for TV distribution and long-distance telephone, but fiber optics are replacing these fixed wireless circuits. In general, terrestrial fixed systems are leaving this band due to the longstanding difficulty of sharing it with satellite services.

Although this band is available to private microwave systems relocating from 2 GHz "Emerging Technologies" spectrum into higher bands, "the Fixed Service has effectively lost the 3.7–4.2 GHz band due to its inability to coordinate with licensed GSO downlinks," according to presentations made at an FCC hearing.

Satellite earth stations in this band also experience interference from defense radars in 2700–3700 MHz. The interference can come from front-end overload of the earth station receiver and from energy from the radar's spurious emissions in 3.7–4.2 GHz. Ways of mitigating the interference include RF filters, antenna shrouding, signal coding in the satellite system, and filtering of the radar output.

The 3.6–4.4 GHz spectrum is one of many restricted bands in which the FCC Part 15 rules permit unlicensed devices to emit only very low level emissions.

4.2–4.4 GHz

FG: S5.440 US261 AERONAUTICAL RADIONAVIGATION

Non-FG: S5.440 US261 AERONAUTICAL RADIONAVIGATION

FCC: Aviation (87)

This band is internationally reserved for radio altimeters. For obvious safety reasons, aviation authorities are especially concerned about interference to operations in this band from adjacent channel or harmonic sources.

These devices determine altitude by emitting signals that reflect from terrain back to the aircraft. For high altitude measurement, some aircraft use pulsed altimeters (see 1610–1610.6 MHz).

The frequencies 4.202 GHz and 6.427 GHz are available, but apparently not used, for two-way satellite time signal communications.

Passive sensors on remote sensing satellites also may use this band to measure the temperature of the sea surface.

The 3.6–4.4 GHz spectrum is one of many restricted bands in which the FCC Part 15 rules permit unlicensed devices to emit only very low level emissions.

4.4–4.5 GHz

FG: FIXED. MOBILE

Non-FG: None

The federal 4 GHz spectrum spans 4.4 GHz to 4.99 GHz. This spectrum is designated in the U.S. and NATO countries for military fixed and mobile communications.

Typical uses include point-to-point microwave, drone vehicle control and telemetry.

Nuclear Fire Department

The Department of Energy uses this band for telemetry, video, and mobile communications in support of the little-known, anti-terrorist *Nuclear Emergency Search Team* (NEST). Its radios tune 4.4–4.99 GHz.

Dubbed the "Nation's Nuclear Fire Department," NEST must search for nuclear materials and respond to nuclear extortion or accidents in any location. The organization is said to be based at Nellis Air Force Base, Nevada, home of the mysterious "Area 51" of UFO lore.

The Treasury Department operates tethered radar balloons (aerostats) in this band, deployed along the southern border and Caribbean areas to detect aircraft suspected of carrying drugs.

4.5–4.8 GHz

FG: US245 FIXED. MOBILE

Non-FG: US245 FIXED-SATELLITE (space-to-Earth) 792A

See 4.4–4.5 GHz for the uses of this band.

By Presidential action under the 1993 *Omnibus Budget Reconciliation Act* (OBRA, see 2390–2400 MHz), the 4.635–4.66 GHz segment was reallocated to the private sector in February 1997 for future commercial uses.

But the President changed his mind. In March 1999, he withdrew the entire 4.635–4.685 GHz spectrum from commercial availability, after conflicts appeared with the Navy's *Cooperative Engagement Capability* (CEC), a radar information distribution network, in this spectrum.

He made a nearby band available to the FCC for commercial use instead (see 4.94–4.99 GHz).

The 4.635–4.66 GHz segment was to have been combined with the 4.66–4.685 GHz band to form a 50 MHz wide *General Wireless Communications Service* (GWCS) with no specific use. The new 4.94–4.99 GHz service will not be called GWCS.

The 4.5–7.5 GHz spectrum is used in microwave radiometry from aircraft to measure ocean wind speed and rain characteristics in hurricanes and storms.

The 4.5–5.15 GHz spectrum is one of many restricted bands in which the FCC Part 15 rules permit unlicensed devices to emit only very low level emissions.

4.8–4.94 GHz

FG: S5.149 US203 FIXED. MOBILE

Non-FG: S5.149 US203

See 4.4–4.5 GHz for the uses of this band.

Footnote US203 provides for *Radio Astronomy Service* (RAS) uses. The 4.8–4.99 GHz spectrum is allocated worldwide to the RAS on a secondary basis, with primary allocations to RAS in some countries.

The 4.5–5.15-GHz spectrum is one of many restricted bands in which FCC Part 15 rules permit unlicensed devices to emit only very low level emissions.

4.94–4.99 GHz

FG: S5.149 S5.339 US257 (proposed to be replaced with US311) FIXED. MOBILE

Non-FG: S5.149 S5.339 US257 (proposed to be replaced with US311)

Note: 4940–4990 MHz became non-Federal government exclusive spectrum in March 1999.

Historically, the Defense Department was the largest user of 4.94–4.99 GHz. The Air Force uses the band to train aircrews in electronic combat, to send radar data over microwave links, and for such uses as missile testing, drone aircraft control, and tropospheric scatter communications— the only part of the spectrum used by the U.S. military for this propagation mode.

Army and Navy uses are similar. An important Navy system in this band is the *Light Airborne Multipurpose System* (LAMPS), a wideband link between helicopters and ships. Other major federal users include the Justice Department, which uses the band in video surveillance, and the Treasury Department in border interdiction.

Footnote US257 recognizes the *Radio Astronomy Service* (RAS) in this spectrum. It authorizes astronomical observations at certain locations and attempts to avoid interference to these passive (non-transmitting) operations, which are used to study the brightness of objects such as supernovas.

The 4.8–4.99 GHz spectrum is allocated worldwide to the RAS on a secondary basis, with primary allocations to RAS in some countries. The 4.95–4.99 GHz segment is allocated worldwide to the *Space Research Service* (SRS) and *Earth Exploration Satellite Services* (EESS), both for passive (reception) uses, on a secondary basis.

Commercial Use

The 4.94–4.99 GHz spectrum was reallocated to the private sector as the FCC Note previously describes. It may become the new home of fixed and mobile services in what was formerly called the *General Wireless Communications Service* (GWCS).

That service has been eliminated as a distinct identity. At this writing, the FCC announced an examination of licensing and service rules for 4.94–4.99 GHz in the context of the *Wireless Communications Service* (WCS, Dockets WT 00-32 and ET 98-237). Importantly, the FCC proposed to pair 4.94–4.99 GHz with 3.65–3.7 GHz.

Mobile to base operations, however, would not be permitted in 3.65–3.7 GHz in order to protect incumbent satellite stations. Mobile to base use would have to be limited to 4.94–4.99 GHz.

Broad Flexibility

The FCC proposed to permit the 4.9 GHz spectrum to be used for any service within the categories of "fixed" and "mobile," including land and maritime mobile. Except for aeronautical mobile (to avoid RAS interference), it is subject to international requirements and frequency coordination.

The *Federal Government* (FG) fixed and mobile allocations, shown previously, would be deleted. All FG frequency assignments, except for RAS observatories, would be withdrawn or limited in the band, with FG agencies expected to relocate their operations to the 4.4–4.94 GHz spectrum.

Complaints

Not all sides were satisfied that the FCC met its statutory responsibilities in allocating this band. The *Association for Maximum Service Television* (AMST), a group of high-power TV stations, said that, "In the 4 GHz band, the Commission failed to discharge its spectrum allocation duty altogether. Instead of determining which service suited the band and served the public, the Commission broadly allocated the spectrum for fixed and mobile services (including one-way, two-way, mobile, fixed, private, commercial, subscription, and non-subscription uses). The Commission left it to auctions to sort out the shifting and probably spotty composition of the band. ...

"Broadcasters had sought an allocation of the 4 GHz band for a non-mutually exclusive, non-subscription application: *electronic news gathering* (ENG) functions that are being crowded out of lower frequencies and provide critical support for live news and special events coverage both currently and in the future digital world. ...

"The amorphous and overbroad allocation the Commission adopted all but forbids the operation of ENG in the 4 GHz band by opening the frequencies to a miscellany of services with vastly different and mutually incompatible interference characteristics, operational modes, spectrum needs, and service traits."

Motorola argued that commercial wireless broadband services could be placed in other spectrum. Public safety agencies (police, fire, and emergency services)—to which Motorola is the leading radio supplier—may be the more appropriate users of 4.9 GHz, for "mission critical broadband applications," it said. "What is the commercial application for the 4.9 GHz band that cannot be met in the existing unlicensed spectrum bands?" Motorola asked.

The 4.5–5.15 GHz spectrum is one of many restricted bands in which the FCC Part 15 rules permit unlicensed devices to emit only very low level emissions.

4.99–5 GHz

FG: US246 RADIO ASTRONOMY US74. Space Research (passive)

Non-FG: US246 RADIO ASTRONOMY US74. Space Research (passive)

No transmissions are permitted in this receive-only scientific band, widely used in radio astronomy.

Astronomers use it to study the distributions of brightness of objects in galaxies and to make radio maps of interstellar clouds and supernova remnants. This portion of the spectrum is desirable for astronomy because of the low level of galactic background continuum radiation.

This is a key band for *Very Long Baseline Interferometry* (VLBI), which links radio telescopes together across long distances. Scientists use this technique to study continental drift, rotation rate of Earth, earthquakes, and spacecraft navigation.

The 4.5–5.15 GHz spectrum is one of many restricted bands in which the FCC Part 15 rules permit unlicensed devices to emit only very low level emissions.

5-5.25 GHz

FG: S5.446 733 796 797 US211 US307 AERONAUTICAL RADIONAVIGATION US260

Non-FG: S5.446 733 796 797 US211 US307 AERONAUTICAL RADIONAVIGATION US260

FCC: Satellite Communications (25). Aviation (87)

Note: The NTIA Manual (footnote G126) states that differential GPS stations may be authorized in the 5000-5150 MHz segment, but the FCC has not yet addressed this footnote.

This aviation band is seeing new uses for satellites and terrestrial broadband communications.

The 1995 World Radiocommunication Conference allocated 5.091-5.250 GHz to the *Fixed Satellite Service* (FSS) on a co-primary basis worldwide, limited to feeder links from gateway stations in the *Non-Geostationary Mobile Satellite Service* (NGSO MSS or Big LEO, see 1610-1610.6 MHz).

The proposed U.S. allocation (Docket ET 98-142) is 5.15-5.25 GHz for use by non-government FSS on a primary basis for NGSO MSS feeder links, along with 6.7-7.075 GHz for downlinks and 15.43-15.63 GHz for both uplinks and downlinks.

Aviation Uses

U.S. NGSO MSS feeders could have access to this band below 5.15 GHz but the FCC cautioned potential users that restrictions relating to aeronautical and FSS sharing would apply.

Government differential *Global Positioning System* (GPS) stations may be authorized in 5-5.15 GHz as well (see 1215-1240 MHz) as mentioned by the FCC Note; however, Footnote G126 is not officially part of this band's allocations at this writing.

The *International Civil Aviation Organization* (ICAO) Standard *Microwave Landing System* (MLS) offers civil and military airport approach and landing services in 5.03-5.091 GHz. A MLS station provides navigation and guidance for aircraft out to a range of 43 kilometers (23 nautical miles) and an altitude of 6,096 meters (20,000 feet).

The requirements of international MLS "took precedence over all other uses of the 5000–5250 MHz band," according to the FCC; and satellite uses in this band must accommodate these aeronautical operations.

MLS is supposed to eventually replace the *Instrument Landing System* (ILS) as the international standard (see 108–117.975 MHz). MLS signals are less affected by terrain, structures, and weather.

Nevertheless, the Federal Aviation Adminstration terminated development of MLS based on good results with GPS as well as financial constraints. The "phase-down" of MLS is expected to start in 2008.

The FAA is developing other services for this band, including *Terminal Doppler Weather Radar* (TDWR, see 5.6–5.65 GHz) and *Automatic Dependent Surveillance* (ADS), part of Data Link (see 136–137 MHz and 1215–1240 MHz), which relays aircraft position to controllers and other aircraft.

Another federal use of this band is for radar-based fuze sensors in air-to-air missiles to detect targets. Scientific uses of the band include scatterometer radar observation of sea conditions.

5 GHz LANs

The *Broadband Radio Access Networks* (BRAN) project in the *European Telecommunications Standards Institute* (ETSI) is developing *High Performance Radio Local Area Network* (Hiperlan) standards in the 5 GHz spectrum. These products offer wireless connections for such applications as printer access, ad hoc portable networks, and video transmission.

Hiperlan 1, a wireless Ethernet-type standard completed in 1996, offers speeds in the 20 Mb/s range. Hiperlan 2 will offer higher speeds, better support for quality-of-service-sensitive applications like voice, video and multimedia, and a technology known as *Orthogonal Frequency Division Multiplexing* (OFDM), used in the European digital audio broadcast system and excellent for multipath environments.

Hiperlan 3 or HiperAccess will be devoted to wireless local loop telephone services. Hiperlan 4 or HiperLink is a point-to-point standard for 17 GHz.

Two petitioners asked the FCC to allocate spectrum for Hiperlan-like wireless LANs in the U.S.: the *Wireless Information Networks Forum* (WINForum) and Apple Computer. WINForum requested spectrum for what it called the SUPERNet or Shared Unlicensed Personal Radio Network.

Apple petitioned for a *National Information Infrastructure* (NII) Band in this spectrum. Apple previously petitioned for, and received, access to 2390–2400 MHz and 1910–1920 MHz for unlicensed Data-PCS (see 1850–1990 MHz).

"The copious bandwidth of the NII Band will permit microcellular frequency re-use for yet more spectrum capacity and efficiency," Apple told the FCC. "This capability translates not only into the ability to do conventional tasks more quickly, but also into the ability to do things not previously possible."

In addition to indoor and short-range data transfer, Apple envisioned NII Band to be used for longer-range, community networking, and wireless Internet. "With NII Band devices, it will be possible to communicate at distances on the order of 10 to 15 km or more. …The capability to communicate on an unlicensed, wireless basis over such distances will open up new ways of communicating. Perhaps most importantly, it will make possible an entirely new application of wireless networking: the community network."

FCC Acts

Responding to these requests, the FCC decided in January 1997 to make available 300 MHz of spectrum for what it calls *"Unlicensed NII"* (U-NII) devices.

U-NII devices are regulated under Part 15 of the FCC rules and do not constitute a radio service. This action was therefore not a spectrum allocation in the legal sense, although the practical result is similar.

The FCC defined the NII as, "a group of networks, including the public switched telecommunications network, radio and television networks, private communications networks, and other networks not yet built, which together will serve the communications and information processing needs of the people of the United States in the future."

U-NII devices, the FCC said, "will provide short range, high speed wireless digital communications on an unlicensed basis. We anticipate that U-NII devices will support the creation of new wireless local area networks and will facilitate wireless access to the National Information Infrastructure."

The Commission gave U-NII devices access to three 100 MHz segments at 5.15–5.25 GHz, 5.25–5.35 GHz, and 5.725–5.825 GHz, with different antenna and power provisions for each segment. The 5.15–5.25 GHz segment, shared with the Mobile Satellite Service, is limited by regulation to indoor use only.

"We believe that U-NII devices can share with proposed and existing services in these bands," the FCC said, "including the Mobile Satellite Service feeder link operations that may use the 5.15–5.25 GHz band.

"On the other hand, U-NII devices will not have access to spectrum used by microwave landing systems operated by the FAA in the 5.0–5.15 GHz band. Additionally, U-NII devices will not have access to the 5.825–5.875 GHz band.

"This will avoid potential interference with low power Part 15 hearing aid devices and potential *Intelligent Transportation Systems* (ITS) operations in the 5.850–5.875 GHz band, Fixed Satellite Service (FSS) operations in the 5.850–5.925 GHz band, and amateur operations in the 5.650–5.725 and 5.825–5.925 GHz bands."

Concurrent with the U-NII action, the FCC removed 5.15–5.35 GHz from the list of restricted bands. It later permitted increased antenna gain in the uppermost U-NII band, as Apple had requested.

Amateur, satellite, and long-distance carrier interests opposed the NII Band petition. AT&T, in particular, argued that expanding the capabilities of U-NII would violate the Commission's "regulatory parity objectives" and that "the increased reach that signals from U-NII devices would enjoy as a result of increased power or more directional antennas would allow unlicensed spectrum to be used for purposes beyond those envisioned by the Commission."

But "issues of regulatory parity and unfair competitiveness raised by AT&T are not a basis for concern at this time…The status of unlicensed devices is not being altered," the FCC said, "in that they continue to operate without protection from interference caused by other devices or authorized services and are not entitled to exclusive use of the spectrum in a given area, as are most licensed services."

The U-NII regulations could be considered broadband-preferred, with power reductions required for bandwidths less than 20 MHz.

The *Institute of Electrical and Electronics Engineers* (IEEE) is developing the 802.11a networking standard for the U-NII and Hiperlan 5 GHz bands. Related IEEE 802 wireless standards concern 2.4 GHz (see 2400–2402 MHz).

Some observers believe that most unlicensed, broadband wireless networking will gravitate to the 5 GHz IEEE standards, with shorter-range, *personal area networks* (PANs) dominant at 2.4 GHz. At this writing, most for-profit and not-for-profit wireless Internet service providers that use unlicensed equipment are focused on 2.4 GHz.

Future of U-NII

With regard to the future of unlicensed systems, a *National Telecommunications and Information Administration* (NTIA) report said that, "In general, one should expect very substantial growth in unlicensed systems of many types, e.g., cordless phones and wireless LANs. Eventually there may be too many additional systems to expect interference-free operation in crowded locations.

"Or, maybe not. ...The situation will be similar in the U-NII band. ...As with other unlicensed applications, the possible growth of interference in this band due to uncoordinated use is a potential problem for which no one has sufficient experience to give a convincing answer yet."

5.25–5.35 GHz

FG: RADIOLOCATION S5.333 US110 G59

Non-FG: RADIOLOCATION S5.333 US110

This is one of the three segments available to *Unlicensed National Information Infrastructure* devices (U-NII, see 5–5.25 GHz).

The 5.25–5.925 GHz spectrum supports airborne weather radars and military systems, such as Air Force test range instrumentation radars, the Army Patriot surface-to-air missile system radar, Navy surface search, and navigation and fire control radars. The largest category of 5 GHz radars are ground-based, including mobile systems.

Non-military government radar operations in this spectrum include range safety, geological surveys, radars on unmanned aerial vehicles, and a category known as "nuclear incident situations."

Radars in this spectrum emit up to 100 billion watts of power. The 5.725–5.875 GHz segment contains the highest concentration of these radars.

The 5.25–5.35 GHz band is one of those used in Earth observation, providing multi-spectral images obtained by spaceborne active microwave sensors.

Radiolocation stations installed on spacecraft may also be employed in this band in the *Earth Exploration Satellite Service* (EESS) and *Space Research Service* (SRS) on a secondary basis.

5.35-5.46 GHz

FG: US48 AERONAUTICAL RADIONAVIGATION S5.449. RADIOLOCATION G56

Non-FG: US48 AERONAUTICAL RADIONAVIGATION S5.449. Radiolocation

FCC: Aviation (87)

This is part of the 5.25-5.925 GHz government radar spectrum (see 5.25-5.35 GHz). Airborne, shipborne, and land-based weather radars may be found throughout 5.35-5.6 GHz.

Military radars in this spectrum are used to track targets and are used for *Identification Friend or Foe* (IFF) interrogation.

The 5.35-5.46 GHz band is one of many restricted bands in which the FCC Part 15 rules permit unlicensed devices to emit only very low level emissions.

5.46-5.47 GHz

FG: US49 US65 RADIONAVIGATION. S5.449. Radiolocation G56

Non-FG: US49 US65 RADIONAVIGATION. S5.449. Radiolocation

This is part of the 5.25-5.925 GHz government radar spectrum (see 5.25-5.35 GHz).

5.47-5.6 GHz

FG: US50 US65 MARITIME RADIONAVIGATION. Radiolocation G56

Non-FG: US50 US65 MARITIME RADIONAVIGATION. Radiolocation

FCC: Maritime (80)

This is part of the 5.25–5.925 GHz government radar spectrum (see 5.25–5.35 GHz).

5.6-5.65 GHz

FG: S5.452 US65 MARITIME RADIONAVIGATION. METEOROLOGICAL AIDS. Radiolocation US51 G56

Non-FG: S5.452 US65 MARITIME RADIONAVIGATION. METEOROLOGICAL AIDS. Radiolocation US51

FCC: Maritime (80)

This is part of the 5.25–5.925 GHz government radar spectrum (see 5.25–5.35 GHz).

Terminal Doppler Weather Radar (TDWR) uses this band to detect and guide aircraft around wind shear—dangerous, rapid downdrafts at low altitudes.

Like the NEXRAD weather radar (2700–2900 MHz), TDWR has become politicized in some areas. Residents, concerned about radiation or aesthetics, have blocked or delayed radar construction in several cities. TDWR systems emit peak power of 250 kW.

Each TDWR covers an area of sky up 24,000 feet around 55 miles of the airport. The radar was developed in the 1980s after air crashes implicated wind shear.

Air traffic controllers drubbed TDWR for susceptibility to storm outages and nicknamed it "Thunderstorms Destroy Weather Radar."

5.65-5.83 GHz

FG: S5.150 US245 RADIOLOCATION G2

Non-FG: S5.150 S5.282 Amateur

FCC: ISM Equipment (18). Amateur (97)

This is part of the 5.25-5.925 GHz government radar spectrum (see 5.25-5.35 GHz). The major federal users of this band are the Defense Department and NASA for surface, missile and rocket tracking radars, aircraft guidance and surveillance radar, telemetry, and ground facilities to develop these systems.

Radar use includes interrogating transponders onboard craft. These devices reply on a different frequency, or they may amplify the signals reflected back to the radar.

Unlicensed Devices

FCC Part 15 rules for unlicensed devices permit field disturbance sensors, such as door openers, to operate in the 5.785-5.815 GHz spectrum. Spread spectrum unlicensed devices may use 5.725-5.85 GHz and non-spread spectrum devices may use 5.725-5.875 GHz.

The FCC eliminated a limit on directional antenna gain for certain spread spectrum systems in 5.725-5.85 GHz. Certain important restrictions apply to such systems that employ high gain directional antennas (greater than 6 dBi).

These systems must be restricted to fixed, point-to-point operation. Point-to-multipoint, omnidirectional and multiple co-located systems transmitting the same information are not allowed to take advantage of the relaxed antenna rules. Although the systems are unlicensed, the FCC holds the operator or installer of the system responsible for ensuring that it is used only for fixed point-to-point operation.

European nations are adopting the 5.8 GHz spectrum as the location for automatic vehicle identification tags. This reportedly enhances prospects for an international standard. In the U.S., most of these tags are at 902-928 MHz.

Wireless local-area networks for computers use 5.725-5.85 GHz for spread spectrum communications (see 902-928 MHz and 2400-2402 MHz).

According to the FCC, the band's uses include "intelligent transportation system communications links; high speed Internet connections for schools, hospitals, and government offices; energy utility applications; PCS and cellular backbone connections; and T-1 common carrier links in rural areas." Also in 5.75–5.825 GHz are Unlicensed National Information Infrastructure devices that can be used for much the same purposes (U-NII, see 5–5.25 GHz).

The Amateur Radio Service 5-cm band shares the 5.65–5.925 GHz spectrum on a secondary, non-interference basis, with 5.65–5.67 GHz and 5.83–5.85 GHz allocated for amateur satellite use.

The FCC defines the *Industrial, Scientific, and Medical* (ISM) frequency in this band as 5800 MHz, ±75 MHz.

5.83–5.85 GHz

FG: S5.150 US245 RADIOLOCATION G2

Non-FG: S5.150 Amateur. Amateur-satellite (space-to-Earth)

FCC: ISM Equipment (18). Amateur (97)

This is part of the 5.25–5.925 GHz government radar spectrum (see 5.25–5.35 GHz).

The major federal users of this band are the Defense Department and NASA for surface, missile and rocket tracking radar, aircraft guidance, interrogation, and surveillance radar, telemetry, and ground facilities to develop these systems. See 5.65–5.83 GHz for additional uses of this band.

5.85–5.925 GHz

FG: S5.150 US245 RADIOLOCATION G2

Non-FG: S5.150 FIXED-SATELLITE (Earth-to-space) US245. MOBILE NG160. Amateur

FCC: ISM Equipment (18). Private Land Mobile (90). Amateur (97).

This is part of the 5.25–5.925 GHz government radar spectrum (see 5.25–5.35 GHz). This band may see widespread use in vehicle applications. See 5.65–5.83 GHz for additional uses of this band.

The main federal uses of this band are military radar and control of unmanned air vehicles, mostly on remote test ranges, and transportable satellite earth station operations in 5.85–6.425 GHz.

As the uplink side of the "extended C-band," it is paired with the *Fixed Satellite Service* (FSS) 3.6–3.65 GHz and 3.65–3.7 GHz bands.

Unlicensed devices and *Industrial, Scientific, and Medical* (ISM) devices have access to 5.725–5.875 GHz. An example ISM product in 5.85–5.875 GHz is an advanced hearing aid that uses radio to communicate between a miniature earpiece and a signal processing unit on the user's belt.

Dedicated Short Range Communications (DSRC)

In October 1999, the FCC allocated 5.85–5.925 GHz to the *Intelligent Transportation Systems* (ITS) Radio Services for use by *Dedicated Short Range Communications* (DSRC, Docket ET 98-95), as part of the international trend toward wireless ITS. We expect DSRC to eventually enjoy large markets in consumer and commercial vehicle operations.

The FCC formally defines DSRC as, "The use of non-voice radio techniques to transfer data over short distances between roadside and mobile radio units, between mobile units, and between portable and mobile units to perform operations related to the improvement of traffic flow, traffic safety, and other intelligent transportation service applications in a variety of public and commercial environments. DSRC systems may also transmit status and instructional messages related to the units involved."

"This is certainly the largest allocation of spectrum that the FCC has given to transportation," said Paul Najarian, telecommunications director at trade association *ITS America* (ITSA). "It truly signals the convergence of the telecommunications and transportation industries in finding solutions that will save lives, time, money, and improve the quality of life."

DSRC already is in wide use in the form of electronic toll collection. "Beacons" and "readers" at tollgates communicate with radio "tag" devices mounted on cars. Computers automatically debit the driver toll account.

Most of these existing systems are in 902–928 MHz. They are supposed to migrate eventually to the 5.8 GHz version of DSRC, which will expand well beyond toll collection applications.

ITS advocates foresee many new uses for DSRC, including electronic commerce, for payment of parking and fast-food charges, radio license plates that identify vehicles to police and inspectors, and collision avoidance systems, which alert drivers when they are on a collision path. Roadway heat sensors are expected to use DSRC to alert truck drivers of hazardous brake conditions. In-vehicle signing, the display of road sign and traffic information inside the vehicle, may enable drivers to avoid hazards and bottlenecks.

An intriguing DSRC application is in Automated Highway Systems, which permit hands-off, computer controlled driving of cars in "platoons" that travel dedicated highway lanes.

ITS America said, "This allocation will enhance the efficiency of use of the transportation infrastructure, improve mobility and reduce traffic congestion, enable quicker emergency incident response from public safety agencies, improve safety inspections of commercial vehicles while reducing costly weigh station and border crossing delays, reduce health care costs attributable to traffic accidents, improve the management and security of the flow of hazardous materials throughout the nation and help realize billions of dollars of gain in economic productivity."

The FCC cited "a need for up to 32 different DSRC transactions, many of which will require two-way capabilities, wideband channels, and the need for multiple channels in a single location. ...[N]ot all channels will be available for DSRC deployment in all areas due to incumbent radar, ISM, and FSS operations. Therefore, we find that 75 megahertz of DSRC spectrum within the United States is warranted due to the scope of the National ITS Architecture, the incumbent operations in this band in the U.S. and consideration [of] DSRC developments domestically and internationally."

Alien to the Industry

Especially important is whether or not DSRC licenses will be auctioned. "DSRC is completely alien to the wireless industry," according to ITSA's Najarian, but it offers that industry an "inestimable benefit," a very large bandwidth. The balance of commercial versus public safety applications of DSRC will have much impact on the possibility of auctions in the service.

Although it has allocated spectrum to DSRC, at this writing, the FCC had not proposed any licensing, channelization, or operational rules for the service. It promised a future proceeding to examine these issues. It is not known, therefore, who will be permitted to deploy DSRC facilities or whether this will be a licensed, unlicensed, or licensed-by-rule service.

In ITU Region 2 (the Americas and Greenland), the 5.65–5.925 GHz spectrum is allocated to the Amateur Radio Service on a secondary basis. Amateur uses of this band include satellite uplinks and downlinks, bouncing signals off of the Moon ("Earth-Moon-Earth" or EME), weak-signal, and beacon operation.

5.925–6.425 GHz

FG: None

Non-FG: FIXED NG41. FIXED-SATELLITE (Earth-to-space)

FCC: International Fixed (23). Satellite Communications (25). Fixed Microwave (101)

This is the C-band satellite uplink, also used by fixed service stations (see 3.7–4.2 GHz).

6.425–6.525 GHz

FG: S5.440 S5.458

Non-FG: S5.440 S5.458 FIXED-SATELLITE (Earth-to-space). MOBILE

FCC: Auxiliary Broadcasting (74). Cable TV Relay (78). Fixed Microwave (101)

The main terrestrial uses at 6.425–6.525 GHz are *Broadcast Auxiliary Services* (BAS) operations such as remote TV pickup and as *Local Television Transmission Service* (LTTS) where common carriers provide service to TV stations, cable systems, and networks.

Coordinating BAS operations in the 6–7 GHz spectrum is increasingly difficult due to congestion in these bands in major markets, according to the *National Association of Broadcasters* (NAB).

Licensees of Ka-band *Geostationary Orbit Fixed Satellite Services* (GSO FSS, see 17.7–17.8 GHz) asked the FCC to designate 6.425–6.525 GHz (uplink) and 3.65–3.7 GHz (downlink) for GSO FSS tracking, telemetry, and control.

The 6.425–7.250 GHz spectrum is available to passive sensors for ocean temperature measurements.

The frequencies 6.427 GHz and 4.202 GHz are available, but apparently not used, for two-way satellite time signal communications.

6.525–6.875 GHz

FG: S5.458

Non-FG: S5.458 FIXED. FIXED-SATELLITE (Earth-to-space) 792A

FCC: Satellite Communications (25). Fixed Microwave (101)

Governmental, industrial, communications carrier, and transportation licensees use this general-purpose fixed microwave band.

It is the main band to which microwave incumbents migrate from the 2 GHz spectrum reallocated for *Personal Communications Services* (PCS) and "Emerging Technologies." The FCC has stated that this spectrum is lightly used by the *Fixed Satellite Service* (FSS), in contrast to the conventional 4/6 GHz and 12/14 GHz bands.

The FCC proposed in Docket ET 98-142 to allocate 6.700–7.075 GHz to non-government FSS downlinks associated with feeder links for *Non-Geostationary Mobile Satellite Services* (NGSO MSS). The docket would divide this band into 6.525–6.700 GHz and 6.700–6.875 GHz.

Radio astronomers have discovered a strong spectral line of the methanol molecule at 6668.518 MHz. The methanol line is an important tracer of star formation activity.

A segment in this band is listed as available for uplinks in the *Radiodetermination Satellite Service* (RDSS, see 1610–1610.6 MHz). No RDSS satellites exist.

6.875–7.075 GHz

FG: S5.458

Non-FG: S5.458 NG118 FIXED. FIXED-SATELLITE (Earth-to-space) 792A. MOBILE

FCC: Auxiliary Broadcasting (74). Cable TV Relay (78)

The 6.875–7.125 GHz spectrum is one of the bands allocated on a primary basis to *Broadcast Auxiliary Services* (BAS) and the *Cable Television Relay Service* (CARS).

Applications include *electronic news gathering* (ENG), remote event coverage, intercity relay, relay associated with television translators, and *studio-transmitter links* (STL).

The advent of Digital TV may exacerbate congestion in TV auxiliary bands (see 470–512 MHz and 6.425–6.525 GHz). "Dual-mode" (analog plus DTV) operations are expected to increase the burden on this spectrum.

Feeder links for Big LEO satellites may use this band (see 1610–1610.6 MHz). Satellite *Digital Audio Radio Services* (DARS) feeder links are in 7.025–7.075 GHz (see 2320–2345 MHz).

7.075–7.125 GHz

FG: S5.458

Non-FG: S5.458 NG118 FIXED. MOBILE

FCC: Auxiliary Broadcasting (74). Cable TV Relay (78)

See 6.875–7.075 GHz for information about this band.

7.125–7.19 GHz

FG: S5.458 US252 G116 FIXED

Non-FG: S5.458 US252

The Federal Aviation Administration uses the 7.125–8.5 GHz spectrum for the FAA *Radio Communications Link* (RCL) network connecting air traffic and radar sites. The footnotes suggest that this spectrum also is used in military satellite communications.

7.19-7.235 GHz

FG: S5.458 FIXED. SPACE RESEARCH (Earth-to-space)

Non-FG: S5.458

This uplink band is for deep space communications in conjunction with 8.4–8.5 GHz downlinks (see 8.4–8.45 GHz).

7.235-7.25 GHz

FG: S5.458 FIXED

Non-FG: S5.458

See 7.125–7.19 GHz for information about this band.

7.25-7.3 GHz

FG: G117 FIXED-SATELLITE (space-to-Earth). MOBILE-SATELLITE (space-to-Earth). Fixed

Non-FG: None

See 7.125–7.19 GHz for information about this band.

The 7.25–7.75 GHz spectrum is one of many restricted bands in which the FCC Part 15 rules permit unlicensed devices to emit only very low level emissions.

7.3-7.45 GHz

FG: G117 FIXED. FIXED-SATELLITE (space-to-Earth). Mobile-Satellite (space-to-Earth)

Non-FG: None

See 7.125–7.19 GHz for information about this band.

The 7.25–7.75 GHz spectrum is one of many restricted bands in which the FCC Part 15 rules permit unlicensed devices to emit only very low level emissions.

7.45–7.55 GHz

FG: G104 G117 FIXED. FIXED-SATELLITE (space-to-Earth).

METEOROLOGICAL-SATELLITE (space-to-Earth). Mobile-Satellite (space-to-Earth)

Non-FG: None

See 7.125–7.19 GHz for information about this band.

The Meteorological-Satellite Service allocation is used to downlink weather data.

The 7.25–7.75 GHz spectrum is one of many restricted bands in which the FCC Part 15 rules permit unlicensed devices to emit only very low level emissions.

7.55–7.75 GHz

FG: G117 FIXED. FIXED-SATELLITE (space-to-Earth). Mobile-Satellite (space-to-Earth)

Non-FG: None

See 7.125–7.19 GHz for information about this band.

The 7.25–7.75 GHz spectrum is one of many restricted bands in which the FCC Part 15 rules permit unlicensed devices to emit only very low level emissions.

7.75–7.9 GHz

FG: FIXED

Non-FG: None

See 7.125–7.19 GHz for information about this band.

7.9–8.025 GHz

FG: G117 FIXED-SATELLITE (Earth-to-space). MOBILE-SATELLITE (Earth-to-space). Fixed

Non-FG: None

Military terrestrial microwave operations are found throughout 7.9–8.45 GHz. The Department of Energy and power agencies such as the Tennessee Valley Authority also use this spectrum for *System Control And Data Acquisition* (SCADA) telemetry in the management of vast power distribution networks.

Military and government satellite uses include the *Defense Satellite Communications System* (DSCS), with service links in 7.9–8.4 GHz and tracking, telemetry, and control links in the 1761–1842 MHz and 2200–2290 MHz *Space-Ground Link Subsystem* (SGLS) bands. Some of the defense satellite downlinks associated with this band are in the VHF spectrum (see 235–267 MHz).

The geostationary DSCS satellites provide the Departments of Defense, State, and other agencies with jam-resistant communications for applications including command and control, crisis management, intelligence, early warning detection, diplomatic traffic, and treaty monitoring.

These bands also host the geostationary *Defense Support Program* (DSP) satellites, which use infrared telescopes to detect missiles. DSP satellites have been in operation since 1970. During the Gulf War, they provided up to five minutes warning of launches from Iraq, in what were the first episodes of ballistic missile defense.

The Air Force will gradually replace DSP satellites with the $22 billion *Space Based Infrared System* (SBIRS). The SBIRS will include up to 30 geostationary and low-Earth orbit spacecraft.

8.025-8.175 GHz

FG: G117 EARTH EXPLORATION-SATELLITE (space-to-Earth).
FIXED. FIXED-SATELLITE (Earth-to-space). Mobile-Satellite
(Earth-to-space) (no airborne transmissions)

Non-FG: None

The *Earth Exploration Satellite Service* (EESS) for land remote sensing uses 8.025–8.4 GHz to deliver images, and for some satellites, telemetry and control downlinks.

EESS is defined as "a radiocommunication service between earth stations and one or more space stations, which may include links between space stations in which: (1) information relating to the characteristics of the Earth and its natural phenomena is obtained from active or passive sensors on earth satellites; (2) similar information is collected from airborne or earth-based platforms; (3) such information may be distributed to earth stations within the system concerned; and (4) platform interrogation may be included." (For information on "platforms," see 401–402 MHz.)

The sale of satellite imagery is approaching a multibillion dollar business. Among authorized U.S. users of this spectrum are Landsat and the AstroVision, EarthWatch, Orbimage, Resource21, and Space Imaging commercial remote sensing satellite providers. These will offer high resolution images, formerly available only from intelligence satellites (see 2200–2290 MHz). EarthWatch (marketed as DigitalGlobe) was the first U.S. EESS licensee. Its two satellites, EarlyBird and QuickBird, failed. Space Imaging's first Ikonos satellite failed; but its second attempt proved successful.

"Land remote-sensing satellites can provide information that can assist in studying and understanding human impacts on the environment, managing resources, and carrying out national security functions," the FCC has stated. "The collected data can be used for mapping, conservation, law enforcement, environmental monitoring, and forecasting change in vegetation and other features of the Earth's surface."

Also using 8.025–8.4 GHz are the satellites in the NASA *Earth Observation System* (EOS). These satellites will survey land surface, atmosphere and ocean characteristics, and will continuously transmit data to users worldwide.

The 8.025–8.4 GHz spectrum is becoming heavily used, particularly as advances in technology result in higher resolution remote sensing instruments that require greater download bandwidths. The 25.5–27.5 GHz spectrum is a candidate expansion band for EESS downlinks (see 25.25–27 GHz).

Communications satellites of the *North Atlantic Treaty Organization* (NATO) SATCOM system also operate in 8.025–8.4 GHz.

The 8.025–8.5 GHz spectrum is one of many restricted bands in which the FCC Part 15 rules permit unlicensed devices to emit only very low level emissions.

8.175–8.215 GHz

FG: US258 G104 G117 EARTH EXPLORATION-SATELLITE (space-to-Earth). FIXED. FIXED-SATELLITE (Earth-to-space). METEOROLOGICAL-SATELLITE (Earth-to-space). Mobile-Satellite (Earth-to-space) (no airborne transmissions)

Non-FG: None

See 8.025–8.175 GHz for the uses of this band.

The 8.025–8.5 GHz spectrum is one of many restricted bands in which the FCC Part 15 rules permit unlicensed devices to emit only very low level emissions.

8.215–8.4 GHz

FG: US258 G117 EARTH EXPLORATION-SATELLITE (space-to-Earth). FIXED. FIXED-SATELLITE (Earth-to-space). Mobile-Satellite (Earth-to-space) (no airborne transmissions)

Non-FG: US258

See 8.025–8.175 GHz for the uses of this band.

The 8.025–8.5 GHz spectrum is one of many restricted bands in which the FCC Part 15 rules permit unlicensed devices to emit only very low level emissions.

8.4-8.45 GHz

FG: FIXED. SPACE RESEARCH (space-to-Earth) (deep space only)

Non-FG: None

The 8.4–8.5 GHz spectrum contains space research downlinks (uplinks at 7.19–7.235 GHz), with 8.4–8.45 GHz limited to craft in deep space (beyond Earth's orbit).

The "deep space only" provision is applied to some bands because of the different radio characteristics at these long distances. Uplinks to deep space receivers must use more power than uplinks to satellites closer to Earth, and downlinks from deep space naturally appear at lower received power than those from closer satellites (see also 2290–2300 MHz).

The 8.025–8.5 GHz spectrum is one of many restricted bands in which the FCC Part 15 rules permit unlicensed devices to emit only very low level emissions.

8.45-8.5 GHz

FG: FIXED. SPACE RESEARCH (space-to-Earth)

Non-FG: SPACE RESEARCH (space-to-Earth)

See 8.4–8.45 GHz for the uses of this band.

The 8.025–8.5 GHz spectrum is one of many restricted bands in which the FCC Part 15 rules permit unlicensed devices to emit only very low level emissions.

8.5-9 GHz

FG: US53 RADIOLOCATION S5.333 US110 G59

Non-FG: US53 Radiolocation S5.333 US110

The 8.5–10.55 GHz spectrum is used for military and civil radar, including meteorological radar, airborne navigation, transportable artillery-locating radar, fire control radar, and ballistic missile defense imaging radar.

Coastal radars use this band to map ocean currents in harbor areas. The 8.55–8.65 GHz band is one of those used to obtain multi-spectral images through use of spaceborne active microwave sensors.

In the Mojave Desert northeast of Los Angeles, NASA's Goldstone Solar System Radar transmits at 8.56 GHz in studies of planetary bodies, asteroids, and orbital debris.

Pulses from the Goldstone radar can take as long as 4.5 hours to return from the target. The radar also operates at 2320 MHz, but this frequency is jeopardized by new radio services (see 2320–2345 MHz).

9–9.2 GHz

FG: US48 US54 G19 AERONAUTICAL RADIONAVIGATION. S5.337 Radiolocation G2

Non-FG: US48 US54 AERONAUTICAL RADIONAVIGATION. S5.337 Radiolocation

FCC: Aviation (87)

Military aviation uses this band for precision approach radar, airborne search and rescue, law enforcement radar, airborne navigation, and surveillance radar.

The band has been identified for new types of *Airport Surface Detection Equipment* (ASDE, see 15.7–16.6 GHz).

The 9–9.2 GHz band is a restricted band in which Part 15 rules permit unlicensed devices to emit only very low level emissions.

9.2–9.3 GHz

FG: S5.474 MARITIME RADIONAVIGATION S5.472. Radiolocation US110 G59

Non-FG: S5.474 MARITIME RADIONAVIGATION S5.472. Radiolocation US110

This is part of the 9.2–10.55 GHz federal radar spectrum.

Maritime navigation radars operate in 9.2–9.5 GHz. The Coast Guard uses these radars to observe harbor and coastal traffic. Besides the navigation function, they work with *search and rescue transponders* (SART) that respond when interrogated by the radar. Evacuees can grab the wand-like SART device and take it onboard survival craft.

The Air Force makes extensive use of this band for the air and ground control of *unmanned aerial vehicles* (UAVs) used in electronic warfare and reconnaissance.

An intriguing type of UAV is the *micro air vehicle* (MAV), measuring less than six inches in any one dimension and flyable indoors and outdoors. Some advanced MAVs are disk-like "flying saucers," and others flap their wings, according to Defense Department testimony.

Head-mounted displays will offer the soldier "see-through, fly-through" visualization of imagery picked up by UAVs.

9.3–9.5 GHz

FG: S5.427 S5.474 US67 US71 RADIONAVIGATION S5.476 US66. Radiolocation US51 G56. Meteorological Aids

Non-FG: S5.427 S5.474 US67 US71 RADIONAVIGATION S5.476 US66. Radiolocation US51. Meteorological Aids

This is part of the 9.2–10.55 GHz federal radar spectrum.

State and local agencies, broadcasters, researchers, and forecasters operate weather radars in this band. Many aircraft carry weather radars using this band as well.

Mobile radars on vehicles are used to investigate coastal weather phenomena, especially hurricanes and tornadoes. Similar radars use other nearby bands including 8.5–9 GHz and 9.5–10 GHz. Marine navigation radars and radar transponder beacons also use this band (see 9.2–9.3 GHz).

Radars in this band are important tools in the study of bird migration. The homing and migration abilities of birds continue to intrigue scientists after decades of research. Visual, auditory, and olfactory cues, and magnetic orientation have failed to explain the phenomenon.

This is a restricted band in which the Part 15 rules permit unlicensed devices to emit only very low level emissions.

9.5-10 GHz

FG: S5.479 RADIOLOCATION S5.333 US110

Non-FG: S5.479 Radiolocation S5.333 US110

This meteorological radar band (see 9.3–9.5 GHz) is part of the 9.2–10.55 GHz federal radar spectrum.

International Space Station radar systems are in this band. Other planned platforms include Earth observation satellites and the Space Shuttle, for use by *synthetic aperture radar* (SAR), altimeters, and scatterometers.

The military services use this band to control unmanned aerial vehicles, for wideband data links for high-performance aircraft, capable of jam-resistant transmission over long distances, intrusion detection, submarine warfare systems, weapons location, battlefield air traffic control, mapping and imaging, terrain avoidance, and many other applications.

Non-military radar in this spectrum includes such exotic uses as insect tracking and the study of fire properties. Active microwave road condition sensors use this band to measure ice, snow, salinity, and traction.

10-10.45 GHz

FG: S5.479 US58 US108 G32 RADIOLOCATION

Non-FG: S5.479 US58 US108 NG42 Radiolocation. Amateur

FCC: Private Land Mobile (90). Amateur (97)

The 10–10.5 GHz spectrum is available to military radar and satellite weather radar.

The radar component of the Army ballistic missile defense system, *Theater High-Altitude Area Defense* (THAAD), uses this spectrum. This radar performs surveillance and tracking, and is the communications link with interceptor missiles directed at incoming enemy missiles.

Sophisticated radars use this spectrum to generate high resolution, real-time images of the ocean surface.

This is the Amateur Radio 3-cm band. Amateur communication range in this spectrum has exceeded 1,000 miles under tropospheric ducting conditions.

The FCC includes 10–10.5 GHz among the bands available for experimental use by students and schools.

The 8–12.5 GHz spectrum is traditionally designated the "X-band."

10.45–10.5 GHz

FG: US58 US108 G32 RADIOLOCATION

Non-FG: US58 US108 NG42 NG134 Radiolocation. Amateur. Amateur-Satellite.

FCC: Private Land Mobile (90). Amateur (97)

See 10–10.45 GHz for the uses of this band.

10.5–10.55 GHz

FG: US59 RADIOLOCATION

Non-FG: US59 RADIOLOCATION

FCC: Private Land Mobile (90)

This is one of the police speed radar bands. X-band units transmit on 10.525 GHz. Other police radar bands include 24.05–24.25 GHz and 33.4–36 GHz.

The FCC also allows non-police entities to obtain radar licenses. Baseball teams, tennis clubs, car and boat racing organizations, railroads, and other transportation firms may be authorized to use radar to measure the speed of objects or vehicles.

Field disturbance sensors for door openers and security systems also use these bands, sometimes causing false responses in police radars and radar detectors. High-security federal facilities are among the users of such sensors in this band.

10.55–10.6 GHz

FG: None

Non-FG: FIXED

FCC: Fixed Microwave (101)

This 10 GHz spectrum is mostly used for fixed microwave links, connecting cell sites with the central switch in a cellular system, for example.

The *Digital Electronic Message Service* (DEMS), a common carrier data service, and the private (but otherwise similar) *Digital Termination System* (DTS) stations were allocated 10.55–10.68 GHz.

Few DEMS or DTS stations were built in this spectrum. The FCC converted this spectrum for conventional fixed use. It permitted DEMS stations in another allocation, 18.82–18.92 GHz, and 19.16–19.26 GHz.

Later, in a controversial action, the FCC relocated DEMS from 18 GHz to 24 GHz, to avoid interference from DEMS to the ground stations of signals intelligence satellites (see 18.8–19.3 GHz and 24.25–24.45 GHz). The FCC has proposed to rename DEMS as the "24 GHz Service."

The *Telecommunications Industry Association* (TIA) petitioned (RM-9418) the FCC to update its rules for the 10.55–10.68 GHz spectrum in order to encourage fixed service use. The changes involve reducing the minimum antenna diameter and associated technical specifications.

10.6–10.68 GHz

FG: US265 US277 EARTH EXPLORATION-SATELLITE (passive). SPACE RESEARCH (passive)

Non-FG: US265 US277 EARTH EXPLORATION-SATELLITE (passive). FIXED. SPACE RESEARCH (passive)

FCC: Fixed Microwave (101)

This is part of the 10.55–10.68 GHz conventional fixed microwave service allocation.

It is part of the spectrum that was intended for the *Digital Electronic Message Service* (DEMS) and *Digital Termination Service* (DTS) that never extensively developed in this allocation (see 10.55–10.6 GHz).

Operation in this band in the vicinity of 20 federal government facilities across the country must be coordinated with the *National Telecommunications and Information Administration* (NTIA).

The *Telecommunications Industry Association* (TIA) petitioned the FCC (RM-9418) to update its rules for the 10.55–10.68 GHz spectrum in order to encourage fixed service use. The changes involve reducing the minimum antenna diameter and associated technical specifications.

The 10.6–10.7 GHz spectrum is used in satellite-based land and ocean surface remote sensing and imagery, including the downlink of synthetic aperture radar data from unmanned aerial vehicles. Passive sensing and radio astronomy use 10.68–10.7 GHz on a primary basis.

The 10.6–12.7 GHz spectrum is one of many restricted bands in which the FCC Part 15 rules permit unlicensed devices to emit only very low level emissions.

10.68–10.7 GHz

FG: US246 US355 EARTH EXPLORATION-SATELLITE (passive). RADIO ASTRONOMY US74. SPACE RESEARCH (passive)

Non-FG: US246 US355 EARTH EXPLORATION-SATELLITE (passive). RADIO ASTRONOMY US74. SPACE RESEARCH (passive)

No transmissions are permitted in this scientific band. Its uses include remote sensing of rainfall and radio astronomy studies of quasars, which are among the most energetic and distant objects that astronomers can detect.

The 10.6–12.7 GHz spectrum is one of many restricted bands in which the FCC Part 15 rules permit unlicensed devices to emit only very low level emissions.

10.7–11.7 GHz

FG: US211 US355

Non-FG: US355 FIXED NG41. FIXED-SATELLITE (space-to-Earth) S5.441 US211 NG104

FCC: International Fixed (23). Satellite Communications (25). Fixed Microwave (101)

This band contains terrestrial fixed links used by utilities, railroads, public safety agencies, telephone networks, and other governmental and industrial licensees. The *Local Television Transmission Service* (LTTS) uses the band for broadcast studio transmitter links.

The 10.95–11.2 GHz (downlink) segment is paired with the 13.75–14 GHz (uplink) segment and the 11.45–11.7 GHz (downlink) segment is paired with the 14–14.2 GHz (uplink) segment to form two of the Ku-band allocations available to the Fixed Satellite Service. The 10.7–11.7 GHz band also is one of several bands for use by *Non-Geostationary Satellite Services* (NGSO FSS, see 11.7–12.2 GHz).

Another use of this band is *telemetry, tracking and control* (TT&C) functions for geostationary satellites. "TT&C communications are used throughout the satellite's life, including the launch and deployment phase," the FCC has explained. "The TT&C function allows the earth station to control both the physical orbital position and internal functioning of the spacecraft."

The 10.6–12.7 GHz spectrum is one of many restricted bands in which the FCC Part 15 rules permit unlicensed devices to emit only very low level emissions.

11.7–12.2 GHz

This band has different frequency boundaries for Federal Government and Non-Federal Government allocations. The FG allocation is divided into the 11.7–12.1 GHz segment, in which S5.486 is the only entry, and the 12.1–12.2 GHz segment, in which no entry exists—meaning no federal use.

The Non-FG allocation is to the entire 11.7–12.2 GHz band as follows: S5.486 S5.488 FIXED-SATELLITE (space-to-Earth) NG143 NG145. Mobile except aeronautical mobile. The FCC use designators are Satellite Communications (25) and Fixed Microwave (101).

The Ku-band generally refers to spectrum in the 12–18 GHz range. The 11.7–12.2 GHz Ku-band *Fixed Satellite Service* (FSS) downlink is associated with uplink spectrum at 14.2–14.5 GHz. *Very Small Aperture Terminal* (VSAT) operations, which distribute data to and from many business locations, use these bands.

The Ku-band serves satellite networks for voice, data, facsimile, video transmission, and satellite control, and for broadcasting to consumers in what the FCC calls the *Direct-to-Home Fixed Satellite Service* (DTH-FSS).

The DTH-FSS is similar to, but should not be confused with, the *Direct Broadcast Satellite* (DBS) or *Broadcasting Satellite Service* (BSS) (see 12.2–12.7 GHz). DTH-FSS originated in the C-band, where many home subscribers still receive programming (see 3.7–4.2 GHz).

The 11.7–12.2 GHz band is another band available for *Broadcast Auxiliary Services* (BAS) studio transmitter links and TV remote pickup.

Massive New Systems

In Docket ET 98-206, the FCC allocated 11.7–12.2 GHz and 12.2–12.7 GHz to *Non-Geostationary Fixed Satellite Service* (NGSO FSS) user downlinks. User uplinks are in the 14–14.5 GHz spectrum, gateway earth station uplinks in 12.75–13.25 GHz and 13.75–14 GHz, and gateway downlinks are in 10.7–11.7 GHz. Gateway stations do not originate or terminate communications traffic, but interconnect the satellites with other networks such as the Internet and the public telephone network.

"The implementation of NGSO FSS systems will allow new advanced services to be provided to the public, as well as provide increased competition to existing satellite and terrestrial services," the FCC said.

"Indeed, the NGSO FSS, because of its ability to serve large portions of the Earth's surface, can bring advanced services to rural areas."

The NGSO FSS is characterized by large numbers of continuously orbiting satellites, providing voice and data service to specialized, tracking-capable user stations. The FCC received proposals for several NGSO FSS systems, following a November 1998 request for license applications.

The FCC based its invitation and NGSO FSS rules on favorable World Radio Conference technical provisions for NGSO FSS operation and on a 1997 application from SkyBridge (formerly Sativod), a subsidiary of Alcatel of France. SkyBridge proposed a $3.5 billion international system of 80 satellites, to provide "bandwidth on demand" Internet and telephone services, and "a variety of substitutes for terrestrial infrastructure links."

SkyBridge bills itself as a "Dumb Network." Unlike the NGSO Iridium system, SkyBridge satellites will perform no on-board processing or switching, therefore enjoying some reduced costs.

The system will require some 200 gateway stations around the world, with two in the U.S. It is projected to accommodate more than 20 million users, with a mix of 70 percent commercial and 30 percent residential. The company estimates that commercial use would cost about $300 for an average 10 GB/month, and residential use $30 for an average 1 GB/month. SkyBridge anticipates a 2003 service start. The company proposed a SkyBridge II system with additional services (see 18.3–18.58 GHz).

Other proposals filed in response to the invitation include Boeing Corp., for a 20 satellite, $4.5 billion system that would include both SkyBridge-type services and *Mobile Satellite Services* (MSS) on an "ancillary" basis, Teledesic Corp., for a 30-satellite, $3.5 billion "KuBS" (Ku-Band Supplement) system, also with MSS capability, two systems from Hughes Communications: *HughesLINK* (H-LINK), with 22 satellites at $2.6 billion, and HughesNET with 70 satellites at $6.95 billion, and the Denali system of 13 satellites to provide both FSS and MSS.

Teledesic said that its proposed KuBS "will function primarily as a high-bandwidth supplement to the Teledesic Network, relieving Ka-band congestion in capacity-constrained areas." Although its basic service was proposed as 10 megabit connections to fixed terminals, KuBS would provide as much as 100 megabit uplinks and 1 GB downlinks.

The system "will also permit Teledesic to offer Mobile Satellite Service on an ancillary, non-interference basis particularly to aeronautical and maritime users operating in the polar regions who currently have no real-time access to broadband communications," the company said. KuBS would be funded by cash flow from the Teledesic Network, intended to start commercial operation in 2003 (see 18.8–19.3 GHz).

According to Hughes, its systems "will enable users to readily update and retrieve applications, create secure IP-multicast sessions and conduct high-speed file transfers; support media streaming; participate in distance learning; and perform database updates and replication to and from personal computers."

An unusual application filed in response to the FCC proposal is Virgo, a $2.64 billion system of 15 satellites in "virtual geostationary" orbits. This type of orbit enables satellites to "appear to hang in the sky, as their rotational velocity more nearly matches that of the Earth than is the case with nongeostationary satellites in circular orbits," the company told the FCC.

The design permits user terminals to be pointed well away from the equator, reducing interference possibilities by ensuring that user antennas for the geostationary fixed satellite services and direct satellite broadcasting services are looking at different parts of the sky. "It is as if the geostationary arc, for all of its congestion, did not exist, and the virtual geostationary orbit fixed satellite service was to be established in fallow spectrum," Virtual Geosatellite said. Virgo is backed by Ellipso, a Big LEO licensee (see 1610–1610.6 MHz).

At this writing, the FCC had not yet proposed licensing and service rules for NGSO FSS, which must be completed in order for the Commission to grant any of the applications it has received.

The 10.6–12.7 GHz spectrum is one of many restricted bands in which the FCC's Part 15 rules permit unlicensed devices to emit only very low level emissions.

12.2–12.7 GHz

FG: S5.490

Non-FG: S5.487A S5.488 S5.490 FIXED. BROADCASTING-SATELLITE

FCC: International Fixed (23). Satellite Communications (25). Direct Broadcast Satellite (100). Fixed Microwave (101)

The FCC reallocated this band to downlinks for the *Direct Broadcast Satellite* (DBS) service from terrestrial fixed microwave services in 1982.

Internationally, DBS is known as the *Broadcasting Satellite Service* (BSS). According to the FCC, "in the United States the term DBS is used interchangeably with BSS."

DBS ventures of the 1980s lost huge sums of money without providing service, notably *Satellite Television Corporation* (STC), a Comsat unit. STC lost $200 million on the project and sold off the satellites it had built, which were later destroyed in launch accidents.

DBS should not be confused with the *Direct-to-Home Fixed Satellite* (DTH-FSS, see 11.7–12.2 GHz), sometimes called "medium-power" service.

DirecTV

The U.S. DBS industry reignited with the launch of DBS satellites by Hughes, for its DirecTV service. DirecTV later acquired the licenses of its competitors *U.S. Satellite Broadcasting* (USSB), a subsidiary of Hubbard Broadcasting, and Tempo Satellite. The acquisitions enabled DirecTV to operate five DBS satellites from three orbital slots. In approving the consolidation, the FCC reasoned that competition between DBS and cable TV is more important to consumers than competition between DBS licensees.

The satellites receive uplink signals at 17.3–17.8 GHz. The wide orbital spacing, high power, and SHF operation of DBS enable consumers to use small, unobtrusive antennas to receive the signals. DBS programming includes the same national networks available on cable TV, plus pay-per-view movies, and CD-quality audio channels.

DirecTV intends to launch a sixth satellite, DIRECTV 5, to be co-located with the former Tempo satellite USABSS-7, and to retain the Tempo satellite as a spare.

DirecTV successfully petitioned the FCC to allocate additional spectrum to BSS and the Fixed Satellite Service for DBS expansion. The company applied for licenses for six "DX" satellites (see 17.3–17.7 GHz).

DirecTV added Internet service with the DirecPC brand. In an interactive version of this service, both sides of the data link will use the satellite, unlike the initial version that required a telephone landline for the return connection.

One of the Hughes DBS satellites, DBS-2, carries an unusual payload: SpaceArc, Archives of Humanity. SpaceArc is a recording of 40,000 messages from people in 45 countries, a documentary on events of 1993, and music of artists including Jimi Hendrix and Peter Gabriel.

EchoStar

After USSB, the third domestic DBS competitor to begin operations was EchoStar's DISH Network. EchoStar won a license for a second satellite at auction for about $52 million. MCI won a nationwide DBS license for an auction bid of $682.5 million, but EchoStar later acquired the unused license from a MCI Worldcom-News Corp. partnership in a $1.1 billion deal.

EchoStar partnered with Microsoft and VSAT provider Gilat Satellite Networks to launch a full two-way Internet service, StarBand, in 2000. At this writing, StarBand uses GE and Telstar satellites in addition to the EchoStar satellites.

Under the allotment scheme of the ITU Region 2 BSS Plan covering the Americas, the FCC granted frequency and orbital assignments to other DBS companies including Directsat and DBS Corp. (both acquired by EchoStar), R/L DBS Co., and Dominion Video Satellite.

R/L DBS (then Continental Satellite) received its license in 1989. After years of disputes over the company's ownership—and a rare admission by the FCC of its own culpability in delaying R/L its frequency assignment—the FCC gave R/L until December 29, 2003 to begin providing service.

R/L DBS is owned by Rainbow DBS Holdings and Loral. The company plans to offer regional programming to TV markets in the eastern U.S. "R/L DBS's planned service is perhaps the last opportunity in the near term for entry by a competitive provider within the DBS service itself," the FCC said.

Controversial Service Created

In late 2000, the FCC found that 12.2–12.7 GHz could be made available on a secondary-to-DBS basis to a new *Multichannel Video Distribution and Data Service* (MVDDS) under the Part 101 Fixed Service rules, and sought further comment on its design and regulation (Docket ET 98-206). At this writing, a formal allocation to MVDDS was pending.

MVDDS will provide consumers with TV programming and Internet connectivity from terrestrial transmitters, under exacting technical arrangements that permit it to share the spectrum with satellites. The main MVDDS proponent is Northpoint Technology.

The company's technology "creates—in essence—a land-based 'satellite' slot. Instead of broadcasting from space with the 9-degree angle of difference used by the current DBS satellites, Northpoint broadcasts are made from a terrestrial base with a very large angle of difference," it said. "This angular separation, combined with directional reception antenna characteristics, allows both satellite and terrestrial signals to co-exist without interfering with one another."

Northpoint partner Broadwave USA filed waiver requests and license applications in 1999 to provide the service through a corporate owned operation and local affiliates. Other potential competitors followed with similar applications, even though the FCC had not yet proposed to create MVDDS.

The creation of MVDDS infuriated the *Satellite Broadcasting and Communications Association* (SBCA), which had vigorously lobbied Congress to delay or kill the new service. "SBCA and several of its member companies have been working hard to prevent this from happening," the association said, arguing that it is "incomprehensible for the FCC to allow sharing of the DBS frequency band by terrestrial wireless services," and "inconceivable that the Commission would risk the success and future growth of DBS by allowing these services to share the spectrum."

Yet "tests conducted in the 12.2–12.7 GHz band by Northpoint under an experimental authorization confirm that the MVDDS could operate without excessively impacting DBS subscribers," the FCC said. "Northpoint has also filed extensive technical studies to demonstrate that any impact on DBS operations would be minimal and could be mitigated using existing engineering techniques."

Prevention of MVDDS interference with DBS involves avoiding large concentrations of DBS consumer antennas in close proximity to MVDDS transmitters. Other measures include shielding, repointing or replacing the consumer DBS antenna (at the MVDDS licensee's expense) and reducing MVDDS transmission power during rainy weather when DBS reception will be weaker. The FCC proposed to require MVDDS receive antennas to be "technically similar to home DBS antennas" and they would have to "generally point southward."

The FCC also was persuaded that MVDDS could function without interference to the *Non-Geostationary Fixed Satellite* (NGSO FSS, see 11.7–12.2 GHz).

Service and License Structure

As is its practice with other auctioned services, the FCC proposed to license MVDDS to geographic areas—rather than to specific transmit sites—perhaps with a scheme that tracks the *Designated Market Areas* (DMAs) used by the Nielsen TV audience measurements. One license would be available in each of the 211 DMAs. Each license would cover 500 MHz of spectrum, enough to provide 100 channels with current technology, although the FCC sought comment on smaller spectrum blocks.

End-users would not be able to transmit in this band back to the network: "For two-way services, licensees could find spectrum in other bands or use telephone lines or other means for the return path," the FCC said.

Other Uses of DBS Spectrum

When DBS was in its early stages, there was no assurance that the business would succeed. To reduce this risk, the FCC permitted DBS licensees to provide other, "non-conforming" services as a source of early revenues.

But "advances in technology, ability [of DBS] to compete with cable services, and new service offerings may warrant revisiting this policy," the FCC said (Docket IB 98-21).

The Commission asked for comment on the nature of non-conforming uses, on whether it should eliminate, relax or maintain restrictions on those uses, and on the interference criteria that should apply to such non-DBS services provided over DBS spectrum.

The 10.6–12.7 GHz spectrum is one of many restricted bands in which the FCC Part 15 rules permit unlicensed devices to emit only very low level emissions.

12.7–12.75 GHz

FG: None

Non-FG: NG53 FIXED NG118. FIXED-SATELLITE (Earth-to-space). MOBILE

FCC: Satellite Communications (25). Auxiliary Broadcasting (74). Cable TV Relay (78). Fixed Microwave (101)

The main terrestrial users of the 12.7–12.75 GHz and 12.75–13.25 GHz bands are cable TV systems, which use them in the *Cable Television Relay Service* (CARS) to convey video signals between points in their networks. This includes the studio-to-headend link to connect an origination site with the site that transmits into the cable.

The FCC officially defines the CARS station as a "fixed or mobile station used for the transmission of television and related audio signals, signals of standard and FM broadcast stations, signals of instructional television fixed stations [ITFS], and cablecasting from the point of reception to a terminal point from which the signals are distributed to the public." CARS has access to 12.7–13.2 GHz.

At this writing, the FCC only assigns CARS licenses to cable TV systems, cable networks, and licensees in the Multipoint Distribution Service and ITFS (see 2500–2655 MHz).

TV stations have access to the 12.7–13.2 GHz and 13.2–13.25 GHz segments for *Broadcast Auxiliary Stations* (BAS), including *studio transmitter* (STL) links and *electronic news gathering* (ENG) operations where the signal is sent from an event to the TV truck and then on to the broadcast station.

"At some future time, it may no longer be in the public interest to allocate this large [12.7–13.25 GHz] band to the fixed service if it is used heavily in only a few cities," NTIA analysts opined in 1995. They believed that as much as half of this spectrum could eventually be reallocated to other purposes.

But NTIA said in 2000 that, "Earlier predictions of CARS license decreases in the 13 GHz band were in error partly because cable operators have not yet returned unused CARS licenses and partly because of the need to distribute additional video programming to fill wider-bandwidth cable systems." Nevertheless, NTIA observed that, "In many current cable systems, including most urban systems, optical fiber networks have mostly replaced the use of CARS microwave links."

Mess up DTV

In Docket CS 99-250, the FCC is considering eligibility for CARS licenses, which would include access to this 12 GHz spectrum, for *Private Cable Operators* (PCOs) who provide video service to multi-unit residences, as well as other so-called *Multichannel Video Programming Distributors* (MVPDs—the FCC's catchall term for non-broadcast TV providers).

Under FCC Part 101 fixed microwave rules, PCOs already use 18.142–18.58 GHz (see 17.7–17.8 GHz) and 21.2–23.6 GHz (see 21.2–21.4 GHz) to provide video programming, but 12 GHz affords longer link distances than those higher bands.

MVPDs would have to bid at auction for access to 12 GHz spectrum, if the FCC received mutually exclusive applications for particular licenses and if it did not adopt methods to avoid mutual exclusivity.

The Commission also is exploring the conditions for possible CARS use of 12 GHz spectrum to deliver information other than video programming, such as data and voice.

The proposals drew fire from TV broadcasters. They dubbed PCO eligibility a "spectrum grab" that would worsen congestion for BAS and CARS users and undermine the transition to digital TV technology by making it harder to telecast sports events (DTV, see 470–512 MHz).

"In short," said the Society of Broadcast Engineers, "mess up sports coverage by making RF (radiofrequency) channels unavailable for RF camera shots, and face the likelihood of consumers messing up all of the Commission's plans by refusing to spend their dollars on new television receivers."

12.75–13.25 GHz

FG: US251

Non-FG: US251 NG53 FIXED NG118. FIXED-SATELLITE (Earth-to-space) S5.441 NG104 MOBILE

FCC: Satellite Communications (25). Auxiliary Broadcasting (74). Cable TV Relay (78). Fixed Microwave (101)

See 12.7–12.75 GHz about this band. This also is one of several bands for use by *Non-Geostationary Orbit Fixed Satellite Services* (NGSO FSS, see 11.7–12.2 GHz).

This is a Space Research Service band, used to communicate with planetary spacecraft and limited to the NASA Deep Space Network facility at Goldstone, California.

13.25–13.4 GHz

FG: AERONAUTICAL RADIONAVIGATION S5.497. Space Research (Earth-to-space)

Non-FG: AERONAUTICAL RADIONAVIGATION S5.497. Space Research (Earth-to-space)

FCC: Aviation (87)

This is part of the 13.25–14.2 GHz federal radar band used by military and scientific services.

Airborne Doppler radar navigation aids use 13.25–13.4 GHz and 8.75–8.85 GHz (see 8.5–9 GHz). Airborne radar in this band also is used in measurement of ocean surface characteristics.

This band also is allocated, on a secondary basis, to satellite uplinks in the Space Research Service. NASA uses this spectrum for satellite-based altimeters, scatterometers, and precipitation radars.

This is one of many restricted bands in which the FCC Part 15 rules permit unlicensed devices to emit only very low level emissions.

13.4–13.75 GHz

FG: RADIOLOCATION S5.333 US110 G59. Space Research. Standard Frequency and Time Signal-Satellite (Earth-to-space)

Non-FG: Radiolocation S5.333 US110. Space Research. Standard Frequency and Time Signal-Satellite (Earth-to-space)

FCC: Private Land Mobile (90)

This is part of the 13.25–14.2 GHz federal radar band, used by military and scientific services.

Shipboard radars operate in 13.4–13.75 GHz and 13.75–14 GHz worldwide. In the U.S., the Navy and Coast Guard are the main users.

The Navy's Geosat program operates satellite radar altimeters in this band. The current *Geosat Follow-On* (GFO) mission transmits oceanography data to Navy ships, government weather and space scientists, and university researchers.

The satellite altimeter, an active sensor, transmits radio energy and measures the reflected energy to obtain wave height, a measure of surface wind speed.

Satellite altimetry is the only known way to precisely measure sea surface topography. The technique reveals changes in sea level prompted by such factors as warming of the global climate.

This band is available for police radars (hence the FCC Part 90 designation) but reportedly is little used for this purpose. Future low-detectability radars are expected to use the band.

No U.S. use of this band, or its companion 20.2–21.2 GHz, for satellite frequency and time communications exists.

13.75-14 GHz

FG: S5.503A US356 US357 RADIOLOCATION US110 G59. Standard Frequency and Time Signal-Satellite (Earth-to-space). Space Research US337

Non-FG: S5.503A US356 US357 FIXED-SATELLITE (Earth-to-space) US337. Radiolocation US110. Standard Frequency and Time Signal-Satellite (Earth-to-space). Space Research

FCC: Satellite Communications (25). Private Land Mobile (90)

The 13.75–14 GHz Ku-band *Fixed Satellite Service* (FSS) uplink band is paired with 10.95–11.2 GHz (downlink).

"At this time, the FSS uplink use is relatively light due to the short time that the FSS has been allocated in the band and the prevalence of the Government operations," the FCC said in 1998. "We have, however, licensed satellites to provide international operations in the band."

This is one of several bands for use by *Non-Geostationary Orbit Fixed Satellite Services* (NGSO FSS, see 11.7–12.2 GHz).

This is part of the 13.25–14.2 GHz federal radar band, used by military and scientific services. High-power shipboard radars, often referred to simply as "mobile radars," operate in 13.4–13.75 GHz and 13.75–14 GHz worldwide. In the U.S., the Navy and Coast Guard are the main users.

Scientific Uses

NASA's *Tracking and Data Relay Satellite System* (TDRSS) uses the entire 13.75–14 GHz secondary Space Research Service allocation to downlink to its earth stations at White Sands, New Mexico, and Guam (see also 2025–2110 MHz).

The geostationary TDRSS relays communications between earth stations and satellites or spacecraft in *low-Earth orbit* (LEO) such as the Space Shuttle, when those objects are not capable of direct communication with the ground.

This link to LEO is between 13.75 GHz and 13.8 GHz, depending on desired bandwidth; the LEO object return link to TDRSS is at 14.9–15.1 GHz.

Other NASA missions use the band for active sensors. These transmit signals to a target and receive reflected signals, which are compared with transmitted signals.

NASA's active sensors used in climate and topography studies include scatterometers, which measure wind velocity, spaceborne altimeters, which measure the height of the satellite over the ocean surface, and space- and airborne precipitation radars. The Space Shuttle uses a rendezvous radar for satellite retrieval in this band.

The location of any Fixed Satellite Service facilities has to be carefully coordinated with the government in order to avoid interference to these various NASA operations. A TDRSS move to 25.25–27.5 GHz is planned.

14–14.2 GHz

FG: RADIONAVIGATION US292. Space Research

Non-FG: FIXED-SATELLITE (Earth-to-space). RADIONAVIGATION US292. Land Mobile-Satellite (Earth-to-space). Space Research

FCC: Satellite Communications (25). Aviation (87). Maritime (80)

The 14–14.2 GHz (uplink) Ku-band *Fixed Satellite Service* (FSS) band is paired with 11.45–11.7 GHz (downlink).

The 14 GHz spectrum is used by *geostationary* (GSO) FSS *Very Small Aperture Terminal* (VSAT) systems for video and data communications, widely deployed at business locations from the largest corporate headquarters to the smallest convenience store. Qualcomm's OmniTracs data communications service for trucks also uses this spectrum.

This is one of several bands for use by *Non-Geostationary Orbit Fixed Satellite Services* (NGSO FSS, see 11.7–12.2 GHz).

This is part of the 13.25–14.2 GHz federal radar band, used by military and scientific services. It is available to maritime radars, and to airborne radionavigation on a secondary basis. The FCC said, however, that there are "no significant" radionavigation uses in 14–14.2 GHz except for handheld navigation devices used at some waterways.

Satellite scatterometers use this spectrum to observe ocean winds by reflecting microwave pulses across broad regions of Earth's surface.

The International Space Station will use 14–14.3 GHz and 14.5–14.89 GHz for communications with spacecraft in its near vicinity.

14.2–14.4 GHz

FG: None

Non-FG: FIXED-SATELLITE (Earth-to-space). Land Mobile-Satellite (Earth-to-space). Mobile except aeronautical mobile

FCC: Satellite Communications (25). Fixed Microwave (101)

The 14.2–14.5 GHz Ku-band *Fixed Satellite Service* (FSS) uplink spectrum is paired with 11.7–12.2 GHz (downlink).

The 14.2–14.4 GHz spectrum is used for some television pickup operations. See 14–14.2 GHz for more information on 14 GHz uses.

14.4–14.47 GHz

FG: Fixed. Mobile

Non-FG: FIXED-SATELLITE (Earth-to-space). Land Mobile-Satellite (Earth-to-space)

FCC: Satellite Communications (25)

The 14.2–14.5 GHz Ku-band *Fixed Satellite Service* (FSS) uplink spectrum is paired with 11.7–12.2 GHz (downlink). This is one of several bands for use by *Non-Geostationary Orbit Fixed Satellite Services* (NGSO FSS, see 11.7–12.2 GHz).

The *Federal Aviation Administration* (FAA) uses 14.4–15.35 GHz for television microwave links that convey radar and video imagery between air facilities; however, the FCC has stated that government use of 14.4–14.5 GHz "appears light."

This band is used to downlink synthetic aperture radar data from unmanned aerial vehicles.

Radio astronomers use this band and 4.83 GHz to detect formaldehyde, a component of the molecular structure of galaxies. See 14–14.2 GHz for more information on 14 GHz uses.

14.47–14.5 GHz

FG: S5.149 US203 Fixed. Mobile

Non-FG: S5.149 US203 FIXED-SATELLITE (Earth-to-space). Land Mobile-Satellite (Earth-to-space)

FCC: Satellite Communications (25)

The 14.2–14.5 GHz Ku-band *Fixed Satellite Service* (FSS) uplink spectrum is paired with 11.7–12.2 GHz (downlink). This is one of several bands for use by *Non-Geostationary Orbit Fixed Satellite Services* (NGSO FSS, see 11.7–12.2 GHz). See 14–14.2 GHz for more information on 14 GHz uses.

This is an important band for radio astronomy spectral line observations.

The 14.47–14.5 GHz band is one of many restricted bands in which the FCC Part 15 rules permit unlicensed devices to emit only very low level emissions.

14.5–14.7145 GHz

FG: FIXED. Mobile. Space Research

Non-FG: None

Military uses of this band include electronic warfare, radar microwave links, cross-section measurement, and threat simulation.

NASA's *Tracking and Data Relay Satellite System* (TDRSS) has uplinks at 14.5–15.35 GHz (see 2025–2110 MHz and 13.75–14 GHz).

This is part of the FAA television microwave link band that relays video between aviation facilities.

14.7145–15.1365 GHz

FG: US310 MOBILE. Fixed. Space Research

Non-FG: US310

NASA's *Tracking and Data Relay Satellite System* (TDRSS) has uplinks at 14.5–15.35 GHz (see 2025–2110 MHz and 13.75–14 GHz).

15.1365–15.35 GHz

FG: S5.339 US211 FIXED. Mobile. Space Research

Non-FG: S5.339 US211

Air traffic control links, electronic warfare, government data and control links, radio astronomy, and satellite-based rain and water vapor sensors use this band.

NASA's *Tracking and Data Relay Satellite System* (TDRSS) has uplinks at 14.5–15.35 GHz (see 2025–2110 MHz and 13.75–14 GHz).

15.35–15.4 GHz

FG: US246 EARTH EXPLORATION-SATELLITE (passive). RADIO ASTRONOMY US74. SPACE RESEARCH (passive)

Non-FG: US246 EARTH EXPLORATION-SATELLITE (passive). RADIO ASTRONOMY US74. SPACE RESEARCH (passive)

No transmissions are permitted in this receive-only scientific band. It is used in radio astronomy observations of continuum radiation and quasars.

The 15.35–16.2 GHz spectrum is one of many restricted bands in which the FCC Part 15 rules permit unlicensed devices to emit only very low level emissions.

15.4-15.7 GHz

FG: 733 797 US211 AERONAUTICAL RADIONAVIGATION US260

Non-FG: 733 797 US211 AERONAUTICAL RADIONAVIGATION US260

FCC: Aviation (87)

Aeronautical radionavigation systems operate in 15.4–15.7 GHz worldwide.

These include aircraft landing systems, aircraft multipurpose radars, radar sensing and measurement systems used in low level operations and helicopter flight, and *surface based radars* (SBR) that monitor aircraft movements in landing areas.

The 1995 and 1997 World Radio Conferences (WRC-95 and WRC-97) allocated 15.4–15.7 GHz to the *Fixed Satellite Service* (FSS) for *Non-Geostationary Satellite Service* (NGSO MSS, see 1610–1610.6 MHz) feeder downlinks on a primary basis worldwide, constrained by protection of aeronautical radionavigation and radio astronomy.

In Docket ET 98-142, the FCC proposed to divide 15.4–15.7 GHz into three bands: 15.4–15.43 GHz, 15.43–15.63 GHz, and 15.63–15.7 GHz, with the first and last of these bands designated for aviation use only and 15.43–15.63 GHz to be used by the FSS for NGSO MSS uplinks and downlinks.

The 15.35–16.2 GHz spectrum is one of many restricted bands in which the FCC Part 15 rules permit unlicensed devices to emit only very low level emissions.

15.7-16.6 GHz

FG: RADIOLOCATION US110 G59

Non-FG: Radiolocation US110

FCC: Private Land Mobile (90)

This is the most intensively used of the bands from 15.7 to 17.7 GHz allocated to radiolocation.

Military uses include ground and airborne radars, telemetering and telecommand of aircraft, airborne weapons control, mortar location, and combat surveillance.

This is an important part of the spectrum for *airborne synthetic aperture radar* (AIRSAR), which generates higher resolution images than generated from conventional radar. Airborne SAR processes radar signals obtained along a flight path to create, in effect, a large antenna.

A SAR system provides its own "illumination"—the transmitted radio energy—and therefore can "see" through conditions such as night, clouds, and dust that foil conventional imaging technology.

SAR has many applications in terrain mapping, geology, and oceanography, treaty verification and weapons nonproliferation, reconnaissance, surveillance, and targeting. SAR systems in various bands are flown onboard the Space Shuttle and satellites as well as on aircraft.

Another airborne radar in this band, part of the Advanced Air Defense System, detects missiles and forwards data to ground personnel over a 420 MHz link.

Airport Radar

The Federal Aviation Administration uses the 15.7–16.2 GHz segment for *Airport Surface Detection Equipment* (ASDE).

These are radars that monitor aircraft, vehicles, and obstructions at ground facilities, and at up to 200 feet in altitude.

ASDE helps ensure that aircraft are on their proper runways or taxiways, particularly when weather restricts visibility. Latest versions of ASDE permit controllers to view maps of the airport that can be zoomed and rotated.

The 15.35–16.2 GHz spectrum is one of many restricted bands in which the FCC Part 15 rules permit unlicensed devices to emit only very low level emissions.

16.6-17.1 GHz

FG: RADIOLOCATION US110 G59. Space Research (deep space) (Earth-to-space)

Non-FG: Radiolocation US110

FCC: Private Land Mobile (90)

See 15.7–16.6 GHz for the uses of this band.

Space frequency authorities advocate use of this band for Space Research Service downlinks for near-Earth communications. They recommend that the deep space limitation be removed at a future World Radio Conference.

17.1-17.2 GHz

FG: RADIOLOCATION. US110 G59

Non-FG: Radiolocation. US110

FCC: Private Land Mobile (90)

See 15.7–16.6 GHz for the uses of this band.

17.2-17.3 GHz

FG: RADIOLOCATION US110 G59. Earth Exploration-Satellite (Active). Space Research (Active)

Non-FG: Radiolocation US110. Earth Exploration-Satellite (Active). Space Research (Active)

FCC: Private Land Mobile (90)

See 15.7–16.6 GHz for the uses of this band.

Besides the government radars in this band, active sensors onboard remote sensing satellites use the band to measure vegetation and snow, a use of the *Earth Exploration Satellite Service* (EESS).

Active sensors transmit signals and then measure "backscatter," the signals reflected from the sensed object. Backscatter from leaves, for example, can reveal internal moisture and physical dimensions. Such information can help determine the health of crops and even the species of trees.

All active sensor frequencies are in radiolocation bands.

17.3–17.7 GHz

FG: Radiolocation US259 G59

Non-FG: US259 FIXED-SATELLITE (Earth-to-space) US271 BROADCASTING-SATELLITE NG163

FCC: Satellite Communications (25). Direct Broadcast Satellite (100)

Fixed Satellite Service (FSS) feeder uplinks devoted to *Direct Broadcast Satellites* (DBS, also called *Broadcasting Satellite Service* or BSS) are authorized in 17.3–17.8 GHz (see 12.2–12.7 GHz). Feeder uplinks supply programming and control commands to spacecraft from special earth stations. Customers do not use these links.

DBS provider DirecTV Enterprises petitioned the FCC to make a downlink (direct to customers) allocation in this band to BSS and a feeder link allocation at 24 GHz to the FSS for use by future DBS systems.

The FCC granted this request in Docket IB 98-172. "This increased amount of spectrum should allow BSS operators to offer an increased variety of programming and services which would enhance competition in the multichannel video programming market," it said.

The FCC allocated 17.3–17.7 GHz for BSS downlinks, and two segments at 24 GHz for BSS feeder uplinks: 24.75–25.05 GHz on a primary basis— for use only with 17 GHz BSS—and 25.05–25.25 GHz for co-primary sharing between FSS and the *24 GHz Service* (TFGS, originally the *Digital Electronic Message Service* or DEMS).

The allocations will become effective April 1, 2007, as provided by ITU regulations (see Footnote NG163).

The use of the same spectrum for uplink and downlink, a practice called "reverse band working" or "reverse band operation," is not supposed to greatly interfere with DBS. Only a small number of DBS uplink stations could be affected They are supposed to be engineered to accommodate any such interference.

The FCC reasoned that FSS and TFGS could share spectrum essentially because FSS feeder uplink stations are few and far between; and because any FSS interference would only be received at TFGS hub stations—not at the many TFGS user receivers (that are in a different band, 24.25–24.45 GHz).

DX Satellites

DirecTV also applied for a license to use this spectrum for a $1.4 billion system of six "DX" DBS satellites with advanced capabilities.

The company said that, "the trend toward higher technical [broadcast] quality, further fueled by the transition of terrestrial broadcast television to digital technology, demands more and more satellite transmission capacity. ...

"An entirely new programming category of Internet-like multimedia has become an important part of broadcasting satellite service delivery plans in the U.S. that will strain the capacity of today's bandwidth-limited broadcasting satellite service systems. ...Deployment of all six expansion satellites will significantly increase DirecTV's capacity for nationwide services," it said. "This dramatic increase is necessary to meet increased multichannel competition and to provide the new services expected by the American public early in the 21st century." But at this writing, the FCC has not established service rules or opened a filing window for applications for the "new" DBS.

High-power military radars are in this band, which share with uplinks to *geostationary* (GSO) satellites by limiting radar emissions in the direction of the GSO arc. The government plans to remove these radars, but removal will take time—a key reason why the FCC is not releasing this spectrum to BSS before 2007.

17.7–17.8 GHz

FG: None

Non-FG: NG144 FIXED. FIXED-SATELLITE (Earth-to-space) US271

FCC: Satellite Communications (25). Auxiliary Broadcasting (74).
Cable TV Relay (78). Fixed Microwave (101)

This is the first band in a portion of the Ka-band that extends from 17.7 GHz to 20.2 GHz, available to fixed stations (in 17.7–18.58 GHz and 19.3–19.7 GHz), and satellites of various types (in 18.3–20.2 GHz), segregated by a bandplan established in International Bureau Docket IB 98-172.

This docket allocated 17.7–17.8 GHz and 17.8–18.3 GHz to terrestrial fixed use on a primary basis. Previously, these two bands were available for shared co-primary use between *Geostationary Fixed Satellite Service* (GSO FSS) and terrestrial fixed stations.

But in IB 98-172, the FCC reallocated these bands to terrestrial fixed use only in order, it said, to avoid any future interference from satellite use. (Nevertheless, the FCC did not remove the FSS from the Table entry shown previously for Non-FG allocations.)

Typical terrestrial uses in this part of the spectrum include *Broadcast Auxillary Service* (BAS), *Cable TV Relay Service* (CARS), public safety, public utility services, and other private fixed operations, such as interconnection of base stations in mobile communications systems, video security links, links associated with the *Multipoint Distribution Service* (MDS), and distribution of *Private Cable Operator* (PCO) TV service to such customers as apartment buildings and hotels.

PCO use of this spectrum differs from public cable TV use in the CARS service (see 12.7–12.75 GHz). PCO facilities serve only individual buildings and do not cover customers throughout a geographic area. The wireless signal from the PCO center is received at the building and distributed within it via coaxial cable.

"As a result of serving only individual buildings, as well as being forbidden to cross public right-of-ways," according to NTIA, "private cable has not developed any associated wide-area fiber optic networks. Therefore, at a time when CATV is abandoning its CARS systems in favor of large multipurpose optical fiber networks, private cable at 18 GHz is still rapidly expanding its microwave distribution network."

"Terrestrial fixed service use of this band is expected to increase as a result of migration of users from the congested lower terrestrial fixed service bands to this band, and from the need for new systems to support the introduction of new services such as digital television broadcasting, *Personal Communications Services* (PCS), and other digital communications systems," the FCC has stated.

The 17.7–21.4 GHz spectrum is one of many restricted bands in which the FCC Part 15 rules permit unlicensed devices to emit only very low level emissions.

17.8-18.3 GHz

FG: S5.519 US334 FIXED-SATELLITE (space-to-Earth) G117

Non-FG: S5.519 US334 NG144 FIXED

FCC: Auxiliary Broadcasting (74). Cable TV Relay (78). Fixed Microwave (101)

See 17.7–17.8 GHz about this band. In Docket IB 98-172, the FCC reallocated this band to the terrestrial fixed service on a primary basis.

Footnote US334 permits certain government satellites to operate on a primary basis outside a specified orbital location.

To avoid interference to these classified eavesdropping satellites (see 18.8–19.3 GHz), the FCC established exclusion and coordination zones around the satellites' earth stations, which are located at Morrison and Denver, Colo. and near Washington, D.C.

Broadcast Auxiliary, Cable TV Relay, and Fixed Microwave stations must stay away from these earth stations or coordinate their operations with the federal government, depending on their proximity to the earth stations.

The 17.7–21.4 GHz spectrum is one of many restricted bands in which the FCC Part 15 rules permit unlicensed devices to emit only very low level emissions.

18.3-18.58 GHz

FG: US334 FIXED-SATELLITE (space-to-Earth) G117

Non-FG: US334 NG144 FIXED. FIXED-SATELLITE (space-to-Earth). NG164

FCC: Satellite Communications (25). Auxiliary Broadcast. (74). Cable TV Relay (78). Fixed Microwave (101)

This part of the Ka-band is allocated on a co-primary shared basis to the *Geostationary Orbit Fixed Satellite Service* (GSO FSS) and to the terrestrial fixed service. The fixed uses in this spectrum are described in 17.7–17.8 GHz.

Worldwide Impact

In 1997, the FCC authorized several broadband satellite systems to use the Ka-band FSS allocations, and accepted applications for additional systems. They will provide high-speed Internet services, at startup costs in the billions of dollars. These systems could have worldwide social and economic impact if they are ever built.

"Within the next five to ten years, we anticipate that these services will be provided to millions of United States businesses and consumers using small antenna Ka-band satellite earth stations," the FCC said. It later expressed concern that consumers could make errors in antenna pointing, bringing harmful interference to satellites.

Representative of these licensees is the $2.3 billion four-satellite Millennium, a project of Motorola subsidiary Comm Inc.

"Motorola intends to transplant the concept of 'any person, anytime, anywhere' communications—the goal of the Iridium system in the area of the Mobile Satellite Service—to broadband communications in the Fixed Satellite Service," the company said.

Motorola trumpeted school, health, consumer, and home-based applications to the FCC: "A wide array of residential services will include work-at-home interconnection between home and office computers; educational services linking students and teachers around the hemisphere; medical information access bringing together patients and doctors; home shopping and service information products; customized news, sports, financial and other information products; on-line and Internet access; point-to-point or multipoint communications with other users including collaborative opportunities; and pay-per-view video, games, magazines, newspapers and other special events. Business services will include: financial transaction processing, collaborative communications, LAN-to-LAN communications, training and education, and health care services."

Millennium, Motorola continued, will "literally shrink the hemisphere by eliminating the factor of physical distance that currently inhibits interaction between people. ... Through small and inexpensive satellite terminals, consumers will have access to data rates ranging from 64 kb/s to over 92 Mb/s."

The FCC assigned Millennium uplinks in 28.35–28.6 GHz and 29.5–30 GHz, with downlink in 19.7–20.2 GHz. The Commission said that it later may be able to grant additional downlink spectrum to the system in 17.7–18.8 GHz after sharing issues with the other services in that spectrum are resolved. They were apparently resolved in Docket IB 98-172, which established allocations for satellite and terrestrial fixed systems in 17.7–20.2 GHz.

Motorola proposed to augment Millennium with other constellations in what it called the Celestri Architecture.

These other components include Celestri, which was to incorporate 63 *Non-Geostationary Orbit* (NGSO) FSS satellites plus five GSO FSS satellites, and M-Star (see 37.6–38.6 GHz).

Millennium emphasized multipoint services to the Americas. Celestri was intended to offer point-to-point services to a global market. Motorola later made a substantial investment in Teledesic Corp., including an "integration" of Celestri and Teledesic. The failure of Iridium LLC and the restructuring of Teledesic undoubtedly impacted these plans.

More Licensees and Applicants

Other GSO FSS licensees included GE American Communications, for a nine-satellite GE-Star system, Orion Network Systems, for a three-satellite system and Orion Atlantic for a single satellite (Loral now owns Orion), WildBlue, for a two-satellite system, EchoStar Satellite Corp. for a two-satellite system, Loral Space & Communications Ltd., for a three-satellite CyberStar system, Lockheed Martin Corp. for a nine-satellite Astrolink system, PanAmSat (now owned by Hughes) for two satellites, and Hughes Communications Galaxy for four Ka-band satellites in the 21-satellite Galaxy/Spaceway system.

Another licensee, VisionStar Inc., emphasized the use of its single $207.5 million GSO FSS satellite for video, in conjunction with the Local Multipoint Distribution Service that its owner co-founded (LMDS, see 27.5–29.5 GHz).

The FCC also accepted a second round of applications. Some companies used the opportunity to apply for additions to their holdings. For example, Lockheed Martin asked for permission to add to Astrolink 17.8–18.3 GHz and 18.8–19.3 GHz for downlinks and 27.85–28.35 GHz and 28.6–29.1 GHz for uplinks, as well as millimeter wave spectrum for *intersatellite* (ISL) links.

The company also requested a license for Astrolink Phase II, a system of five GSO FSS satellites to "interconnect with and augment the capabilities of" its already licensed Astrolink. Phase II would use 17.8–19.3 GHz and 19.7–20.2 GHz for downlinks, and 27.85–29.1 GHz and 29.25–30 GHz for uplinks.

Separately, Lockheed Martin proposed a $6.8 billion LM-MEO (medium Earth orbit) system of 32 satellites in 28.35–29.1 GHz and 29.5–30 GHz for uplinks, and 18.05–18.3 GHz, 18.8–19.3 GHz, and 19.7–20.2 GHz for downlinks. A separate V-band component of LM-MEO would use 47.2–50.2 GHz for uplinks and 37.5–42.5 GHz for downlinks.

LM-MEO would provide "seamless broadband connectivity to public and private networks" using small, inexpensive Ka-band user terminals, while its V-band service would provide "telecommunications trunking among geographic regions with high traffic demands."

If these satellite operators can't avoid interfering with the terrestrial fixed service, they may have to pay to relocate terrestrial systems to other frequencies or media, unless the problem occurs in bands where terrestrial use is primary or co-primary with satellite use.

Hughes Communications applied to license the $2.3 billion Spaceway EXP system, which would use eight geostationary satellites to provide very high data rate transoceanic service, inter- and intracontinental long haul data transport and high capacity backbone service with multicasting capability. Other second-round applicants included CAI Data Systems, DirectCom Networks, Celsat America, KaStarcom, Pacific Century Group, Pegasus Development Corp., Loral Orion, and TRW.

SkyBridge (see 11.7–12.2 GHz) proposed a $7.3 billion SkyBridge II System of 96 LEO satellites to provide a global "Transit Mission" using spot beams for high data rate point-to-point service, and a "Service Mission" for "bandwidth on demand" to consumers. SkyBridge II proposed to use downlinks in the 18 GHz spectrum and uplinks in 28.35–29.1 GHz and 29.5–30 GHz.

Footnote US334 pertains to classified intelligence-collection satellites (see 18.8–19.3 GHz).

The 17.7–21.4 GHz spectrum is one of many restricted bands in which the FCC Part 15 rules permit unlicensed devices to emit only very low level emissions.

18.58–18.6 GHz

FG: US334 FIXED-SATELLITE (space-to-Earth) G117

Non-FG: US334 NG144 FIXED-SATELLITE (space-to-Earth) NG164

FCC: Satellite Communications (25)

International Bureau Docket IB 98-172 allocated 18.58–18.6 GHz and 18.6–18.8 GHz to Fixed Satellite Service stations in Geostationary Orbit (GSO FSS). See 18.3–18.58 GHz for examples of GSO FSS systems licensed and applied for in this part of the spectrum.

To avoid interference between terrestrial fixed and satellite systems, that docket eliminated a previous, co-primary allocation in these bands to the terrestrial fixed service. Existing fixed microwave stations in 18.58–19.3 GHz can continue to operate on a co-primary basis; but they are "subject to the overriding right of satellite providers to require them to relocate," the FCC said.

Footnote US334 pertains to classified intelligence-collection satellites (see 18.8–19.3 GHz).

The 17.7–21.4 GHz spectrum is one of many restricted bands in which the FCC Part 15 rules permit unlicensed devices to emit only very low level emissions.

18.6–18.8 GHz

FG: US254 US334 EARTH EXPLORATION-SATELLITE (passive). FIXED-SATELLITE (space-to-Earth) US255 G117. SPACE RESEARCH (passive)

Non-FG: US254 US334 NG144 EARTH EXPLORATION-SATELLITE (passive). FIXED-SATELLITE (space-to-Earth) US255 NG164. SPACE RESEARCH (passive)

FCC: Satellite Communications (25)

International Bureau Docket IB 98-172 allocated 18.58–18.6 GHz and 18.6–18.8 GHz to Fixed Satellite Service stations in Geostionary Orbit (GSO FSS). See 18.3–18.58 GHz for examples of GSO FSS systems licensed and applied for in this part of the spectrum.

To avoid interference between terrestrial fixed and satellite systems, that docket eliminated a previous, co-primary allocation in these bands to the terrestrial fixed service, and established procedures for coordination and relocation of existing fixed microwave stations operating here.

Remote sensing satellites use passive sensors in 18.6–18.8 GHz to gather data on rain and snowfall, soil moisture, ocean salinity, temperature, and oil spills.

Footnote US334 pertains to classified intelligence-collection satellites (see 18.8–19.3 GHz).

The 17.7–21.4 GHz spectrum is one of many restricted bands in which the FCC Part 15 rules permit unlicensed devices to emit only very low level emissions.

18.8–19.3 GHz

FG: US334 FIXED-SATELLITE (space-to-Earth) G117

Non-FG: US334 NG144 FIXED-SATELLITE (space-to-Earth) NG165

FCC: Satellite Communications (25)

FCC Docket IB 98-172 allocated this band to the *Non-Geostationary Fixed Satellite Service* (NGSO FSS).

A global NGSO FSS project, Calling Communications, was announced in 1993 to little public attention. Microsoft's Bill Gates and telecom magnate Craig McCaw acquired Calling Communications and renamed it Teledesic Corp. The company filed a formal license application at the FCC on March 21, 1994, thrusting the project into the limelight. The FCC granted the application in 1997.

At a cost of $9 billion, Teledesic planned to orbit 288 satellites plus spares, in 12 orbital planes, although the company has indicated that the number of satellites could change.

With primary downlinks in 18.8–19.3 GHz and uplinks at 28.6–29.1 GHz, this international system will provide an "Internet in the sky" (a Teledesic trademark), with voice channels, broadband channels supporting videoconferencing and interactive multimedia, and real-time two-way data services offered by resellers to the public.

Each satellite is a node in a packet network, with optical communication links to eight adjacent satellites. A separate set of Teledesic services will use, on a secondary basis, 17.8–18.6 GHz (downlink) and 27.6–28.4 GHz (uplink).

Teledesic plans to become operational by 2004. The company anticipates expansion in the Q-band (33–50 GHz).

Teledesic already operates T1, an experimental satellite, to study issues including atmospheric drag, power control, propagation delay, spacecraft attitude sensing, GPS synchronization, and stability of laser *intersatellite links* (ISL).

At this writing, Teledesic had requested FCC permission to reorganize its ownership structure. Teledesic LLC, the FCC licensee, would become 100 percent owned by Teledesic Corp., itself 100 percent owned by ICO-Teledesic Global Ltd. that also owns New ICO (see 1990–2025 MHz) and is controlled by Craig McCaw through his Eagle River Investments company.

"The financial difficulties of NGSO satellite systems such as Iridium and ICO have created an unfavorable atmosphere within the investment community," Teledesic told the FCC. "Nonetheless, Mr. McCaw and his companies have committed over a billion dollars of their own money to provide globally ubiquitous communications capabilities—a feat that only a constellation of NGSO satellites can accomplish. But despite the mammoth investment made to date by Mr. McCaw, billions more will be needed in order to realize this vision." Competing with Teledesic will be other NGSO FSS and *Geostationary Fixed Satellite Service* (GSO FSS) ventures (see 11.7–12.2 GHz and 18.3–18.58 GHz) and Q/V-Band ventures (see 37.6–38.6 GHz).

Teligent

Parts of the 18 GHz spectrum are available to radio networks formerly known as *Digital Termination Systems* (DTS), and now considered part of the Private Operational Fixed Point-to-Point Microwave Service. The band was also available to common-carrier operations in the *Digital Electronic Message Service* (DEMS, see also 10.55–10.6 GHz). DEMS was allocated 18.82–18.92 GHz and 19.16–19.26 GHz in 1984.

DEMS was little used and virtually unknown, however, until 1993, when cellular and cable TV entrepreneurs Rajendra Singh and Myles Berkman began obtaining DEMS licenses without competitive bidding and without the knowledge of FCC policymakers (see the following).

Teligent LLC now holds those DEMS licenses. Teligent is a *competitive local exchange carrier* (CLEC), providing fixed wireless phone and data communications to small and medium size businesses in major markets.

At this writing, the FCC is preparing to auction any remaining DEMS licenses at 24 GHz, where DEMS has been relocated. The FCC *General Counsel* (GC) told an appeals court that the Commission would not auction the spectrum unless opponents to the DEMS proceeding prevailed in court and if the FCC decided on remand to hold an auction—"both of which are highly conjectural at best," he said.

The FCC proposed to rename DEMS as the "24 GHz Service" and announced plans for the auction in November, 1999 (Docket WT 99-327). For information on this auction, see 24.25–24.45 GHz.

FCC *Wireless Telecommunications Bureau* (WTB) staff were unable to explain to us the discrepancy between the GC's pleadings and the auction plans, except to suggest that the GC should have checked with the WTB before making those claims.

Spysat Connection

Footnote US334 provides for the operation, in 17.8–20.2 GHz, of classified government systems in the Fixed Satellite Service. These mysterious spacecraft are *signals intelligence* (SIGINT) satellites that are among the largest manmade objects in orbit.

They sport exotic names such as Vortex, Jumpseat, Trumpet, Orion, Mercury, Magnum, and Mentor, and may also use links in 30–31 GHz.

The satellites eavesdrop on massive amounts of telecommunications traffic and aerospace telemetry. They are reportedly part of "Echelon," the cooperative global electronic intelligence network operated by the U.S., Canada, United Kingdom, Australia, and New Zealand.

Officially, the *National Telecommunications and Information Administration* (NTIA) has said only that satellites in this band "involve military functions, as well as specific sensitive national security interests of the United States." The prospect of nationwide deployment of DEMS systems in this band raised concerns that DEMS might interfere with the satellites' Colorado and D.C.-area ground stations.

According to Commerce Department inspectors, the FCC's Gettysburg, Pa. licensing facility "issued DEMS licenses without the knowledge of FCC headquarters or [NTIA] officials, even though the DEMS technology interfered with existing licenses for federal government satellite operations in two geographic areas." This prompted the Defense

Department, through NTIA, to ask the FCC to move DEMS to 24 GHz spectrum that NTIA had made available.

The Commission in March 1997 reallocated the 24.25–24.45 GHz and 25.05–25.25 GHz bands to DEMS.

Permitted Fixed Stations

Non-DEMS fixed operations in 18 GHz, such as systems in the *Cable Television Relay Service* (CARS), are not permitted in zones near the ground stations but are allowed to continue elsewhere.

These "point-to-point and low power operations that will be permitted to continue to operate differ significantly from DEMS operations, and do not pose the same risk to Government operations," according to FCC chairman William Kennard.

In Docket IB 98-172, the FCC said that existing low power fixed systems will be permitted to continue in 18.82–18.87 GHz and 19.16–19.21 GHz, on a co-primary basis with the FSS, but that new stations will be authorized only for use indoors.

Separately, Teledesic had raised concerns that DEMS could interfere with downlinks of its planned NGSO FSS satellites, operating in 18.8–19.3 GHz. The FCC's reallocation of DEMS to 24 GHz has resolved this issue, with some physical relocation of DEMS stations funded by Teledesic.

Spectrum Fiasco

These FCC actions, taken under the shield of national security and without opportunity for public comment as required by the *Administrative Procedures Act* (APA), made competing wireless carriers furious and triggered Congressional inquiries. The reallocation quadrupled the size of the DEMS allocation (from five 20 MHz channels at 18 GHz to five 80 MHz channels at 24 GHz), ostensibly because of increased rain attenuation and link reliability concerns in 24 GHz.

The FCC prohibited public comment on alternative system designs, service models, or regulatory schemes that might have required less spectrum. Indeed, the Commission conceded years later that the technical parameters, including channelization, it selected for 24 GHz DEMS "were derived, for purposes of expedience, from those applied to operations at 18 GHz, and may not have been exactly suited to operations at the higher 24 GHz band."

The vacation of 18 GHz by DEMS was a national security matter. Yet the sudden supply of a generously sized, virtually empty 24 GHz band, and its free availability to DEMS and Teligent, were not security matters. "It appears that in this instance," House Commerce Committee Chairman Tom Bliley (R-Va.) said, "the Commission put the cart before the horse—first deciding to bypass notice and comment, and then trying to squeeze this case within one of the APA's statutory exemptions. I do not believe that is any way to run an agency charged with the public trust."

Bliley said that the FCC's "collective memory lapse" on how and why it took these actions "is as troubling as it is unacceptable, and highlights a shocking lack of accountability and transparency in agency decision making." He demanded that FCC and NTIA turn over detailed records relating to DEMS and Teligent to Congressional investigators.

Bliley later lauded Teligent at a ceremonial service launch, describing the company as "proof that the Telecommunications Act is working."

Altair RIP

A former inhabitant of this spectrum was Altair, a Motorola line of local area network radios (wireless LANs) at 18 GHz and 19 GHz. The company obtained hundreds of licenses across the U.S. for this purpose.

Observers cited the product's lack of features such as mobility, low cost, interoperability, or interior building penetration as reasons for its poor reception in the market. Motorola exited the Altair business in 1995.

Motorola had earlier attempted to place wireless LANs in meteorology spectrum. This effort, aggressively promoted by the FCC, failed after scientific interests in the Commerce Department defended the frequencies (see 1700–1710 MHz).

The 17.7–21.4 GHz spectrum is one of many restricted bands in which the FCC Part 15 rules permit unlicensed devices to emit only very low level emissions.

19.3–19.7 GHz

FG: US334 FIXED-SATELLITE (space-to-Earth) G117

Non-FG: US334 NG144 FIXED. FIXED-SATELLITE (space-to-Earth) NG166

FCC: Satellite Communications (25). Auxiliary Broadcast (74). Cable TV Relay (78). Fixed Microwave (101)

The FCC reallocated this band in Docket IB 98-172 for use by feeder links in the Mobile Satellite Service. For example, the Iridium system uses gateway earth stations in this band. Terrestrial fixed links are co-primary in this band.

The 17.7–21.4 GHz spectrum is one of many restricted bands in which the FCC Part 15 rules permit unlicensed devices to emit only very low level emissions.

19.7–20.1 GHz

FG: US334 FIXED-SATELLITE (space-to-Earth) G117

Non-FG: S5.525 S5.526 S5.527 S5.528 S5.529 US334 FIXED-SATELLITE (space-to-Earth). MOBILE-SATELLITE (space-to-Earth)

FCC: Satellite Communications (25)

This band and 20.1–20.2 GHz are allocated to the *Geostationary Orbit Fixed Satellite Service* (GSO FSS). See 18.3–18.58 GHz for examples of GSO FSS systems licensed and applied for in this part of the spectrum.

The international Footnotes place a variety of special conditions on the use of this band by the Mobile Satellite Service.

The 17.7–21.4 GHz spectrum is one of many restricted bands in which the FCC Part 15 rules permit unlicensed devices to emit only very low level emissions.

20.1–20.2 GHz

FG: US334 FIXED-SATELLITE (space-to-Earth) G117

Non-FG: S5.525 S5.526 S5.527 S5.528 US334 FIXED-SATELLITE (space-to-Earth). MOBILE-SATELLITE (space-to-Earth)

FCC: Satellite Communications (25)

This band is allocated to the *Geostationary Orbit Fixed Satellite Service* (GSO FSS) along with 19.7–20.1 GHz. See 18.3–18.58 GHz for examples of GSO FSS systems for this part of the spectrum.

The 17.7–21.4 GHz spectrum is one of many restricted bands in which the FCC Part 15 rules permit unlicensed devices to emit only very low level emissions.

20.2–21.2 GHz

FG: G117 FIXED-SATELLITE (space-to-Earth). MOBILE-SATELLITE (space-to-Earth). Standard Frequency and Time Signal-Satellite (space-to-Earth)

Non-FG: Standard Frequency and Time Signal-Satellite (space-to-Earth).

No U.S. use of this band, or its companion band 13.4–14 GHz, for two-way satellite frequency and time communications exists.

The Pentagon's Milstar has a downlink in this band, and an uplink at 43.5–45.5 GHz.

Milstar is the Military, Strategic and Tactical Relay system, intended to provide the National Command Authorities (the President and the Secretary of Defense) and other military commanders with a means of communication operable nearly worldwide and in all levels of military conflict, and which is resistant to electronic jamming and some effects of nuclear detonations. Expenditures on Milstar are pegged at $17 billion through 2002.

Controlled from Falcon Air Force Base in Colorado Springs, Colorado, Milstar provides communications for users in the field, in missile silos, submarines, bombers, and special trucks that travel the highways during periods of high military alert. Its services include voice conferencing, missile warning, and emergency action message dissemination.

Milstar terminals are interoperable across all military services. Features such as *Extremely High Frequency* (EHF) operation, time, frequency permutation, and frequency hopping keep Milstar robust against enemy jamming or propagation disturbances (scintillation) from nuclear detonations.

EHF allows use of physically small, high-gain antennas, permitting mobility and low transmit power. The beamwidth of a Milstar EHF uplink is narrow, making it difficult for an opponent to place interceptor aircraft in the beam.

Milstar incorporates a lower-frequency UHF service (350 MHz uplink, 250 MHz downlink) compatible with older military satellite systems.

The satellites contain on-board message switching and 60 GHz crosslinks between satellites. The satellites themselves route messages, through crosslinks if necessary, to the appropriate downlink to make a connection. This permits communications without the need for ground relays or facilities on foreign soil.

Disclosure of Milstar

Government Accounting Office investigations into Milstar revealed a massive program riddled with delays, poor or nonexistent testing, and facing unacceptable system degradation even as U.S. forces become more dependent on it. Huge cost overruns, technical setbacks, and whistleblowers' charges eventually brought the ostensibly classified program undesired scrutiny.

An Air Force colonel told Congressional committees his views about Milstar cost him his job evaluating the space budget. He described Milstar as a costly monument to Cold War technology that should be terminated.

America's Defense Monitor called Milstar "another expensive weapon system made obsolete by the collapse of Communism." The Defense Department claims that nuclear warfighting capability was removed from Milstar and replaced with conventional components. The system is still supposed to warn of and assist in retaliation for ballistic missile attacks.

Roger and the Box

In 1993, a Milstar power amplifier on its way to McClellan Air Force Base in California mysteriously ended up on the unclaimed freight market. Not knowing exactly what the device was, Roger Spillman of Cooleemee, N.C. bought the $363,735 device from a salvage dealer for $75.00. He had hoped to use it with a CB radio in his truck.

The Air Force seized the unit after a friend of Spillman's, a radio amateur, telephoned the manufacturer to ask for an owner's manual. The friend told us that Air Force agents said the unit was for the "President's launch codes." The national press and TV widely reported the story, pushing Milstar into the public eye.

The first Milstar Development Flight Satellite, DFS-1, was placed in orbit February 7, 1994, and blocks of Milstar satellites continue to be launched.

Milstar's successor, the Advanced EHF Satellite System, is being developed. The first launch of this system is expected in 2006.

The 17.7–21.4 GHz spectrum is one of many restricted bands in which FCC Part 15 rules permit unlicensed devices to emit only very low level emissions.

21.2–21.4 GHz

FG: US263 EARTH EXPLORATION-SATELLITE (passive). FIXED. MOBILE. SPACE RESEARCH (passive)

Non-FG: US263 EARTH EXPLORATION-SATELLITE (passive). FIXED. MOBILE. SPACE RESEARCH (passive)

FCC: Fixed Microwave (101)

The *Telecommunications Industry Association* (TIA) has called the 21.2–23.6 GHz spectrum (the "23 GHz band") an "untapped resource for fixed service users. ...It is allocated for fixed service use, is shared between non-government and government users, and is especially suitable for medium or high capacity, short-range systems."

According to Harris Corp., the 23 GHz band "will be increasingly used to interconnect cell sites in urban areas or in campus-type systems, where the antennas would more than likely be mounted on rooftops, monopoles, and other structures that cannot support large microwave dishes."

The 23 GHz band is suitable for general-purpose fixed links up to about 10 miles, and is used in links for telephone bypass, aviation, TV stations, PBX telephone systems, computer networks, security cameras, and video teleconferencing.

The band also is shared with spacecraft in the *Earth Exploration Satellite Service* (EESS, see 13.4–13.75 GHz), for measurement of Earth's water vapor. The FCC is reviewing technical standards for 23 GHz in Docket WT 00-19.

The 17.7–21.4 GHz spectrum is one of many restricted bands in which the FCC Part 15 rules permit unlicensed devices to emit only very low level emissions.

21.4-22 GHz

FG: FIXED. MOBILE

Non-FG: FIXED. MOBILE

FCC: Fixed Microwave (101)

See 21.2–21.4 GHz for the uses of this band.

NASA's *Microwave Anisotropy Probe* (MAP) will use the 22 GHz, 30 GHz, 40 GHz, 60 GHz, and 94 GHz bands to measure the distribution of cosmic background radiation, the faint radio afterglow of the Big Bang (see also 37.5–38 MHz and 217–231 GHz).

22-22.21 GHz

FG: S5.149 FIXED. MOBILE except aeronautical mobile

Non-FG: S5.149 FIXED. MOBILE except aeronautical mobile

FCC: Fixed Microwave (101)

See 21.2–21.4 GHz for the uses of this band.

The 22.01–23.12 GHz spectrum is one of many restricted bands in which the FCC Part 15 rules permit unlicensed devices to emit only very low level emissions.

22.21-22.5 GHz

FG: S5.149 US263 EARTH EXPLORATION-SATELLITE (passive).
FIXED. MOBILE except aeronautical mobile. RADIO ASTRONOMY. SPACE RESEARCH (passive)

Non-FG: S5.149 US263 EARTH EXPLORATION-SATELLITE (passive). FIXED. MOBILE except aeronautical mobile. RADIO ASTRONOMY. SPACE RESEARCH (passive)

FCC: Fixed Microwave (101)

Atmospheric gases emit and absorb energy at discrete resonance frequencies. Molecules of water resonate at 22.235 GHz. Measurements around this frequency indicate the amount of water in an atmosphere.

Radio astronomers use this data for such purposes as measurement of distances to far galaxies. The band also is used for *Very Long Baseline Interferometry* (VLBI) by connected radio telescopes.

These passive (receive-only) uses are not protected against interference from active (transmitting) uses as part of the 21.2–23.6 GHz fixed spectrum (see 21.2–21.4 GHz).

The 22.01–23.12 GHz spectrum, however, is one of many restricted bands in which the FCC Part 15 rules permit unlicensed devices to emit only very low level emissions.

22.5–22.55 GHz

FG: US211 FIXED. MOBILE

Non-FG: US211 FIXED. MOBILE

FCC: Fixed Microwave (101)

See 21.2–21.4 GHz for the uses of this band.

The 22.01–23.12 GHz spectrum is one of many restricted bands in which the FCC Part 15 rules permit unlicensed devices to emit only very low level emissions.

22.55–23.55 GHz

FG: S5.149 US278 FIXED. INTER-SATELLITE. MOBILE

Non-FG: S5.149 US278 FIXED. INTER-SATELLITE. MOBILE

FCC: Satellite Communications (25). Fixed Microwave (101)

This is part of the 21.2–23.6 GHz fixed microwave band (see 21.2–21.4 GHz).

This band is allocated for intersatellite (satellite-to-satellite or ISL, also called Inter-Satellite Service or ISS) links in both non-geostationary and geostationary satellite systems.

To be compatible with planned European and Japanese relay satellite systems, NASA targets 22.55–23.55 GHz and 25.25–27.5 GHz for future *Tracking and Data Relay Satellites* (TDRSS, see 2025–2110 MHz).

The 22.01–23.12 GHz spectrum is one of many restricted bands in which the FCC Part 15 rules permit unlicensed devices to emit only very low level emissions.

23.55–23.6 GHz

FG: FIXED. MOBILE

Non-FG: FIXED. MOBILE

FCC: Fixed Microwave (101)

See 21.2–21.4 GHz for the uses of this band.

23.6–24 GHz

FG: US246 EARTH EXPLORATION-SATELLITE (passive). RADIO ASTRONOMY US74. SPACE RESEARCH (passive)

Non-FG: US246 EARTH EXPLORATION-SATELLITE (passive). RADIO ASTRONOMY US74. SPACE RESEARCH (passive)

No transmissions are permitted in this scientific band, used in radio astronomy and Earth observations.

Molecular lines in this band indicate ammonia in the interstellar medium. Satellite sensors use the band to measure water vapor.

This is one of many restricted bands in which the FCC Part 15 rules permit unlicensed devices to emit only very low level emissions.

Figure 15
Intrusion detection systems like this one use the 24 GHz band to protect military assets. (Southwest Microwave photo)

24–24.05 GHz

FG: S5.150 US211

Non-FG: S5.150 US211 AMATEUR. AMATEUR-SATELLITE

FCC: ISM Equipment (18). Amateur (97)

This band and 24.05–24.25 GHz are available to unlicensed devices. The center *Industrial, Scientific, and Medical* (ISM) frequency is 24.125 GHz.

This part of the spectrum contains active satellite-based sensors that measure vegetation and snow, radars that measure fluid levels in storage and processing facilities, and field disturbance sensors that open doors and sound alarms.

Unlicensed, high speed, fixed radios are now available for 24–24.25 GHz under the FCC Part 15 rules. The FCC proposed to improve conditions for unlicensed 24 GHz operation (see 24.05–24.25 GHz).

24.05–24.25 GHz

FG: S5.150 RADIOLOCATION US110 G59. Earth Exploration-Satellite (Active)

Non-FG: S5.150 Radiolocation US110. Amateur. Earth Exploration-Satellite (Active)

FCC: ISM Equipment (18). Private Land Mobile (90). Amateur (97)

Government radiolocation is primary in this band. K-band police speed radars share this band, at 24.15 GHz, Spaceborne instruments for precipitation and ocean measurement also use the band.

This band and 24–24.05 GHz are available to unlicensed devices under the FCC Part 15 rules. The center *Industrial, Scientific, and Medical* (ISM) frequency is 24.125 GHz.

The FCC proposed in Docket ET 98-156 to permit unlicensed, fixed point-to-point operation at field strengths as high as 2500 mV/m in 24.05–24.25 GHz provided that high-gain antennas are used. This would allow use of 24 GHz to better avoid the expense and delays of the conventional microwave licensing process.

Key applications for such operations are backbone links in Internet services provided in other unlicensed bands such as the *Unlicensed National Informational Infrastructure* (U-NII, see 5–5.25 GHz) and 2.4 GHz (see 2400–2402 MHz). Paths as long as 2.5 miles could use 2500 mV/m 24 GHz radios at 99.99 percent reliability, the FCC was told.

Especially interested in this band are makers of traffic light control systems that were formerly sold in what is now Local Multipoint Distribution Service spectrum (see 31–31.3 GHz).

Several manufacturers supported the proposal, which was still pending at this writing. But FCC staff members were concerned that Part 15 devices may not be appropriate for high-reliability applications; that interference to Part 15 links might adversely affect traffic lights; that systems in this band might interfere with the 24 GHz Service in adjacent spectrum (see 24.25–24.45 GHz); and that the Amateur Radio Service might experience interference from the proposed systems.

The supporters responded that they want to make equipment available that can be set up quickly for temporary use, to be replaced by licensed equipment later if necessary; and that unlicensed 24 GHz equipment would be clearly labeled to indicate that the user must accept interference if it occurs and not interfere with licensed services. They told the FCC that traffic signals will operate normally in case of communications failure, using internal timing instead of synchronizing with other intersections.

The Amateur Radio 1.2-cm band is 24–24.25 GHz. Radio amateurs opposed the proposal. The 24–24.05 GHz band is included in the capabilities of the ambitious Phase 3D amateur satellite. For this reason, the FCC declined to permit point-to-point unlicensed operations in 24–24.05 GHz.

"Predictably, however, amateur opposition has continued nonetheless," the proponents told the FCC. Manufacturers told the Commission that actual 24 GHz interference from a high-gain (33 dBi, as proposed) antenna into an amateur receiver is "astronomically unlikely" and that the user must stop transmitting if interference does occur.

Controversial Warning System

The *Safety Warning System* (SWS) was developed to provide in-vehicle warning of road hazards. SWS transmits alerts in this band to consumer receivers, including radar detectors.

Fixed SWS transmitters will be installed at construction sites, flooded areas, and other hazardous places, while mobile SWS transmitters can be mounted on emergency vehicles. In this application, the transmitter illuminates the receiver from behind, alerting the motorist to an approaching vehicle.

A corporation formed to license the technology, SWS, L.C., became a Part 5 Experimental Radio Service licensee and deployed SWS transmitters in 26 states, well before the FCC considered allocating spectrum to SWS on a permanent basis.

Extensive experimental investments of this type often play a key role in persuading the FCC to allocate spectrum to a proposed service.

Eventually, after receiving a petition from the Radio Association for Defending Airwave Rights (a radar detector advocacy outfit representing the same manufacturers who make SWS consumer devices), the FCC decided to allocate 24.1 GHz to SWS and to permit licensees in the Public Safety frequency pool (see 30.56–32 MHz) and railroad licensees in the Industrial/Business pool to broadcast safety alerting signals on this frequency.

In taking this action, the Commission bucked a formidable army of SWS opponents, including the U.S. *Department of Transportation* (DOT), the *International Association of Chiefs of Police* (IACP) and the *National Association of Governors' Highway Safety Representatives* (NAGHSR).

DOT said that SWS is "unlikely to enhance the safety of motorists, and undercuts that safety by promoting the widespread deployment of a device whose primary use is to facilitate unlawful speeds without detection."

"These devices, when not receiving emergency notification of road hazards and approaching emergency vehicles, will be used to alert the driver of speed enforcement activities of police," the IACP said. NAGHSR similarly argued that SWS "is a way to legitimize the use of radar detectors by motorists" and claimed that more sophisticated driver alerting systems are being developed with "millions of Federal dollars."

In response, the FCC pointed to the millions of federal dollars being spent on SWS. The money, appropriated in the massive 1998 Transportation Equity Act, is for grants to state and local governments "to permit further studies and testing of SWS."

Proponents of the system conceded that SWS works with more than 20 million "early-generation safety warning system receivers...already in use by consumers" which are "basically radar detectors." Current and future SWS products "will not incorporate the circuitry required for the device to function as a radar detector," however. Radar detectors are not allowed by law in some states, and federal law bans radar detectors in interstate commercial vehicle operations.

The Commission said that SWS "will provide immediate benefits without the motoring public having to wait possibly 10–15 years for the development and implementation of future ITS (Intelligent Transportation System) technology." It added railroad companies to the group of licensees eligible to transmit SWS signals, in a program to reduce the high number of accidents that occur at rail crossings.

Dedicated Short Range Communications (DSRC), part of ITS to be deployed nationwide, is supposed to provide safety warning, in-vehicle signing, mobile electronic commerce, and collision avoidance among its functions (see 5.85–5.925 GHz). DSRC could be considered an advanced successor to SWS. We expect DSRC to penetrate commercial transportation markets well before consumer markets.

At this writing, yet another proposal for an in-vehicle warning system was pending at the FCC: the *Emergency Radio Data System* (ERDS, see 88–108 MHz).

24.25–24.45 GHz

FG: None

Non-FG: RADIONAVIGATION. FIXED

FCC: Aviation (87). Fixed Microwave (101)

No aviation-related operations are in this band. This is the hub-to-user band, with 25.05–25.25 GHz the user-to-hub band, allocated to the *Digital Electronic Message Service* (DEMS), dominated by Teligent and used to provide *competitive local exchange carrier* (CLEC) telephone and data service. For the peculiar history of this service, see 18.8–19.3 GHz and 10.55–10.6 GHz.

At this writing, Docket WT 99-327 proposed allocation and auction rules for the DEMS bands. It proposed to delete the non-government radionavigation allocations in 24.25–24.45 GHz and 25.05–25.25 GHz, which were never used.

The DEMS would be renamed the *"24 GHz Service"* (TFGS) and defined broadly as "A radiocommunication service that may encompass any digital fixed service." TFGS licenses would be auctioned in 175 *Economic Areas* (EAs) defined by the Department of Commerce. Existing facilities, licensed under an older geographic scheme, would be "excluded from the applicable EAs."

Each EA would accommodate five licenses, with each license consisting of a pair of 40 MHz channels. As is now usually the case, the Commission will permit licensees to partition and disaggregate their spectrum and geographic service areas to other parties if desired.

Although the FCC proposed to retain these allocations for fixed service use, it requested comment on whether mobile service should be added to the permissible uses. "Based on the information currently available, it appears that, in the near term, equipment may not be available for mobile use in the 24 GHz band," the FCC said.

"[N]o one has requested the opportunity to provide mobile service at 24 GHz. If, contrary to our assumption, equipment is available for mobile use in this band, and interference problems can be resolved, we know of no reason why we would not allow mobile operations."

24.45–24.65 GHz

FG: S5.333 INTER-SATELLITE. RADIONAVIGATION

Non-FG: S5.333 INTER-SATELLITE. RADIONAVIGATION

FCC: Satellite Communications (25)

The spectrum from 24.45 GHz to 25.05 GHz "remains available" for use by the aeronautical radionavigation service, according to the FCC in Docket WT 99-327, which was its last pronouncement on the status of this spectrum at this writing. We believe that this spectrum is unused.

24.65–24.75 GHz

FG: INTER-SATELLITE. RADIOLOCATION-SATELLITE (Earth-to-space)

Non-FG: INTER-SATELLITE. RADIOLOCATION-SATELLITE (Earth-to-space)

FCC: Satellite Communications (25)

24.75–25.05 GHz

FG: RADIONAVIGATION

Non-FG: FIXED-SATELLITE (Earth-to-space) NG167. RADIONAVIGATION

FCC: Satellite Communications (25). Aviation (87)

In Docket IB 98-172, the FCC granted a petition by DirecTV to allocate this band and 25.05–25.25 GHz to the *Fixed Satellite Service* (FSS) for use in feeder uplinks to a future generation of *Direct Broadcast Satellite* (DBS, also called Broadcasting Satellite Service or BSS). For more information on this allocation, see 17.3–17.7 GHz.

25.05–25.25 GHz

FG: None

Non-FG: FIXED-SATELLITE (Earth-to-space) NG167. FIXED. RADIONAVIGATION

FCC: Satellite Communications (25). Aviation (87). Fixed Microwave (101)

This is the user-to-hub band, associated with the hub-to-user band 24.25–24.45 GHz, in the *Digital Electronic Message Service* (DEMS), used for *competitive local exchange carrier* (CLEC) services (see 18.8–19.3 GHz). The FCC has proposed to rename DEMS the *24 GHz Service* (TFGS).

This band and 24.75–25.05 GHz were allocated in Docket IB 98-172 to the *Fixed Satellite Service* (FSS) for use exclusively in uplinks to a future generation of *Direct Broadcast Satellites* (DBS, also called *Broadcasting Satellite Service* or BSS). For more information on this allocation, see 17.3–17.7 GHz.

At this writing, in Docket WT 99-327, the FCC proposed auction and allocation rules for DEMS/TFGS and asks for comment on its interaction with DBS.

25.25–27 GHz

FG: FIXED. MOBILE. Standard Frequency and Time Signal-Satellite (Earth-to-space) (Author's note: These *federal government* (FG) allocations apply to 25.25—25.5 GHz and 25.5—27 GHz. In 25.5—27 GHz, the Earth Exploration-Satellite Service (space-to-space) is also allocated to the FG. The official FCC Note in the following gives further detail, and points out differences between the FCC and NTIA allocations.)

Non-FG: Standard Frequency and Time Signal-Satellite (Earth-to-space). Earth Exploration-Satellite (space-to-space)

Note: In its Manual, NTIA has added primary intersatellite service allocations to the bands comprising 25.25–27.5 GHz, limited the use of these allocations by adopting footnote S5.536, and has changed the directional indicator for the Earth exploration-satellite service allocation in the 25.5–27 GHz band from space-to-space to space-to-Earth.

International agreements prescribe 25.25–27.5 GHz for links between satellites in *the Space Research Service* (SRS) and *Earth Exploration Satellite Service* (EESS, see 8.025–8.175 GHz), and for data from industrial and medical activities in space.

Future changes to this spectrum are likely. For example, NTIA has stated that it intends to upgrade the EESS FG allocation to primary in 25.5–27 GHz and to delete the secondary Non-FG EESS allocation from 25.25–25.5 GHz and 27–27.5 GHz. Forthcoming FCC proceedings on *World Radio Conference* (WRC) implementation will consider this spectrum.

The Air Force is developing the *Wireless Broadband System* (WBS) in 25.25–27.5 GHz. WBS is a one- and two-way voice, video and data distribution network for use at theatre air bases.

NASA targets the 25.25–27.5 GHz and 22.55–23.55 GHz bands for future *Tracking and Data Relay Satellites* (TDRSS, see 2025–2110 MHz and 13.75–14 GHz).

Also in development for 25.25–25.55 GHz and 27.225–27.5 GHz is NASA's *Proximity Operations Communication System* (POCS) used to relay data, video and voice between space vehicles operating in close proximity, for example when docking or operating telerobots.

Although allocated for such a purpose, there reportedly is no U.S. use of this band yet, or its companion band 30–31.3 GHz, for two-way satellite frequency and time communications. Moreover, according to NTIA the 25.525–27.5 GHz spectrum is "virtually unused" as far as fixed services are concerned. No commercial systems have been installed in the band either.

The 26.5–40 GHz spectrum is traditionally designated the "Ka-band," although the FCC places the lower limit of the "Q-band" at 33 GHz.

27-27.5 GHz

FG: FIXED. MOBILE

Non-FG: Earth Exploration-Satellite (space-to-space)

Note: In its Manual, NTIA has added primary intersatellite service allocations to the bands comprising 25.25–27.5 GHz, limited the use of these allocations by adopting footnote S5.536, and has changed the directional indicator for the Earth exploration-satellite service allocation in the 25.5–27 GHz band from space-to-space to space-to-Earth.

See 25.25–27 GHz for information on this band.

27.5-29.5 GHz

FG: None

Non-FG: FIXED. FIXED-SATELLITE (Earth-to-space). MOBILE

FCC: Satellite Communications (25). Fixed Microwave (101)

This portion of the Ka-band contains segments allocated to the *Local Multipoint Distribution Service* (LMDS), a broadband fixed service (with mobile operation permissible) and to satellite uplinks, paired with downlinks lower in the spectrum (see 17.7–17.8 GHz).

LMDS was envisioned as a multichannel wireless cable TV medium. It is now considered a *competitive local exchange carrier* (CLEC) service, to provide voice and data connectivity to business customers. LMDS uses a cellular infrastructure, with multiple base stations and small customer transceivers able to return communications.

At this writing, few if any LMDS stations were operating. The advent of a national testbed for LMDS standards, sponsored by the *National Institute of Standards and Technology* (NIST), will likely speed development of this service, as will the establishment by the *Institute of Electrical and Electronics Engineers* (IEEE) of its Working Group 802.16 on Broadband Wireless Access Standards. Its charter is "to develop standards and recommended practices to support the development and deployment of fixed broadband wireless access systems."

The FCC auctions two LMDS licenses per market. The "A" license, substantial in size, covers a 1,150 MHz bandwidth consisting of 27.5–28.35 GHz, 29.1–29.25 GHz, and 31.075–31.225 GHz. The 150 MHz wide "B" license covers 31–31.075 GHz and 31.225–31.3 GHz.

LMDS History

In 1991, New York real estate developers Suite 12 Group and Hye Crest Management persuaded the FCC to waive certain rules in order to permit the company to offer multipoint video programming in 27.5–28.5 GHz, using technology developed by radio engineer Bernard Bossard. This was the earliest deployment of millimeter wave communications systems in the consumer marketplace.

The Commission received hundreds of additional waiver applications, which it dismissed, leaving CellularVision (successor to Suite 12 and Hye Crest) for a time the sole company authorized to provide wireless video service directly to the public in the millimeter waves.

In 1993, the FCC started lengthy proceedings to formalize the creation of a new radio service nationwide. The FCC eventually settled on the bandplan described previously, constituting one of the largest allocations made to any service.

Mobile LMDS

The deregulation-minded FCC also decided to permit mobile LMDS: "Although LMDS is allocated as a fixed service, we know of no reason why we would not allow mobile operations if they are proposed and we obtain a record in support of such an allocation."

According to former FCC wireless bureau chief Michele Farquhar, "We have learned the hard way that limiting flexibility at the birth of a new service will only cause us problems in the future. In nearly every service we authorize, licensees have come back to the Commission to ask that we change the services they can provide, the technical parameters governing how they provide service, or the amount of spectrum they seek to use. ...In LMDS, I think we got it right the first time. We are permitting the maximum amount of flexibility, based on the wide variety of services that companies told us they wanted to provide." No mobile LMDS services have been proposed or publicly discussed at this writing.

LMDS licenses are assigned on a *Basic Trading Area* (BTA) basis. The FCC allotted two licenses in each of 493 BTAs and BTA-like areas, based on geographic regions published by Rand McNally.

The LMDS auction in March 1998 raised about $578 million and a second auction in May 1999 raised about $45 million, well short of the billions predicted. The largest and second largest LMDS auction winners were WNP Communications and Nextband Communications, respectively. Both were acquired by Craig McCaw's NextLink Communications, a fiber-based CLEC. NextLink—now XO Communications—holds LMDS licenses covering 95 percent of the population of the 30 largest U.S. cities.

The Commission imposed controversial restrictions on the ability of cable and local exchange telephone companies to hold "A" block LMDS licenses in their principal operating regions. But these restrictions expired in June 2000.

LMDS licensees are permitted to partition portions of their license area, and disaggregate portions of their licensed spectrum, to others. Licensees also will benefit from the FCC's liberal new build-out regime. They will not be required to build certain numbers of stations, but must only demonstrate "substantial service" to their license area within 10 years.

CellularVision later sold some of its capacity to WinStar Communications and to XO Communications, and discontinued the analog wireless cable TV service it had provided from 1992–1998.

CellularVision changed its name to SpeedUS.com, "an emerging facilities-based wireless broadband super high-speed Internet company." Its New York LMDS license continues into 2006. It uses the license in "Speed," a pilot Internet Service Provider business. "A full marketing effort will not commence until new LMDS equipment becomes commercially available with cost and performance that allow implementation of Speed service on an economically attractive basis," the company said.

SpeedUS.com principals own key patents in LMDS. Its CEO, Shant Hovnanian, owns VisionStar, a Ka-band satellite licensee (see 18.3–18.58 GHz).

Flying LMDS

A novel alternative for LMDS may become available from Angel Technologies Corp., which proposes to deliver LMDS and other local broadband services from *High Altitude Long Operation* (HALO) aircraft.

Designed by famed aviation pioneer Burt Rutan, the HALO aircraft will slowly circle over a metropolitan area at stratospheric altitude above 50,000 feet. Customers' antennas will track the aircraft, which will operate in shifts to provide continuous 24-hour service. Angel does not anticipate obtaining commercial FCC licenses; rather, its strategy is to partner with spectrum auction winners.

LMDS shares the 27.5–28.35 GHz segment with uplinks, allocated on a secondary basis, in both the *Non-Geostationary Fixed Satellite Service* (NGSO FSS) and the *Geostationary Fixed Satellite Service* (GSO FSS, see 17.7–17.8 GHz).

"As a practical matter," the FCC remarked, "it is unlikely that FSS can operate ubiquitous terminals on an unprotected non-interference basis to LMDS." Nevertheless, some FSS licensees have requested FCC authority to modify their licenses to include uplinks in this spectrum.

In the 28.35–28.6 GHz segment, GSO FSS has an exclusive primary allocation with NGSO FSS secondary.

The 28.6–29.1 GHz segment contains the primary NGSO FSS uplink; GSO FSS is secondary in the segment.

The 29.1–29.25 GHz segment is allocated on a co-primary basis to both LMDS and feeder links from earth stations in the *Mobile Satellite Service* (MSS), namely Big LEOs (see 1610–1610.6 MHz). This is the LMDS "hub-to-sub" segment, devoted to forward transmissions from the LMDS operator to subscribers.

The 29.25–29.5 GHz segment is also allocated to GSO FSS, co-primary with MSS.

29.5–29.9 GHz

FG: None

Non-FG: S5.525 S5.526 S5.527 S5.529 FIXED-SATELLITE (Earth-to-space). Mobile-Satellite (Earth-to-space)

FCC: Satellite Communications (25)

The 29.5–30 GHz portion of the 27.5–30 GHz Ka-band uplink spectrum is devoted to the *Geostationary Fixed Satellite Service* (GSO FSS) on a primary basis and the *Non-Geostationary Fixed Satellite Service* (NGSO FSS) on a secondary basis (see 27.5–29.5 GHz).

29.9-30 GHz

FG: None

Non-FG: S5.525 S5.526 S5.527 S5.543

FIXED-SATELLITE (Earth-to-space). MOBILE-SATELLITE (Earth-to-space)

FCC: Satellite Communications (25)

See 29.5–29.9 GHz for information on this band.

Frequency Allocations

EHF

30-31 GHz

FG: G117 FIXED-SATELLITE (Earth-to-space). MOBILE-SATELLITE (Earth-to-space). Standard Frequency and Time Signal-Satellite (space-to-Earth)

Non-FG: Standard Frequency and Time Signal-Satellite (space-to-Earth)

The 30–31-GHz band is one of several allocated to military fixed and mobile satellites (see 18.8–19.3 GHz). There is no use in the U.S. of this band yet, or its companion band 25.25–27 GHz, for two-way satellite frequency and time communications.

The *National Radio Astronomy Observatory* (NRAO) is developing the $26 million *Atacama Large Millimeter Array* (ALMA) radio telescope to operate at frequencies from 28 GHz to 720 GHz. Millimeter waves enable observations through the dust and haze that can obscure astronomical objects.

The ALMA will use 64, 12-meter transportable dish antennas with carefully manufactured reflective surfaces. It will be the largest and most sensitive instrument in the world to use millimeter and submillimeter wavelengths.

The telescope will be built in a high-altitude location in Chile, and is expected to become operational by 2005. The telescope will help scientists understand the age of the universe, image distant galaxies and pre-planetary structures, measure the properties of subsurface layers of asteroids, and probe the outflows from thousands of stars, among its many possibilities.

Another facility, the joint U.S.-Mexican Large Millimeter Wave Telescope, will be located on a mountain in Mexico. The $50 million, 50-meter telescope will be able to point to a cosmic source with an accuracy of 1/1000th of a degree—the size of a dime at a distance of 10 miles.

31-31.3 GHz

FG: S5.149 US211 Standard Frequency and Time Signal-Satellite (space-to-Earth)

Non-FG: S5.149 US211 FIXED. MOBILE. Standard Frequency and Time Signal-Satellite (space-to-Earth)

FCC: Fixed Microwave (101)

This band had a brief existence as a novel FCC technical and regulatory experiment. Most manufacturers and potential users were unaware of the band. The FCC eventually reallocated it to broadband fixed access.

The band offered a low cost alternative to leased telephone lines, fiber or coaxial cable, or microwave links. The conventional frequency coordination procedures and costs associated with other bands did not apply to 31–31.3 GHz.

In addition to virtually every category of terrestrial fixed service, in 1985, the FCC made this band available to private individuals for personal use in the *General Mobile Radio Service* (GMRS, see 462.5375–462.7375 MHz).

"The Commission is trying to find spectrum for those who do not fall within the traditional service categories, but yet have unfulfilled communication requirements. We continue to feel that this 300 megahertz at 31 GHz is an appropriate location for use by individuals," the FCC said at the time.

The action was widely ignored. No 31-GHz GMRS stations were ever licensed. Those operators who did license in 31 GHz were mostly local and state governments, using the band to control timed traffic signals as required by pollution control laws.

"The number of entities operating under the existing rules for 31-GHz services is small and the locations are very few and confined," the FCC concluded years later, finding only what it termed "simplified communication functions" in the band.

Reallocated to LMDS

The FCC concluded that the *Local Multipoint Distribution Service* (LMDS, see 27.5–29.5 GHz) had a greater need for this band, especially to support interactive services; it also rejected the protests of incumbent licensees. It reallocated the entire 31–31.3-GHz band to LMDS, dividing the band into two 75-MHz segments and one 150-MHz-wide segment: 31–31.075 GHz, 31.075–31.225 GHz, and 31.225–31.3 GHz.

The 31.075–31.225-GHz segment is one of three segments that make up the 1150-MHz-wide "A" LMDS license block, the principal allocation to LMDS. The 31–31.075-GHz and 31.225–31.3-GHz segments comprise the LMDS "B" block.

LMDS may operate anywhere in 31–31.3 GHz, and is required to avoid interfering with incumbent users in the outer two segments. LMDS may operate in the middle 150-MHz segment without concern of any interference it may create to non-LMDS operators there. No new non-LMDS operations will be allowed at 31 GHz.

At this writing, 31-GHz LMDS licensees are planning systems, service rollout, and organizing in a forum of the *Wireless Communications Association* (WCA, see 2500–2655 MHz).

Although allocated here, no time signal satellite operations exist in this band.

Manufacturers of traffic light control links are now focused on another band, to become available on an unlicensed basis (see 24.05–24.25 GHz).

The 31.2–31.8 GHz spectrum is one of many restricted bands in which the FCC Part 15 rules permit unlicensed devices to emit only very low level emissions.

31.3–31.8 GHz

FG: US246 EARTH EXPLORATION-SATELLITE (passive). RADIO ASTRONOMY US74. SPACE RESEARCH (passive)

Non-FG: US246 EARTH EXPLORATION-SATELLITE (passive). RADIO ASTRONOMY US74. SPACE RESEARCH (passive)

No transmissions are permitted in this receive-only scientific band. Satellite-based sensors use the band to measure cloud water, precipitation, and changes in ice flows. Radio astronomy continuum observations are conducted in this band.

The 31.2–31.8-GHz spectrum is one of many restricted bands in which the FCC Part 15 rules permit unlicensed devices to emit only very low level emissions.

31.8-32 GHz

FG: S5.548 US211 RADIONAVIGATION US69. SPACE RESEARCH (deep space) (space-to-Earth) US262

Non-FG: S5.548 US211 SPACE RESEARCH (deep space) (space-to-Earth) US262

The 31.8–32.3-GHz spectrum is allocated to the *Space Research Service* (SRS) for deep space downlinks on a primary basis worldwide (see 2290–2300 MHz). In the U.S., footnote US262 limits this SRS allocation to NASA's Goldstone, California *Deep Space Network* (DSN) station. The 31.8–33-GHz spectrum is allocated to radionavigation on a primary basis worldwide.

According to Docket ET 98-197, the non-government space research allocation in this band is intended for commercial space probes that would also use the Goldstone station. In making the allocation, the FCC quoted a NASA spectrum program manager, who stated that "the realities of deep space communications is that you require very large, very low noise earth stations such as at Goldstone. The likelihood of the commercial sector being able to independently develop a station to support deep space missions is highly remote due to the costs involved."

Military uses in this spectrum include airborne precision ground mapping radar in 31.8–32-GHz and aircraft carrier landing systems in 32–32.3 GHz.

The FCC deleted a former non-government radionavigation allocation from 31.8–32.3 GHz, based on concern by NASA that the SRS would be susceptible to interference from commercial air radionavigation in this spectrum.

NASA's DSN receiving station is designed and located to avoid interference from terrestrial radio sources in the same band. But other than infrequent military operations, NASA cannot protect the station from radio sources emanating from aircraft.

Any future non-government radionavigation in this area of the spectrum will be limited to 32.3–33.4 GHz.

32-32.3 GHz

FG: S5.548 INTER-SATELLITE US278. RADIONAVIGATION US69. SPACE RESEARCH (deep space) (space-to-Earth) US262

Non-FG: S5.548 INTER-SATELLITE US278. SPACE RESEARCH (deep space) (space-to-Earth) US262

Military and non-military radar services make extensive use of 31.8–36 GHz for aviation, weather, and mapping, among many applications. See 31.8–32 GHz for information about this band.

32.3-33 GHz

FG: S5.548 INTER-SATELLITE US278. RADIONAVIGATION US69

Non-FG: S5.548 INTER-SATELLITE US278. RADIONAVIGATION US69

FCC: Aviation (87)

This band and the adjacent 33–33.4-GHz band are mainly devoted to future air radionavigation systems (see also 31.8–32 GHz).

33-33.4 GHz

FG: RADIONAVIGATION US69

Non-FG: RADIONAVIGATION US69

FCC: Aviation (87)

The 26.5–40-GHz spectrum is traditionally designated the "Ka-band," but the FCC identifies the 33–50-GHz range as the "Q-band" and the 40–60 GHz range as the "U-band." See 31.8–32-GHz for information about this band.

33.4-36 GHz

FG: S5.551 US252 RADIOLOCATION US110 G34

Non-FG: S5.551 US252 Radiolocation US110

FCC: Private Land Mobile (90)

Synthetic vision radar research is conducted around 35 GHz. NASA uses this band for deep space uplinks. Ka-band police radars also operate here under FCC Part 90 rules.

Meteorology researchers use radars in this band to study thunderstorm and cloud physics. The radars operate from land vehicles or aircraft, including unmanned drone aircraft. These measurements are also conducted at 95 GHz.

36-37 GHz

FG: US263 US342 EARTH EXPLORATION-SATELLITE (passive). FIXED. MOBILE. SPACE RESEARCH (passive)

Non-FG: US263 US342 EARTH EXPLORATION-SATELLITE (passive). FIXED. MOBILE. SPACE RESEARCH (passive)

Regardless of its allocation to both government and non-government uses, the FCC stated in Docket IB 97-95 that this band "has no commercial designation" and is to be used by federal agencies "to satisfy their fixed and mobile requirements."

North Atlantic Treaty Organization (NATO) joint frequency agreements provide for fixed and mobile military systems in this band.

Passive sensors onboard spacecraft such as the Tropical Rainfall Measurement Mission (see 13.75-14 GHz) use 36-37 GHz to observe clouds, rain rate, snow water content, ocean ice, and oil spills.

The 36.43-36.5-GHz segment is one of many restricted bands in which the FCC Part 15 rules permit unlicensed devices to emit only very low level emissions.

37-37.6 GHz

FG: FIXED. MOBILE. SPACE RESEARCH (space-to-Earth)

Non-FG: FIXED. MOBILE

The FCC designated this and other nearby bands for commercial "wireless services" to be considered later (Docket IB 97-95) (see also 40.5–41 GHz).

"We currently anticipate that the bands being designated for wireless services will be used primarily for terrestrial fixed and mobile services," the FCC said. "Further proceedings to develop service and auction rules for these bands will be necessary, however, before operations are authorized. We note that such future proceedings may further define the specific services to be assigned by auction and to operate in particular bands consistent with the U.S. Table of Frequency Allocations."

NASA plans to use 37–37.5 GHz for manned mission return links, including exploration of Mars, the establishment of a lunar colony, and wideband links that return observation data from radio astronomy satellites. The companion uplink band for manned exploration missions is 40–40.5 GHz.

NASA believes it could order manned lunar missions as early as 2010, and manned Mars missions in the second decade of the 21st century.

North Atlantic Treaty Organization (NATO) joint frequency agreements provide for fixed military systems in 37–39.5 GHz.

The 37-GHz spectrum is used, in a receive-only mode, for imaging radiometry in the aerial surveillance of Earth's features and in environmental studies of incidents, such as oil spills.

37.6-38.6 GHz

FG: FIXED. MOBILE (Author's note: The SPACE RESEARCH (space-to-Earth) Service also is allocated to the FG in 37–38 GHz.)

Non-FG: FIXED. FIXED-SATELLITE (space-to-Earth). MOBILE

FCC: Satellite Communications (25)

The FCC "designated" 37.6–38.6 GHz and 40–41 GHz for use on a primary basis for downlinks in the *Fixed Satellite Service* (FSS). It designated an uplink band at 48.2–50.2 GHz.

Massive systems

Several companies have applied for licenses for massive systems in this area of the spectrum, providing FSS and, in some cases, *Direct Broadcast Satellite Service* (DBS or BSS) and *Mobile Satellite Service* (MSS).

TRW applied for a $3.4 billion *Global EHF Satellite Network* (GESN). The GESN will employ four *geostationary* (GSO) satellites and 15 satellites in medium-Earth orbit, connected by optical intersatellite links.

GESN will provide data services ranging from T1 to OC-30, or the equivalent of more than 850,000 T1 circuits. "Market analysts have predicted that worldwide demand for T1 equivalent circuits will more than double in the next decade to a value of $810 billion by 2005," TRW said.

Motorola proposed M-Star, a $6.15 billion system of 72 non-GSO satellites. The M-Star system is intended for telecommunications infrastructure uses, such as links connecting base stations in cellular, PCS, or wireless local loop telephone networks.

"The M-Star system will enable the vision of lightweight devices that provide varied forms of communications including voice, data, images, and even video, to become a reality by reducing the required infrastructure investment and enhancing competition among wireless service providers," the company said.

Motorola acknowledged that elements of M-Star are similar to its Celestri system (see 17.7–17.8 GHz): "It is possible that in the future these two communications payloads could be combined into one integrated satellite constellation."

Other applicants for similar millimeter wave satellites include Denali Telecom, for the Pentriad system of 13 *highly elliptical orbit* (HEO) satellites; Orbital Sciences, for a seven-satellite system called OrbLink; Globalstar, for the GS-40 system of 80 NGSO satellites; GE American Communications, for the GE*StarPlus system of 11 GSO satellites; and Lockheed Martin, for nine GSO satellites.

Hughes Communications also is an applicant for this spectrum, proposing 56 satellites in four separate systems. Hughes Expressway will use 14 GSO satellites. The SpaceCast system of six GSO satellites will focus on digital video broadcasting, with emphasis on small uplink dish antennas.

"Whereas Expressway is primarily a symmetrical system for the provision of two-way telecommunications," Hughes said, "SpaceCast is primarily a one-way video and multimedia distribution system."

Hughes also applied for the StarLynx system of four GSO satellites for initial service, with 20 in *medium Earth orbit* (MEO) to be added later. StarLynx would offer two-way broadband service to "small user terminals for use in conjunction with personal computers and other electronic devices as well as terminals that can be mounted on vehicles." Meanwhile, Hughes subsidiary PanAmSat applied for the V-Stream system of 12 GSO satellites.

Loral Space and Communications proposed the CyberPath system higher in the millimeter wave spectrum (see 40.5–41 GHz). Teledesic Corp. applied for a $1.8 billion 72-satellite NGSO addition to its currently licensed NGSO system (see 18.8–19.3 GHz).

North Atlantic Treaty Organization (NATO) joint frequency agreements provide for fixed military systems in 37–39.5 GHz.

38.6-39.5 GHz

FG: US291

Non-FG: US291 FIXED. FIXED-SATELLITE (space-to-Earth). MOBILE.

FCC: Auxiliary Broadcasting (74). Fixed Microwave (101).

This is part of the 38.6–40-GHz (39 GHz) Wireless Services spectrum, used largely by Fixed Service carriers to provide business voice and data communications in competition with local exchange telephone companies. Major licensees in this business include WinStar and ART.

The FCC described the uses of this spectrum as follows: "In addition to providing support for existing services (e.g. broadband PCS, cellular, and other commercial and private mobile radio operations), 39 GHz band providers plan to use this spectrum to satisfy needs for a host of other fixed services such as (1) wireless local loops, (2) call termination or origination services to long distance companies, (3) connection of the customers of a *competitive access provider* (CAP) or a *local exchange carrier* (LEC) to its fiber rings, (4) connection and interconnection services to private networks operated by business and government as well as other institutions, (5) Internet access, and (6) cable headend applications." In contrast to the traditional scheme in which incumbent licensees in this spectrum designed their own service areas, the FCC decided to authorize all additional 38.6–40-GHz licenses in pre-defined, standardized geographic areas. This process began in 1995 in Docket ET 95-183 and fostered contention regarding the treatment of applications already on file.

In a July 29, 1999 order in that docket, the Commission settled on *Economic Areas* (EAs) as the geographic unit for this spectrum. EAs, defined by the Bureau of Economic Analysis in the U.S. Department of Commerce, are larger than the *Basic Trading Areas* (BTAs) and smaller than the *Metropolitan Trading Areas* (MTAs) used in other services.

Licensees who want service areas smaller than EAs can, with FCC approval, partition portions of their areas to others, and disaggregate portions of their spectrum to others.

The FCC offered 175 licenses (172 EAs and three EA-like areas covering Guam and Northern Mariana Islands, Puerto Rico and the U.S. Virgin Islands, and American Samoa) for each of 14 paired 50-MHz blocks (designated A through N), each consisting of channels 1-A through 14-A (in 38.6–39.3 GHz) and 1-B through 14-B (in 39.3–40 GHz), with mutually exclusive applications resolved by auction. The May 2000 auction raised more than $410 million from 29 bidders who won 2,173 of the 2,450 available licenses.

Licensees in 38.6–40 GHz may offer fixed point to point and point to multipoint operations. They will not be required to construct a specific number of stations. Instead, upon license renewal, they must show the FCC that they offer "substantial service" to their license area. Auction winners must protect the service areas of incumbent licensees.

In an unexpected development, the FCC also decided to allow 38.6–40-GHz licensees to offer mobile operations. "Permitting such [mobile] flexibility will enable providers to modify their offerings quickly and efficiently to provide the services that consumers demand and that technology makes possible," the FCC said. "[S]ome parties may be developing, or planning to develop, mobile services technology capable of operating without interference to fixed facilities in this band." *North Atlantic Treaty Organization* (NATO) joint frequency agreements provide for fixed military systems in 37–39.5 GHz.

The FCC Part 15 rules permit unlicensed devices to emit only very low level emissions in the spectrum above 38.6 GHz.

39.5-40 GHz

FG: US291 G117 FIXED-SATELLITE (space-to-Earth). MOBILE-SATELLITE (space-to-Earth)

Non-FG: US291 FIXED. FIXED-SATELLITE (space-to-Earth). MOBILE. MOBILE-SATELLITE (space-to-Earth)

FCC: Auxiliary Broadcasting (74). Fixed Microwave (101)

This band contains the 50-MHz-wide channels 5-B through 14-B of the 39-GHz Fixed Services spectrum. See 38.6–39.5 GHz for information about this band.

North Atlantic Treaty Organization (NATO) joint frequency agreements provide for military fixed satellite and mobile satellite downlinks in 39.5–40.5 GHz, paired with 50.4–51.4 GHz.

40-40.5 GHz

FG: G117 EARTH EXPLORATION-SATELLITE (Earth-to-space). FIXED-SATELLITE (space-to-Earth). MOBILE-SATELLITE (space-to-Earth). SPACE RESEARCH (Earth-to-space). Earth Exploration-Satellite (space-to-Earth).

Non-FG: FIXED-SATELLITE (space-to-Earth). MOBILE-SATELLITE (space-to-Earth).

FCC: Satellite Communications (25)

The FCC "designated" 37.6–38.6 GHz and 40–41 GHz for use on a primary basis for downlinks in the Fixed Satellite Service. It designated an uplink band at 48.2–50.2 GHz.

Among its scientific uses, this band is expected to support manned space exploration. *Space Research Service* (SRS) and *Earth Exploration Satellite Service* (EESS) users at 40.0–40.5 GHz will have to share with commercial *Fixed Satellite Service* (FSS) users.

North Atlantic Treaty Organization (NATO) joint frequency agreements provide for military fixed satellite and mobile satellite downlinks in 39.5–40.5 GHz, paired with 50.4–51.4 GHz.

The FCC identifies the "Q-band" as 33–50 GHz and the "U-band" as 40–60 GHz.

40.5-41 GHz

FG: US211

Non-FG: US211 FIXED-SATELLITE (space-to-Earth). BROADCASTING. BROADCASTING-SATELLITE. Mobile. Fixed

FCC: Satellite Communications (25)

The FCC "designated" 37.6–38.6 GHz and 40–41 GHz for use on a primary basis for downlinks in the *Fixed Satellite Service* (FSS). It designated an uplink band at 48.2–50.2 GHz.

One of the first applicants for satellite spectrum in this range is Loral Space and Communications. Its $1.2 billion CyberPath system of ten geostationary satellites will offer "wideband interactive and multimedia one-way and two-way services on demand...anywhere in the world covered by the spot beams."

The Broadcasting Service has long been allocated in the U.S. on a rare "permitted" basis in 40.5–42.5 GHz. The FCC changed its allocation status to primary, in view of action by the 1995 World Radio Conference that granted all previously "permitted" services primary status.

In any case, no broadcasting—defined as transmission for the direct and free receipt by the general public—is allowed in the U.S. millimeter wave spectrum. This is mainly in our view because no one has proposed it.

Challenges to radio wave propagation in this part of the spectrum include the effects of rain, attenuation by water vapor and oxygen, the effects of clouds and clear air, and tropospheric scintillation.

Nosebleed Bands

The FCC is engaged in a long-term, sweeping examination of the millimeter wave bands above 40 GHz—the so-called "nosebleed" portion of the spectrum (Dockets ET 94-124, IB 97-95 and WT 98-136). "Until recently, commercial use of this spectrum was not economically viable," the FCC said. "However, recent technological advances make this spectrum increasingly usable for commercial services and products."

It proposed to allocate several of these bands to unlicensed devices, non-communication devices such as vehicle radars, and undefined commercial "wireless services."

"Millimeter wave technology applications could include transmission of high resolution video images, access to large databases, and communication system backbones," the FCC said. "While spectrum to accommodate wide bandwidth applications is becoming scarce below 40 GHz, the millimeter wave region of the spectrum is largely unused and can accommodate those bandwidths."

The *Telecommunications Industry Association* (TIA) has stated that the "most important technology for supporting new wireless networks in the bands above 30 GHz will be the *High Density Fixed Services* (HDFS)." TIA cited HDFS applications including local access, inter-cell links for mobile and wireless local loop networks, fiber backdrop, local TV distribution, broadband Global Information Infrastructure access, intelligent transport, radio local area networks and Asynchronous Transfer Mode transport, among others.

"The fundamental premise of HDFS systems is that they would provide a cheaper means than optical fiber for connecting medium bandwidth customers," according to the *National Telecommunications and Information Administration* (NTIA). "It is too soon," NTIA said in 2000, "to predict which, if any, of the proposed HDFS systems will be commercially successful. Most of them would assume the role of a *Competitive Local Access Provider* (CLEC), competing with the local phone system, but additionally supplying broadband Internet access and possibly TV/video programming."

HDFS is a somewhat "fuzzy category," NTIA added, noting that a reason for defining a category of HDFS separate from the Fixed Services is that the density and high power of HDFS stations suggests that it would be impractical for HDFS to share spectrum with satellite services as conventional fixed stations often do.

41-42.5 GHz

FG: US211

Non-FG: US211 FIXED. BROADCASTING. BROADCASTING-SATELLITE. MOBILE.

The FCC "designated" this and other nearby bands for commercial "wireless services" to be defined more specifically in future proceedings (Docket IB 97-95) (see 40.5–41 GHz).

42.5-43.5 GHz

FG: US342 FIXED. FIXED-SATELLITE (Earth-to-space). MOBILE except aeronautical mobile. RADIO ASTRONOMY

Non-FG: US342 RADIO ASTRONOMY

Major radio astronomy observatories are active in this band, including the ten *Very Large Baseline Array* (VLBA) sites across the continental U.S., Hawaii, and the U.S. Virgin Islands.

The band is used for sensitive observations of the extended continuum of radio sources. The band is important for astronomy because of its one GHz width and its location in the spectrum.

The continuum emission observed in this band includes emission from ionized gas around newly born stars. This emission provides information on the state of the interstellar medium associated with star formation. The band also contains molecular spectral lines used in studies of the birth and death of stars.

NASA is considering this band and 39.5-40 GHz as uplinks to pair with 37.5-38 GHz.

43.5-45.5 GHz

FG: G117 FIXED-SATELLITE (Earth-to-space). MOBILE-SATELLITE (Earth-to-space)

Non-FG: None

The Milstar uplink is in this exclusive government band (see 20.2-21.2 GHz). This band falls within the 42.5-46.7-GHz spectrum that is undesignated for any commercial use, according to the FCC (in Docket IB 97-95) (see 40.5-41 GHz).

45.5-46.9 GHz

FG: S5.554 MOBILE. MOBILE-SATELLITE (Earth-to-space). RADIONAVIGATION-SATELLITE

Non-FG: S5.554 MOBILE. MOBILE-SATELLITE (Earth-to-space). RADIONAVIGATION-SATELLITE

FCC: RF Devices (15)

Portions of this band fall within the 42.5–46.7-GHz spectrum that is "undesignated" for any commercial use, according to the FCC (in Docket IB 97-95) (see 40.5–41 GHz).

The FCC has, however, designated 46.7–46.9 GHz for anti-collision vehicle radar, which also is allowed in 76–77 GHz where most international development occurs.

Vehicle radar includes forward-looking applications such as radar cruise control, which adjusts vehicle speed based on traffic ahead; collision warning; and collision avoidance that affects vehicle control and braking in response to an impending collision. Side- and rear-looking radars detect obstacles by monitoring objects less visible to the driver. They alert the driver with visual displays or sounds.

Fearful of product-liability suits from accidents involving radar-equipped cars, auto manufacturers reportedly are positioning vehicle radar as a "comfort feature" and not a "safety feature."

FCC Part 15 rules permit the use of 46.7–46.9 GHz and 76–77 GHz for data communications as long as the primary mode of operation is vehicle radar. Operation under the Part 15 vehicle radar rules in these bands is not allowed on aircraft or satellites.

46.9-47 GHz

FG: S5.554 MOBILE. MOBILE-SATELLITE (Earth-to-space). RADIONAVIGATION-SATELLITE

Non-FG: S5.554 MOBILE. MOBILE-SATELLITE (Earth-to-space). RADIONAVIGATION-SATELLITE. FIXED

The FCC designated this and other nearby bands for commercial "wireless services" to be defined more specifically in future proceedings (Docket IB 97-95) (see 40.5–41 GHz).

47-47.2 GHz

FG: None

Non-FG: AMATEUR. AMATEUR-SATELLITE

FCC: Amateur (97)

This is the 6-mm Amateur Radio band. Most amateur millimeter wave experiments occur in this band. The world record for amateur communication at 47 GHz is 286 km, held by French operators.

47.2-48.2 GHz

FG: None

Non-FG: S5.555 FIXED. FIXED-SATELLITE (Earth-to-space) US297. MOBILE US264

The FCC "designated" this and other nearby bands for commercial "wireless services" (Docket IB 97-95 and Docket WT 98-136; see also Docket ET 94-124, 40.5–41 GHz).

However, in December 2000 it closed a *Notice of Proposed Rulemaking* (NPRM) in Docket WT 98-136, canceling momentum toward rules and licenses in 47.2–48.2 GHz until some future date. The cancellation is a good example of what can happen when a docket suffers from rapidly changing market conditions, insufficient advocacy, and an industry unenthusiastic about exotic technology.

Blimps in Proto-space

In July 1997, the FCC allocated 47.2–48.2 GHz to point-to-multipoint services, including wireless services from "airships" (blimps, balloons, or "platforms") as proposed by Sky Station International.

The $4.2 billion Sky Station *Global Stratospheric Telecommunications Service* (GSTS) would consist of hundreds of geostationary, airborne *High Altitude Platform Stations* (HAPS) supported by 500-foot long balloons and positioned over major geographic markets worldwide. Each 37-ton craft would operate from about 100,000 feet altitude.

"The stratosphere is a region of proto-space that the United States and other countries have steadfastly refused to define as part of any nation's sovereign airspace," Sky Station asserted.

Few position-disturbing forces exist in the stratosphere, which is above 99 percent of the oxygen-containing atmosphere.

GSTS would compete with broadband satellite offerings and provide "global and fully portable dollar a day high speed" Internet connectivity. Each airship could cover 400,000 square miles and would be capable of creating 2,100 separate geographical cells.

The FCC found HAPS to be the "dominant use" of the 47-GHz spectrum, a conclusion that satellite companies vehemently disputed. This was an explicit reversal of the FCC's previous finding that the dominant use would be similar to the Local Multipoint Distribution Service (LMDS, 27.5–29.5 GHz).

The Commission said that "the assertedly lower capital requirements, compared to satellite systems, and the flexibility to sequentially activate stratospheric platforms as demand and revenue warrant, present clear, if not yet demonstrated, benefits for a variety of applications."

It divided the 47.2–48.2-GHz segment into five blocks of 200 MHz each for licensing, with each block consisting of a pair of 100-MHz channels separated by 500 MHz. The blocks would be licensed in each of 12 *Regional Economic Area Groupings* (REAGs).

Safety Concerns

"The possibility that these platforms, or parts of them, could fail may present a significant safety concern," the FCC said, adding that "launching and retrieving the platforms may present dangers to aviation."

However, the airships will use "multiple redundant safety features that will eliminate the risk of injury or harm to airborne vehicles and Earth inhabitants," Sky Station said. "These safety features includes multiply-redundant balloon modules (only half of which are sufficient to ensure a gentle and controlled landing), back-up parachute systems, multiple motion sensors (radar altimeters, accelerometers, and GPS devices), and compatibility with helicopter recovery operations. ...Damage on Earth is no more likely to occur than from satellite launch and de-orbit operations."

Nevertheless, NTIA concluded that "There are substantial vehicle engineering challenges and the feasibility of such platforms remains to be demonstrated."

Other types of services, including terrestrial multipoint communications, conventional satellites, and even mobile services were to be allowed in the 47-GHz band.

Poor Participation

Sky Station later asked the FCC to terminate the proceeding because it is redesigning its system "in response to the rapid introduction of new broadband telecommunications technologies and evolving market requirements." Without technical information on the redesign, the FCC said, it is unable to establish workable interference rules.

Moreover, the agency was unsatisfied with input it had received: "Of those comments the Commission received, the majority were brief and addressed few of the issues on which we sought comment. Most commenters did not address the relevant technical issues. ...It is unclear at this time that there is sufficient interest in the 47-GHz band to proceed with promulgating service, licensing, and operating rules based on the current record."

The FCC terminated the NPRM and directed its staff to "further explore with any interested parties possible uses of the 47-GHz spectrum, including use by stratospheric-based platforms and other technologies."

ITU Developments

The 2000 *World Radiocommunication Conference* (WRC-2000) resolved to authorize HAPS as a part of IMT-2000 (see 1850–1990 MHz). Another resolution called for study of HAPS in terrestrial frequencies above 3 GHz. Sky Station said that it has "elevated HAPS to a significant agenda issue" for WRC-2003.

48.2–50.2 GHz

FG: S5.555 US342 FIXED. FIXED-SATELLITE (Earth-to-space) US297. MOBILE US264

Non-FG: S5.555 US342 FIXED. FIXED-SATELLITE (Earth-to-space) US297. MOBILE US264

FCC: Satellite Communications (25)

The FCC "designated" this band on a primary basis for uplinks in the *Fixed-Satellite Service* (FSS), when it designated 37.6–38.6 GHz and 40–41 GHz for downlinks, in Docket IB 97-95 (see also 40.5–41 GHz).

The FCC also designated 49.7–50.2 GHz as a candidate segment for *Intelligent Transportation Systems* (ITS). It could support vehicle radar and links used for the automated, remote control of groups of vehicles ("platooning") and for coordination of evasive action between vehicles about to collide.

At least some of these ITS applications will be served *by Dedicated Short Range Communications* (DSRC) in another band (see 5.85–5.925 GHz).

Radio astronomers use 48.94–49.04 GHz to study molecular material in galaxies. Observatories active in the band include *Very Large Baseline Array* (VLBA) sites across the continental U.S., Hawaii, and the U.S. Virgin Islands.

50.2–50.4 GHz

FG: US246 EARTH EXPLORATION-SATELLITE (passive). SPACE RESEARCH (passive)

Non-FG: US246 EARTH EXPLORATION-SATELLITE (passive). SPACE RESEARCH (passive)

No stations may transmit in this scientific band. Passive (non-transmitting) sensors on remote sensing satellites use this spectrum to measure atmospheric temperature at different altitudes, in weather forecasting, and in climate studies. This is a use of the *Earth Exploration-Satellite Service* (EESS), allocated to this band and several nearby bands (Docket ET 99-261).

According to NASA, 50.2–50.4 GHz is the single most important passive sensing band in 50–60 GHz because it is used as a reference for all measurements in that spectrum. It has the most stringent interference criteria, as the band's passive-only allocation suggests.

The FCC defines "passive sensor" as "a measuring instrument in the earth exploration-satellite service (EESS) or in the space research service by means of which information is obtained by reception of radio waves of natural origin."

50.4–51.4 GHz

FG: G117 FIXED. FIXED-SATELLITE (Earth-to-space). MOBILE. MOBILE-SATELLITE (Earth-to-space)

Non-FG: FIXED. FIXED-SATELLITE (Earth-to-space). MOBILE. MOBILE-SATELLITE (Earth-to-space)

The FCC designated this and other nearby bands for commercial "wireless services" to be defined more specifically in future proceedings (Docket IB 97-95) (see 40.5–41 GHz).

Future U.S. satellites in this band are limited to military uplinks, paired with 39.5–40.5 GHz. Agreements to implement military fixed satellite and mobile satellite operations in this band have been coordinated among the *North Atlantic Treaty Organization* (NATO) nations.

As of this writing, however, the U.S. appears to have no firm plans to launch military satellites in 50.4–51.4 GHz and 39.5–40.5 GHz.

51.4–52.6 GHz

FG: FIXED. MOBILE

Non-FG: FIXED. MOBILE

In Docket ET 99-261, the FCC allocated 51.4–52.6 GHz and 58.2–59 GHz to government and non-government fixed and mobile services on a primary basis, and adjusted the allocations in this spectrum to protect scientific operations from active (transmitting) uses. See 50.2–50.4 GHz for information on scientific uses of this spectrum.

52.6–54.25 GHz

FG: US246 EARTH EXPLORATION-SATELLITE (passive). SPACE RESEARCH (passive)

Non-FG: US246 EARTH EXPLORATION-SATELLITE (passive). SPACE RESEARCH (passive)

No transmissions are permitted in this band. See 50.2–50.4 GHz for information on scientific uses of this spectrum.

54.25-55.78 GHz

FG: US263 EARTH EXPLORATION-SATELLITE (passive). INTER-SATELLITE S5.556A. SPACE RESEARCH (passive)

Non-FG: US263 EARTH EXPLORATION-SATELLITE (passive). INTER-SATELLITE S5.556A. SPACE RESEARCH (passive)

No transmissions are permitted in this environmental remote sensing band except for *Intersatellite Service* (ISS) links.

To protect passive sensor operations in the 54.25–59.3-GHz spectrum, ISS links are generally limited to satellites in geostationary orbit. Such links are less likely to affect remote sensing satellites, which operate closer to the Earth.

Several adjoining bands are allocated to ISS links between satellites. ISS is not widely used at this writing, but the FCC has declared a pressing need for it in both federal and non-federal systems.

55.78-56.9 GHz

FG: US263 US353 EARTH EXPLORATION SATELLITE (passive). FIXED. INTER-SATELLITE S5.556A. MOBILE S5.558. SPACE RESEARCH (passive).

Non-FG: US263 US353 EARTH EXPLORATION SATELLITE (passive). FIXED. INTER-SATELLITE S5.556A. MOBILE S5.558. SPACE RESEARCH (passive).

See 50.2–50.4 GHz for information on scientific uses of this spectrum. Footnote US353 reflects that radio astronomy may be conducted in this band from space.

56.9–57 GHz

FG: US263 EARTH EXPLORATION-SATELLITE (passive). FIXED. INTER-SATELLITE G128. MOBILE S5.558. SPACE RESEARCH (passive)

Non-FG: US263 EARTH EXPLORATION-SATELLITE (passive). FIXED. MOBILE S5.558. SPACE RESEARCH (passive)

See 50.2–50.4 GHz for information on scientific uses of this spectrum.

57–58.2 GHz

FG: US263 EARTH EXPLORATION-SATELLITE (passive). FIXED. INTER-SATELLITE S5.556A. MOBILE S5.558. SPACE RESEARCH (passive)

Non-FG: US263 EARTH EXPLORATION-SATELLITE (passive). FIXED. INTER-SATELLITE S5.556A. MOBILE S5.558. SPACE RESEARCH (passive)

FCC: RF Devices (15)

Docket ET 99-261 made 57–58.2 GHz and 58.2–59 GHz available to unlicensed devices under FCC Part 15 rules, in addition to these bands' allocations to scientific and conventional terrestrial services.

This action opened the entire 57–64-GHz spectrum to unlicensed devices. The Commission declined, however, to extend this unlicensed spectrum to include 64–66 GHz out of concern for interference to other users in that band and given the different propagation characteristics outside of 57–64 GHz.

The wide bandwidth available in this portion of the spectrum can offer capacity approaching that of coaxial cable and fiber optics. It is "most suitable for high re-use, short range communications with a correspondingly low probability of co-channel interference," according to a FCC engineering bulletin.

"The gain of an antenna is proportional to its effective aperture, or size with respect to wavelength. As the wavelength at 58 GHz is in the order of 5 millimeters, an antenna of relatively small physical size is large compared to wavelength and will have correspondingly good gain, narrow beamwidth, and be highly directional. In fixed point-to-point systems, this means the likelihood of co-channel interference is even further reduced."

For these reasons, the FCC rejected claims by the licensed fixed wireless industry that unlicensed, "unregulated" spectrum is unsuitable for such uses as connecting mobile base stations.

"Unlicensed use provides a multitude of users with a high level of flexibility in how they use the spectrum," the Commission said. "Making this spectrum available to unlicensed Part 15 devices harmonizes U.S. spectrum applications with international spectrum applications and allows the U.S. market to benefit from technology currently being used in Europe."

The FCC authorized Gigalink, the first unlicensed millimeter wave fixed radio product, at this writing (see 59.3–64 GHz).

This 57–58.2-GHz band contains the coordination channel used by unlicensed millimeter wave systems: 57–57.05 GHz, as part of the "spectrum etiquette" for such devices (see 59.3–64 GHz). The channel is for techniques to reduce interference that may occur between unlicensed transmitters operating in the same band. This channel was formerly 59–59.05 GHz.

See 50.2–50.4 GHz for information on scientific uses of this spectrum.

58.2–59 GHz

FG: US353 US354 EARTH EXPLORATION-SATELLITE (passive). FIXED. MOBILE. SPACE RESEARCH (passive)

Non-FG: US353 US354 EARTH EXPLORATION-SATELLITE (passive). FIXED. MOBILE. SPACE RESEARCH (passive)

FCC: RF Devices (15)

Footnote US353 reflects that radio astronomy may be conducted in part of the 58.2–59-GHz band from space, as it is one of the bands that cannot be used for terrestrial radio astronomy due to atmospheric absorption of radio energy. Footnote US354 helps to prevent interference to radio astronomy observations. See 57–58.2 GHz for information about this band.

59-59.3 GHz

FG: US353 EARTH EXPLORATION-SATELLITE (passive). FIXED. INTER-SATELLITE S5.556A. MOBILE S5.558. RADIOLOCATION S5.559. SPACE RESEARCH (passive)

FG: US353 EARTH EXPLORATION-SATELLITE (passive). FIXED. MOBILE S5.558. RADIOLOCATION S5.559. SPACE RESEARCH (passive)

FCC: RF Devices (15)

See 50.2–50.4 GHz for information on scientific uses of this spectrum and 57–58.2 GHz regarding unlicensed use.

59.3-64 GHz

FG: S5.138 US353 FIXED. INTER-SATELLITE. MOBILE S5.558. RADIOLOCATION S5.559

Non-FG: S5.138 US353 FIXED. MOBILE S5.558. RADIOLOCATION S5.559

FCC: RF Devices (15). ISM Equipment (18)

Propagation in this band is extremely limited by oxygen in the atmosphere. This attenuation is significant throughout the millimeter wave spectrum, but increases around 60 GHz and 120 GHz. The limited transmission range enables a high concentration of transmitters to be used in an area.

The FCC designated this band for unlicensed Part 15 devices: "Possible uses of this band include the development of short-range, high capacity wireless radio systems that could be used for educational and medical applications, and for wireless access to libraries or other information databases," it said (see 40.5–41 GHz).

FCC rule section 15.255 for this band implements a "spectrum etiquette," or procedure for spectrum access, developed by the industry *Millimeter Wave Communications Working Group* (MWCWG).

Transmitter ID

The etiquette includes specifications for an *automatic transmitter identification system* (ATIS), power limits, and a coordination channel (formerly 59–59.05 GHz, now 57–57.05 GHz) to be used exclusively with techniques that transmitters could use to help reduce interference.

The transmitter ID requirement is limited to transmissions that emanate from inside of a building. "Indoor equipment will be required to have the ID because indoor equipment is under the control of the system operator. The system operator knows its equipment and thus can decode the ID information and find out which transmitter is interfering with the rest of its system," the FCC said. ATIS proposals flopped in other bands (see 154–156.2475 MHz).

Figure 16
GigaLink by Harmonix Corp. is the first, license-free 60 GHz millimeter wave broadband radio. (Harmonix photo)

First Unlicensed 60 GHz Radio

The FCC authorized the first unlicensed, Part 15, 60-GHz radio in late 2000: the Harmonix Corp. GigaLink, which provides up to 622 megabits per second data transmission rate.

"Due to the unique oxygen absorption properties of the 60-GHz spectrum, many GigaLink radios can be co-located without interference problems, with the ability to deploy up to 100,000 radios in a ten square kilometer area," the company said. "Additionally, GigaLink 60-GHz radios are immune to adverse weather conditions, unlike other high-speed fiber alternatives."

Motorola plans to offer a 60-GHz broadband local-area network codenamed "Piano," for short range connections between personal devices such as wireless phones and portable computers. However, Piano devices will first appear in more conventional, lower frequency unlicensed bands.

Not all uses of this spectrum are for communications. Industrial safety and control sensors use this part of the spectrum, as do collision avoidance radars and satellite sounding instruments that observe sea conditions and measure temperature at different altitudes.

"Atmospheric temperature profiles are among the essential parameters which are routinely used by meteorological services for operational weather forecasting, and by the scientific community involved in climate and environmental monitoring studies," the FCC said. "Atmospheric temperature profiles are currently obtained from spaceborne sounding instruments working in the infrared spectrum and in the microwave spectrum, including oxygen absorption around 60 GHz."

ISM Applications

The center frequency 61.25 GHz, ±250 MHz, is available to *Industrial, Scientific and Medical* (ISM) devices under FCC Part 18 rules.

The ISM frequency is used in magnetically contained plasmas in linear accelerators, cyclotron heating, and other exotic applications.

The *International Microwave Power Institute* (IMPI) reported that most current and future ISM applications in this band are "still in their research phase and, as a result, are highly confidential," but involve millimeter wave fusion and extremely high power levels.

The FCC identifies the "V-band" as 50–75 GHz and the "E-band" as 60–90 GHz.

64-65 GHz

FG: FIXED. INTER-SATELLITE. MOBILE except aeronautical mobile

Non-FG: FIXED. MOBILE except aeronautical mobile

65-66 GHz

FG: EARTH EXPLORATION-SATELLITE. FIXED. MOBILE except aeronautical mobile. SPACE RESEARCH

Non-FG: EARTH EXPLORATION-SATELLITE. FIXED. INTER-SATELLITE. MOBILE except aeronautical mobile. SPACE RESEARCH

66-71 GHz

FG: S5.554 MOBILE S5.553 S5.558. MOBILE-SATELLITE. RADIONAVIGATION. RADIONAVIGATION-SATELLITE

Non-FG: S5.554 INTER-SATELLITE. MOBILE S5.553 S5.558. MOBILE-SATELLITE. RADIONAVIGATION. RADIONAVIGATION-SATELLITE

The footnotes provide for feeder link use of this band in the *Mobile Satellite Services* (MSS).

Aviation radar in this band includes an inclement weather guidance system. The 68.9–72.7-GHz spectrum is a candidate for *Intelligent Transportation Systems* (ITS, see 48.2–50.2 GHz).

71-74 GHz

FG: US270 FIXED. FIXED-SATELLITE (Earth-to-space). MOBILE. MOBILE-SATELLITE (Earth-to-space)

Non-FG: US270 FIXED. FIXED-SATELLITE (Earth-to-space). MOBILE. MOBILE-SATELLITE (Earth-to-space)

74-75.5 GHz

FG: FIXED. FIXED-SATELLITE (Earth-to-space) US297. MOBILE

Non-FG: FIXED. FIXED-SATELLITE (Earth-to-space) US297. MOBILE

The FCC identifies the "E-band" as 60–90 GHz and the "W-band" as 75–110 GHz.

75.5-76 GHz

FG: None

Non-FG: AMATEUR. AMATEUR-SATELLITE

FCC: AMATEUR (97)

The 4-mm Amateur Radio band extends from 75.5–81 GHz.

76-77 GHz

FG: S5.560 RADIOLOCATION

Non-FG: RADIOLOCATION. Amateur

FCC: RF Devices (15)

This is an important band for anti-collision vehicle radar (see 45.5–46.9 GHz). Most international development in vehicle radar is in 76–77 GHz.

In July 1998 the FCC suspended Amateur Radio use of this band to ensure against interference to vehicle radar.

The FCC noted, however, that this suspension "will not have an immediate impact on amateur operators because there is little or no use of this band." The Commission said it plans to revisit this issue "in about five years," or when spectrum sharing standards are developed.

FCC Part 15 rules permit the use of 46.7–46.9 GHz and 76–77 GHz for data communications as long as the primary mode of operation is vehicle radar (see also 40.5–41 GHz).

77-77.5 GHz

FG: S5.560 RADIOLOCATION

Non-FG: RADIOLOCATION. Amateur. Amateur-Satellite

FCC: Amateur (97)

This allocation is similar to the 77.5–78-GHz band, except that Amateur Radio is secondary in this band.

77.5-78 GHz

FG: S5.560 RADIOLOCATION

Non-FG: RADIOLOCATION. AMATEUR. AMATEUR-SATELLITE

FCC: Amateur (97)

In July 1998, the FCC allocated this band to the Amateur Radio Service on a co-primary basis "to ensure that future amateur station access to spectrum near 77 GHz is maintained without the threat of preemption by higher priority services," and to continue to foster amateur experimentation with millimeter waves.

78-81 GHz

FG: S5.560 RADIOLOCATION

Non-FG: S5.560 RADIOLOCATION. Amateur. Amateur-Satellite

FCC: Amateur (97)

The Defense Department is developing exotic systems in 80–120 GHz, including quasi-optical gyrotrons for materials processing and traveling wave tubes for countermeasures.

An 80-GHz microscope has been developed. It measures reflected energy from the surface of a material to determine its electrical characteristics such as resistivity.

81-84 GHz

FG: FIXED. FIXED-SATELLITE (space-to-Earth). MOBILE. MOBILE-SATELLITE (space-to-Earth)

Non-FG: FIXED. FIXED-SATELLITE (space-to-Earth). MOBILE. MOBILE-SATELLITE (space-to-Earth)

84-86 GHz

FG: S5.561 US211 FIXED. MOBILE

Non-FG: S5.561 US211 FIXED. MOBILE. BROADCASTING. BROADCASTING-SATELLITE

Satellite remote sensing instruments use this band and the adjacent 86–92-GHz band.

The FCC proposed to allocate spectrum in this band to unlicensed devices (see 40.5–41 GHz).

86-92 GHz

FG: US246 EARTH EXPLORATION-SATELLITE (passive). RADIO ASTRONOMY US74. SPACE RESEARCH (passive)

Non-FG: US246 EARTH EXPLORATION-SATELLITE (passive). RADIO ASTRONOMY US74. SPACE RESEARCH (passive)

No transmissions are permitted in this widely used scientific band.

Because there is little energy absorption by atmospheric water and oxygen, this band is excellent for observations of celestial objects. It is also used by satellite remote sensing instruments that observe weather conditions including water vapor, clouds, ice, and snow.

TRW is developing a passive (non-transmitting) millimeter wave real-time video camera in this band (see also 92–95 GHz). Applications include aids to pilot vision, vehicle collision avoidance systems, airport traffic monitoring, and sentry operations.

The FCC identifies the "W-band" as 75–110 GHz and the "F-band" as 90–170 GHz.

92-95 GHz

FG: S5.149 FIXED. FIXED-SATELLITE (Earth-to-space). MOBILE. RADIOLOCATION

Non-FG: S5.149 FIXED. FIXED-SATELLITE (Earth-to-space). MOBILE. RADIOLOCATION

According to the FCC, "the 94 GHz band is employed for radio astronomy, U.S. Government passive imaging systems, and Department of Defense classified applications."

The reference to passive imaging probably refers to joint Federal Aviation Administration and Defense Department research in synthetic vision in this band, and also at 35 GHz, for use in inclement weather guidance and imaging of the airport runway, even if obscured by fog or smoke. Active radar in missiles is another application being developed in this band.

Millimeter wave imaging can display hidden contraband or weapons, including nonmetal objects, for such uses as airport security. It detects thermal emissions and reflections from materials. The object appears in the image as a shadow against the body. The technology sees through clothing.

The U.S. has internationally proposed to allocate 94–94.1 GHz for spaceborne cloud radars, which determine the vertical profile of clouds and their global distribution.

95–100 GHz

FG: S5.149 S5.554 MOBILE S5.553. MOBILE-SATELLITE. RADIONAVIGATION. RADIONAVIGATION-SATELLITE. Radiolocation

Non-FG: S5.149 S5.554 MOBILE S5.553. MOBILE-SATELLITE. RADIONAVIGATION. RADIONAVIGATION-SATELLITE. Radiolocation

The FCC proposed to allocate 94.7–95.7 GHz to vehicle collision-avoidance radars (see 45.5–46.9 GHz and 40.5–41 GHz).

Radar profilers use this band in studies of cloud distribution, internal motion and structure. The radars operate from land vehicles or aircraft, including unmanned drone aircraft. These measurements also are conducted at 33 GHz.

100–102 GHz

FG: S5.341 US246 EARTH EXPLORATION-SATELLITE (passive). SPACE RESEARCH (passive)

Non-FG: S5.341 US246 EARTH EXPLORATION-SATELLITE (passive). SPACE RESEARCH (passive)

No transmissions are permitted in this receive-only scientific band, used in satellite remote sensing and radio astronomy.

102-105 GHz

FG: S5.341 US211 FIXED. FIXED-SATELLITE (space-to-Earth)

Non-FG: S5.341 US211 FIXED. FIXED-SATELLITE (space-to-Earth)

The FCC proposed to allocate spectrum in this band to unlicensed devices (see 40.5–41 GHz).

105-116 GHz

FG: S5.341 US246 EARTH EXPLORATION-SATELLITE (passive). RADIO ASTRONOMY US74. SPACE RESEARCH (passive)

Non-FG: S5.341 US246 EARTH EXPLORATION-SATELLITE (passive). RADIO ASTRONOMY US74. SPACE RESEARCH (passive)

No transmissions are permitted in this receive-only scientific band, used in satellite remote sensing, and in radio astronomy for observations of the carbon monoxide spectral line and other spectral lines.

The FCC identifies the "F-band" as 90–170 GHz and the "D-band" as 110–170 GHz.

116-119.98 GHz

FG: S5.341 US211 US263 EARTH EXPLORATION-SATELLITE (passive). FIXED. INTER-SATELLITE. MOBILE S5.558. SPACE RESEARCH (passive)

Non-FG: S5.341 US211 US263 EARTH EXPLORATION-SATELLITE (passive). FIXED. INTER-SATELLITE. MOBILE S5.558. SPACE RESEARCH (passive)

The FCC proposed to allocate spectrum in this band to unlicensed devices (see 40.5–41 GHz).

Satellite remote sensing instruments use this band to measure atmospheric temperature and rainfall. Astronomers use the band in observations of ozone, carbon monoxide, and nitrous oxide.

Scientists are concerned about FCC proposals to allocate the band to new uses because of possible interference to remote sensing.

119.98-120.02 GHz

FG: S5.341 US211 US263 EARTH EXPLORATION-SATELLITE (passive). FIXED. INTER-SATELLITE. MOBILE S5.558. SPACE RESEARCH (passive). Amateur

Non-FG: S5.341 US211 US263 EARTH EXPLORATION-SATELLITE (passive). FIXED. INTER-SATELLITE. MOBILE S5.558. SPACE RESEARCH (passive). Amateur

This is another satellite remote sensing band used in atmospheric measurements and radio astronomy.

The 2.5 mm Amateur Radio band covers 119.98–120.02 GHz on a secondary basis.

120.02-126 GHz

FG: S5.138 US211 US263 EARTH EXPLORATION-SATELLITE (passive). FIXED. INTER-SATELLITE. MOBILE S5.558. SPACE RESEARCH (passive)

Non-FG: S5.138 US211 US263 EARTH EXPLORATION-SATELLITE (passive). FIXED. INTER-SATELLITE. MOBILE S5.558. SPACE RESEARCH (passive)

126-134 GHz

FG: FIXED. INTER-SATELLITE. MOBILE 909. RADIOLOCATION S5.559

Non-FG: FIXED. INTER-SATELLITE. MOBILE 909. RADIOLOCATION. S5.559

The FCC proposed to allocate spectrum in this band to unlicensed devices (see 40.5–41 GHz).

134-142 GHz

FG: S5.149 S5.554 S5.555 917 MOBILE S5.553 MOBILE-SATELLITE. RADIONAVIGATION. RADIONAVIGATION-SATELLITE. Radiolocation

Non-FG: S5.149 S5.554 S5.555 917 MOBILE S5.553 MOBILE-SATELLITE. RADIONAVIGATION. RADIONAVIGATION-SATELLITE. Radiolocation

The FCC proposed to allocate some of this spectrum to vehicle anti-collision radars (see 40.5–41.5 GHz). Military tracking and navigation radars operate around 140 GHz.

The FCC identifies the "D-band" as 110–170 GHz and the "G-band" as 140–220 GHz.

142-144 GHz

FG: None

Non-FG: AMATEUR. AMATEUR-SATELLITE

FCC: AMATEUR (97)

The 2.5-mm Amateur Radio band covers 142–149 GHz.

144-149 GHz

FG: S5.149 S5.555 RADIOLOCATION

Non-FG: S5.149 S5.555 RADIOLOCATION. Amateur. Amateur-Satellite

FCC: Amateur (97)

The 148.5–151.5-GHz spectrum and nearby bands are used for satellite remote sensing of atmospheric moisture.

The 2-mm Amateur Radio band covers 142–149 GHz.

149-150 GHz

FG: FIXED. FIXED-SATELLITE (space-to-Earth). MOBILE

Non-FG: FIXED. FIXED-SATELLITE (space-to-Earth). MOBILE

150-151 GHz

FG: S5.149 S5.385 US263 EARTH EXPLORATION-SATELLITE (passive). FIXED. FIXED-SATELLITE (space-to-Earth). MOBILE. SPACE RESEARCH (passive)

Non-FG: S5.149 S5.385 US263 EARTH EXPLORATION-SATELLITE (passive). FIXED. FIXED-SATELLITE (space-to-Earth). MOBILE. SPACE RESEARCH (passive)

151-164 GHz

FG: US211 FIXED. FIXED-SATELLITE (space-to-Earth)

Non-FG: US211 FIXED. FIXED-SATELLITE (space-to-Earth)

The FCC proposed to allocate spectrum in this band to unlicensed devices (see 40.5–41 GHz).

The Big Bang produced cosmic background radiation that peaks around 160 GHz (see 217–231 GHz).

164-168 GHz

FG: US246 EARTH EXPLORATION-SATELLITE (passive). RADIO ASTRONOMY. SPACE RESEARCH (passive)

Non-FG: US246 EARTH EXPLORATION-SATELLITE (passive). RADIO ASTRONOMY. SPACE RESEARCH (passive)

No transmissions are permitted in this receive-only scientific band. Remote sensing satellites use it to measure terrestrial water vapor and chlorine oxide.

168-170 GHz

FG: FIXED. MOBILE

Non-FG: FIXED. MOBILE

The 168–168.3-GHz segment is a candidate for *Intelligent Transportation Systems* (ITS, see 48.2–50.2 GHz).

170-174.5 GHz

FG: S5.149 S5.385 FIXED. INTER-SATELLITE. MOBILE 909

Non-FG: S5.149 S5.385 FIXED. INTER-SATELLITE. MOBILE 909

174.5-176.5 GHz

FG: S5.149 S5.385 US263 EARTH EXPLORATION-SATELLITE (passive). FIXED. INTER-SATELLITE. MOBILE 909. SPACE RESEARCH (passive)

Non-FG: S5.149 S5.385 US263 EARTH EXPLORATION-SATELLITE (passive). FIXED. INTER-SATELLITE. MOBILE 909. SPACE RESEARCH (passive)

Remote sensing satellites measure nitrous oxide in this band.

176.5-182 GHz

FG: S5.149 S5.385 US211 FIXED. INTER-SATELLITE. MOBILE 909

Non-FG: S5.149 S5.385 US211 FIXED. INTER-SATELLITE. MOBILE 909

182-185 GHz

FG: US246 EARTH EXPLORATION-SATELLITE (passive). RADIO ASTRONOMY. SPACE RESEARCH (passive)

Non-FG: US246 EARTH EXPLORATION-SATELLITE (passive). RADIO ASTRONOMY. SPACE RESEARCH (passive)

No transmissions are permitted in this receive-only scientific band.

Remote sensing satellites use it to measure atmospheric water vapor, clouds, precipitation, and ozone.

185-190 GHz

FG: S5.149 S5.385 US211 FIXED. INTER-SATELLITE. MOBILE 909

Non-FG: S5.149 S5.385 US211 FIXED. INTER-SATELLITE. MOBILE 909

190-200 GHz

FG: S5.341 S5.554

MOBILE S5.553. MOBILE-SATELLITE. RADIONAVIGATION. RADIONAVIGATION-SATELLITE

Non-FG: S5.341 S5.554 MOBILE S5.553. MOBILE-SATELLITE. RADIONAVIGATION. RADIONAVIGATION-SATELLITE

200-202 GHz

FG: S5.341 US263 EARTH EXPLORATION-SATELLITE (passive). FIXED. MOBILE. SPACE RESEARCH (passive)

Non-FG: S5.341 US263 EARTH EXPLORATION-SATELLITE (passive). FIXED. MOBILE. SPACE RESEARCH (passive)

Satellite remote sensing instruments measure nitrous oxide in this band.

202-217 GHz

FG: S5.341 FIXED. FIXED-SATELLITE (Earth-to-space). MOBILE

Non-FG: S5.341 FIXED. FIXED-SATELLITE (Earth-to-space). MOBILE

217-231 GHz

FG: S5.341 US246 EARTH EXPLORATION-SATELLITE (passive). RADIO ASTRONOMY US74. SPACE RESEARCH (passive)

Non-FG: S5.341 US246 EARTH EXPLORATION-SATELLITE (passive). RADIO ASTRONOMY US74. SPACE RESEARCH (passive)

No transmissions are permitted in this receive-only scientific band, one of the most important to millimeter wave radio astronomy.

This band is used to detect carbon monoxide, nitrous oxide, and other complex molecules found in gas within galaxies.

It also is used in studies of *cosmic background radiation* (CBR)—the broadband radio noise remaining from the Big Bang—the purported origin of the universe. A new spacecraft, the *Microwave Anisotropy Probe* (MAP), will measure nonuniformity in the CBR (see 21.4–22 GHz).

For Earth remote sensing, a 220-GHz receiver, part of a millimeter wave imaging radiometer, measures atmospheric water vapor, clouds, and precipitation in the *Defense Meteorological Satellite Program* (DMSP, see 1675–1700 MHz).

231-235 GHz

FG: US211 FIXED. FIXED-SATELLITE (space-to-Earth). MOBILE. Radiolocation

Non-FG: US211 FIXED. FIXED-SATELLITE (space-to-Earth). MOBILE. Radiolocation

235-238 GHz

FG: US263 EARTH EXPLORATION-SATELLITE (passive). FIXED. FIXED-SATELLITE (space-to-Earth). MOBILE. SPACE RESEARCH (passive)

Non-FG: US263 EARTH EXPLORATION-SATELLITE (passive). FIXED. FIXED-SATELLITE (space-to-Earth). MOBILE. SPACE RESEARCH (passive)

Remote sensing satellites measure ozone in this band.

238-241 GHz

FG: FIXED. FIXED-SATELLITE (space-to-Earth). MOBILE. Radiolocation

Non-FG: FIXED. FIXED-SATELLITE (space-to-Earth). MOBILE. Radiolocation

241-248 GHz

FG: S5.138 RADIOLOCATION

Non-FG: S5.138 RADIOLOCATION. Amateur. Amateur-Satellite

FCC: ISM Equipment (18). Amateur (97)

Industrial, Scientific and Medical (ISM) devices may use the 244–246-GHz segment. The 1-mm Amateur Radio band covers 241–250 GHz.

248-250 GHz

FG: None

Non-FG: AMATEUR. AMATEUR-SATELLITE

FCC: AMATEUR (97)

The 1-mm Amateur Radio band covers 241–250 GHz.

250-252 GHz

FG: S5.149 S5.555 EARTH EXPLORATION-SATELLITE (passive). SPACE RESEARCH (passive)

Non-FG: S5.149 S5.555 EARTH EXPLORATION-SATELLITE (passive). SPACE RESEARCH (passive)

Remote sensing satellites use this band to detect nitrous oxide.

252-265 GHz

FG: S5.149 S5.385 S5.554 S5.555 US211 MOBILE S5.553.
MOBILE-SATELLITE. RADIONAVIGATION. RADIONAVIGATION-SATELLITE

Non-FG: S5.149 S5.385 S5.554 S5.555 US211 MOBILE S5.553.
MOBILE-SATELLITE. RADIONAVIGATION. RADIONAVIGATION-SATELLITE

265-275 GHz

FG: S5.149 FIXED. FIXED-SATELLITE (Earth-to-space). MOBILE. RADIO ASTRONOMY

Non-FG: S5.149 FIXED. FIXED-SATELLITE (Earth-to-space). MOBILE. RADIO ASTRONOMY

275-300 GHz

FG: S5.565 FIXED. MOBILE

Non-FG: S5.565 FIXED. MOBILE

Frequencies above 300 GHz are not allocated to any radio service in the United States.

Officially, however, the FCC has applied Footnote S5.565 to the band 300–400 GHz and designated it for the Amateur Radio Service.

Research is taking place above 300 GHz into radio technology, especially for applications such as remote sensing, radiolocation, and radio astronomy.

NASA's *Submillimeter Wave Astronomy Satellite* (SWAS), launched in December 1998, investigates the astrochemical phenomena in the 450–550-GHz area (for example, the water spectral line at 556.936 GHz) leading to star formation.

The goal of the mission is to gain a greater understanding of star formation by determining the composition of interstellar clouds and establishing the means by which these clouds cool as they collapse to form stars and planets.

In an unusual physics combination, the SWAS instrument converts incoming radiofrequency energy to acoustic waves within a crystal illuminated by a laser. The dispersed light is imaged onto a sensor that outputs data corresponding to RF intensity over a portion of the spectrum.

The *National Radio Astronomy Observatory* (NRAO) is developing the Millimeter Array telescopes to observe phenomena as high as 600 GHz (see 30–31 GHz).

INTERNATIONAL FOOTNOTES

The following International Footnotes are those cited in the U.S. Table of Allocations as they pertain to bands above 30 MHz. These are a subset of all International Footnotes drawn from the Radio Regulations of the *International Telecommunication Union* (ITU).

Footnotes that may apply to the three ITU Regions, but that do not appear in the U.S. portion of the Table of Allocations, are not shown. Footnotes may refer to various other ITU materials that are not included in this book.

The ITU has switched to new Simplified Radio Regulations, which use the S numbering scheme for international footnotes.

This book endeavors to include those S footnotes that the FCC has adopted, or proposed to adopt, in the domestic allocations table above 30 MHz.

Some international footnotes in the U.S. Table still use the older, three-digit scheme. These are shown following the S footnotes.

S5.111 The carrier frequencies 2182 kHz, 3023 kHz, 5680 kHz, 8364 kHz and the frequencies 121.5 MHz, 156.8 MHz and 243 MHz may also be used, in accordance with the procedures in force for terrestrial radiocommunication services, for search and rescue operations concerning manned space vehicles. The conditions for the use of the frequencies are prescribed in Article S31 and in Appendix S13.

The same applies to the frequencies 10,003 kHz, 14,993 kHz and 19,993 kHz, but in each of these cases emissions must be confined in a band of ± 3 kHz about the frequency.

S5.138 The following bands:

6765–6795 kHz (centre frequency 6780 kHz),

433.05–434.79 MHz (centre frequency 433.92 MHz) in Region 1 except in the countries mentioned in No. S5.280,

61–61.5 GHz (centre frequency 61.25 GHz),

122–123 GHz (centre frequency 122.5 GHz), and

244—246 GHz (centre frequency 245 GHz

are designated for industrial, scientific and medical (ISM) applications. The use of these frequency bands for ISM applications shall be subject to special authorization by the administration concerned, in agreement with other administrations whose radiocommunication services might be affected. In applying this provision, administrations shall have due regard to the latest relevant ITU-R Recommendations.

S5.149 In making assignments to stations of other services to which the bands:

13,360–13,410 kHz,	36.43–36.5 GHz*,
25,550–25,670 kHz,	42.5–43.5 GHz,
37.5–38.25 MHz,	42.77–42.87 GHz*,
73–74.6 MHz in Regions 1 and 3,	43.07–43.17 GHz*,
150.05–153 MHz in Region 1,	43.37–43.47 GHz*,
322–328.6 MHz*,	48.94–49.04 GHz*,
406.1–410 MHz,	72.77–72.91 GHz*,
608–614 MHz in Regions 1 and 3,	93.07–93.27 GHz*,
1330–1400 MHz*,	97.88–98.08 GHz*,
1610.6–1613.8 MHz*,	140.69–140.98 GHz*,
1660–1670 MHz,	144.68–144.98 GHz*,
1718.8–1722.2 MHz*,	145.45–145.75 GHz*,
2655–2690 MHz,	146.82–147.12 GHz*,
3260–3267 MHz*,	150–151 GHz*,
3332–3339 MHz*,	174.42–175.02 GHz*,
3345.8–3352.5 MHz*,	177–177.4 GHz*,
4825–4835 MHz*,	178.2–178.6 GHz*,
4950–4990 MHz,	181–181.46 GHz*,
4990–5000 MHz,	186.2–186.6 GHz*,
6650–6675.2 MHz*,	250–251 GHz*,
10.6–10.68 GHz,	257.5–258 GHz*,
14.47–14.5 GHz*,	261–265 GHz,
22.01–22.21 GHz*,	262.24–262.76 GHz*,
22.21–22.5 GHz,	265–275 GHz,
22.81–22.86 GHz*,	265.64–266.16 GHz*,
23.07–23.12 GHz*,	267.34–267.86 GHz*,
31.2–31.3 GHz,	271.74–272.26 GHz*
31.5–31.8 GHz in Regions 1 and 3,	

are allocated (* indicates radio astronomy use for spectral line observations), administrations are urged to take all practicable steps to protect the radio astronomy service from harmful interference. Emissions from spaceborne or airborne stations can be particularly serious sources of interference to the radio astronomy service (see Nos. S4.5 and S4.6 and Article S29).

S5.150 The following bands:

13,553–13,567 kHz (centre frequency 13,560 kHz),
26,957–27,283 kHz (centre frequency 27,120 kHz),
40.66–40.70 MHz (centre frequency 40.68 MHz),
902–928 MHz in Region 2 (centre frequency 915 MHz),
2400–2500 MHz (centre frequency 2450 MHz),
5725–5875 MHz (centre frequency 5800 MHz), and
24–24.25 GHz (centre frequency 24.125 GHz)

are also designated for industrial, scientific and medical (ISM) applications. Radiocommunication services operating within these bands must accept harmful interference which may be caused by these applications. ISM equipment operating in these bands is subject to the provisions of No. S15.13.

S5.180 The frequency 75 MHz is assigned to marker beacons. Administrations shall refrain from assigning frequencies close to the limits of the guardband to stations of other services which, because of their power or geographical position, might cause harmful interference or otherwise place a constraint on marker beacons.

Every effort should be made to improve further the characteristics of airborne receivers and to limit the power of transmitting stations close to the limits 74.8 MHz and 75.2 MHz.

S5.198 Additional allocation: the band 117.975–136 MHz is also allocated to the aeronautical mobile-satellite (R) service on a secondary basis, subject to agreement obtained under No. S9.21.

S5.199 The bands 121.45–121.55 MHz and 242.95–243.05 MHz are also allocated to the mobile-satellite service for the reception on board satellites of emissions from emergency position-indicating radiobeacons transmitting at 121.5 MHz and 243 MHz (see Appendix S13).

S5.200 In the band 117.975–136 MHz, the frequency 121.5 MHz is the aeronautical emergency frequency and, where required, the frequency 123.1 MHz is the aeronautical frequency auxiliary to 121.5 MHz. Mobile stations of the maritime mobile service may communicate on these frequencies under the conditions laid down in Article S31 and Appendix S13 for distress and safety purposes with stations of the aeronautical mobile service.

S5.208 The use of the band 137–138 MHz by the mobile-satellite service is subject to coordination under No. S9.11A.

S5.208A In making assignments to space stations in the mobile-satellite service in the bands 137–138 MHz, 387–390 MHz and 400.15–401 MHz, administrations shall take all practicable steps to protect the radio astronomy service in the bands 150.05–153 MHz, 322–328.6 MHz, 406.1–410 MHz and 608–614 MHz from harmful interference from unwanted emissions. The threshold levels of interference detrimental to the radio astronomy service are shown in Table 1 of Recommendation ITU-R RA.769-1.

S5.209 The use of the bands 137–138 MHz, 148–150.05 MHz, 399.9–400.05 MHz, 400.15–401 MHz, 454–456 MHz and 459–460 MHz by the mobile-satellite service is limited to non-geostationary-satellite systems.

S5.218 Additional allocation: the band 148–149.9 MHz is also allocated to the space operation service (Earth-to-space) on a primary basis, subject to agreement obtained under No. S9.21. The bandwidth of any individual transmission shall not exceed ± 25 kHz.

S5.219 The use of the band 148–149.9 MHz by the mobile-satellite service is subject to coordination under No. S9.11A. The mobile-satellite service shall not constrain the development and use of the fixed, mobile and space operation services in the band 148–149.9 MHz.

S5.220 The use of the bands 149.9–150.05 MHz and 399.9–400.05 MHz by the mobile-satellite service is subject to coordination under No. S9.11A. The mobile-satellite service shall not constrain the development and use of the radionavigation-satellite service in the bands 149.9–150.05 MHz and 399.9–400.05 MHz.

S5.223 Recognizing that the use of the band 149.9–150.05 MHz by the fixed and mobile services may cause harmful interference to the radionavigation-satellite service, administrations are urged not to authorize such use in application of No. S4.4.

S5.226 The frequency 156.8 MHz is the international distress, safety and calling frequency for the maritime mobile VHF radiotelephone service. The conditions for the use of this frequency are contained in Article S31 and Appendix S13.

In the bands 156–156.7625 MHz, 156.8375–157.45 MHz, 160.6–160.975 MHz and 161.475–162.05 MHz, each administration shall give priority to the maritime mobile service on only such frequencies as are assigned to stations of the maritime mobile service by the administration (see Articles S31 and S52, and Appendix S13).

Any use of frequencies in these bands by stations of other services to which they are allocated should be avoided in areas where such use might cause harmful interference to the maritime mobile VHF radiocommunication service.

However, the frequency 156.8 MHz and the frequency bands in which priority is given to the maritime mobile service may be used for radiocommunications on inland waterways subject to agreement between interested and affected administrations and taking into account current frequency usage and existing agreements.

S5.227 In the maritime mobile VHF service the frequency 156.525 MHz is to be used exclusively for digital selective calling for distress, safety and calling. The conditions for the use of this frequency are prescribed in Articles S31 and S52, and Appendices S13 and S18.

S5.241 In Region 2, no new stations in the radiolocation service may be authorized in the band 216–225 MHz. Stations authorized prior to 1 January 1990 may continue to operate on a secondary basis.

S5.256 The frequency 243 MHz is the frequency in this band for use by survival craft stations and equipment used for survival purposes (see Appendix S13).

S5.258 The use of the band 328.6–335.4 MHz by the aeronautical radionavigation service is limited to Instrument Landing Systems (glide path).

S5.260 Recognizing that the use of the band 399.9–400.05 MHz by the fixed and mobile services may cause harmful interference to the radionavigation satellite service, administrations are urged not to authorize such use in application of No. S4.4.

S5.261 Emissions shall be confined in a band of ± 25 kHz about the standard frequency 400.1 MHz.

S5.263 The band 400.15–401 MHz is also allocated to the space research service in the space-to-space direction for communications with manned space vehicles. In this application, the space research service will not be regarded as a safety service.

S5.264 The use of the band 400.15–401 MHz by the mobile-satellite service is subject to coordination under No. S9.11A. The power flux-density limit indicated in Annex 1 of Appendix S5 shall apply until such time as a competent world radiocommunication conference revises it.

S5.266 The use of the band 406–406.1 MHz by the mobile-satellite service is limited to low power satellite emergency position-indicating radiobeacons (see also Article S31 and Appendix S13).

S5.267 Any emission capable of causing harmful interference to the authorized uses of the band 406–406.1 MHz is prohibited.

S5.282 In the bands 435–438 MHz, 1260–1270 MHz, 2400–2450 MHz, 3400–3410 MHz (in Regions 2 and 3 only) and 5650–5670 MHz, the amateur-satellite service may operate subject to not causing harmful interference to other services operating in accordance with the Table (see No. S5.43). Administrations authorizing such use shall ensure that any harmful interference caused by emissions from a station in the amateur-satellite service is immediately eliminated in accordance with the provisions of No. S25.11. The use of the bands 1260–1270 MHz and 5650–5670 MHz by the amateur-satellite service is limited to the Earth-to-space direction.

S5.286 The band 449.75–450.25 MHz may be used for the space operation service (Earth-to-space) and the space research service (Earth-to-space), subject to agreement obtained under No. S9.21.

S5.287 In the maritime mobile service, the frequencies 457.525 MHz, 457.550 MHz, 457.575 MHz, 467.525 MHz, 467.550 MHz and 467.575 MHz may be used by on-board communication stations. Where needed, equipment designed for 12.5 kHz channel spacing using also the additional frequencies 457.5375 MHz, 457.5625 MHz, 467.5375 MHz and 467.5625 MHz may be introduced for on-board communications. The use of these frequencies in territorial waters may be subject to the national regulations of the administration concerned. The characteristics of the equipment used shall conform to those specified in Recommendation ITU-R M.1174 (see Resolution 341 (WRC-97)).

S5.288 In the territorial waters of the United States and the Philippines, the preferred frequencies for use by on-board communication stations shall be 457.525 MHz, 457.550 MHz, 457.575 MHz and 457.600 MHz paired, respectively, with 467.750 MHz, 467.775 MHz, 467.800 MHz and 467.825 MHz. The characteristics of the equipment used shall conform to those specified in Recommendation ITU-R M.1174.

S5.289 Earth exploration-satellite service applications, other than the meteorological-satellite service, may also be used in the bands 460–470 MHz and 1690–1710 MHz for space-to-Earth transmissions subject to not causing harmful interference to stations operating in accordance with the Table.

S5.328 The band 960–1215 MHz is reserved on a worldwide basis for the use and development of airborne electronic aids to air navigation and any directly associated ground-based facilities.

S5.333 In the bands 1215–1300 MHz, 3100–3300 MHz, 5250–5350 MHz, 8550–8650 MHz, 9500–9800 MHz and 13.4–14.0 GHz, radiolocation stations installed on spacecraft may also be employed for the earth exploration-satellite and space research services on a secondary basis. (SUP–WRC-97)

S5.334 Additional allocation: in Canada and the United States, the bands 1240–1300 MHz and 1350–1370 MHz are also allocated to the aeronautical radionavigation service on a primary basis.

S5.335 In Canada and the United States in the band 1240–1300 MHz, active spaceborne sensors in the earth exploration-satellite and space research services shall not cause interference to, claim protection from, or otherwise impose constraints on operation or development of the aeronautical radionavigation service.

S5.337 The use of the bands 1300–1350 MHz, 2700–2900 MHz and 9000–9200 MHz by the aeronautical radionavigation service is restricted to ground-based radars and to associated airborne transponders which transmit only on frequencies in these bands and only when actuated by radars operating in the same band.

S5.339 The bands 1370–1400 MHz, 2640–2655 MHz, 4950–4990 MHz and 15.20–15.35 GHz are also allocated to the space research (passive) and earth exploration-satellite (passive) services on a secondary basis.

S5.340 All emissions are prohibited in the following bands:

1400–1427 MHz,
2690–2700 MHz, except those provided for by Nos. S5.421 and S5.422,
10.68–10.7 GHz, except those provided for by No. S5.483,
15.35–15.4 GHz, except those provided for by No. S5.511,
23.6–24 GHz,
31.3–31.5 GHz,
31.5–31.8 GHz, in Region 2,
48.94–49.04 GHz, from airborne stations,
50.2–50.4 GHz [2], except those provided for by No. S5.555A,
52.6–54.25 GHz,
86–92 GHz,
105–116 GHz,
140.69–140.98 GHz, from airborne stations and from space stations in the space-to-Earth direction,
182–185 GHz, except those provided for by No. S5.563,
217–231 GHz.

[2] The allocation to the earth exploration-satellite service (passive) and the space research service (passive) in the band 50.2–50.4 GHz should not impose undue constraints on the use of the adjacent bands by the primary allocated services in those bands.

S5.341 In the bands 1400–1727 MHz, 101–120 GHz and 197–220 GHz, passive research is being conducted by some countries in a programme for the search for intentional emissions of extraterrestrial origin.

S5.351 The bands 1525–1544 MHz, 1545–1559 MHz, 1626.5–1645.5 MHz and 1646.5–1660.5 MHz shall not be used for feeder links of any service. In exceptional circumstances, however, an earth station at a specified fixed point in any of the mobile-satellite services may be authorized by an administration to communicate via space stations using these bands.

S5.356 The use of the band 1544–1545 MHz by the mobile-satellite service (space-to-Earth) is limited to distress and safety communications (see Article S31).

S5.364 The use of the band 1610–1626.5 MHz by the mobile-satellite service (Earth-to-space) and by the radiodetermination-satellite service (Earth-to-space) is subject to coordination under No. S9.11A. A mobile earth station operating in either of the services in this band shall not produce a peak e.i.r.p. density in excess of -15 dB(W/4 kHz) in the part of the band used by systems operating in accordance with the provisions of No. S5.366 (to which No. S4.10 applies), unless otherwise agreed by the affected administrations. In the part of the band where such systems are not operating, the mean e.i.r.p. density of a mobile earth station shall not exceed -3 dB(W/4 kHz). Stations of the mobile-satellite service shall not claim protection from stations in the aeronautical radionavigation service, stations operating in accordance with the provisions of No. S5.366 and stations in the fixed service operating in accordance with the provisions of No. S5.359. Administrations responsible for the coordination of mobile-satellite networks shall make all practicable efforts to ensure protection of stations operating in accordance with the provisions of No. S5.366.

S5.365 The use of the band 1613.8–1626.5 MHz by the mobile-satellite service (space-to-Earth) is subject to coordination under No. S9.11A.

S5.366 The band 1610–1626.5 MHz is reserved on a worldwide basis for the use and development of airborne electronic aids to air navigation and any directly associated ground-based or satellite-borne facilities. Such satellite use is subject to agreement obtained under No. S9.21.

S5.367 Additional allocation: The bands 1610–1626.5 MHz and 5000–5150 MHz are also allocated to the aeronautical mobile-satellite (R) service on a primary basis, subject to agreement obtained under No. S9.21.

S5.368 With respect to the radiodetermination-satellite and mobile-satellite services the provisions of No. S4.10 do not apply in the band

1610–1626.5 MHz, with the exception of the aeronautical radionavigation-satellite service.

S5.372 Harmful interference shall not be caused to stations of the radio astronomy service using the band 1610.6–1613.8 MHz by stations of the radiodetermination-satellite and mobile-satellite services (No. S29.13 applies).

S5.375 The use of the band 1645.5–1646.5 MHz by the mobile-satellite service (Earth-to-space) and for inter-satellite links is limited to distress and safety communications (see Article S31).

S5.384 Additional allocation: in India, Indonesia and Japan, the band 1700–1710 MHz is also allocated to the space research service (space-to-Earth) on a primary basis.

S5.385 Additional allocation: the bands 1718.8–1722.2 MHz, 150–151 GHz, 174.42–175.02 GHz, 177–177.4 GHz, 178.2–178.6 GHz, 181–181.46 GHz, 186.2–186.6 GHz and 257.5–258 GHz are also allocated to the radio astronomy service on a secondary basis for spectral line observations.

S5.391 In making assignments to the mobile service in the bands 2025–2110 MHz and 2200–2290 MHz, administrations shall not introduce high-density mobile systems, as described in Recommendation ITU-R SA.1154, and shall take this Recommendation into account for the introduction of any other type of mobile system.

S5.392 Administrations are urged to take all practicable measures to ensure that space-to-space transmissions between two or more non-geo-stationary satellites, in the space research, space operations and Earth exploration-satellite services in the bands 2 025–2 110 MHz and 2 200–2 290 MHz, shall not impose any constraints on Earth-to-space, space-to-Earth and other space-to-space transmissions of those services and in those bands between geostationary and non-geostationary satellites.

S5.393 Additional allocation: in the United States, India and Mexico, the band 2310–2360 MHz is also allocated to the broadcasting-satellite service (sound) and complementary terrestrial sound broadcasting service on a primary basis. Such use is limited to digital audio broadcasting and is subject to the provisions of Resolution 528 (WARC-92).

S5.396 Space stations of the broadcasting-satellite service in the band 2310–2360 MHz operating in accordance with No. S5.393 that may affect the services to which this band is allocated in other countries shall be coordinated and notified in accordance with Resolution 33 (Rev.WRC-97). Complementary terrestrial broadcasting stations shall be subject to bilateral coordination with neighbouring countries prior to their bringing into use.

S5.402 The use of the band 2483.5–2500 MHz by the mobile-satellite and the radiodetermination-satellite services is subject to the coordination under No. S9.11A. Administrations are urged to take all practicable steps to prevent harmful interference to the radio astronomy service from emissions in the 2483.5–2500 MHz band, especially those caused by second-harmonic radiation that would fall into the 4990–5000 MHz band allocated to the radio astronomy service worldwide.

S5.423 In the band 2700–2900 MHz, ground-based radars used for meteorological purposes are authorized to operate on a basis of equality with stations of the aeronautical radionavigation service.

S5.427 In the bands 2900–3100 MHz and 9300–9500 MHz, the response from radar transponders shall not be capable of being confused with the response from radar beacons (racons) and shall not cause interference to ship or aeronautical radars in the radionavigation service, having regard, however, to No. S4.9.

S5.440 The standard frequency and time signal-satellite service may be authorized to use the frequency 4202 MHz for space-to-Earth transmissions and the frequency 6427 MHz for Earth-to-space transmissions. Such transmissions shall be confined within the limits of 2 MHz of these frequencies, subject to agreement obtained under No. S9.21.

S5.446 Additional allocation: in the countries listed in Nos. S5.369 and S5.400, the band 5150–5216 MHz is also allocated to the radiodetermination-satellite service (space-to-Earth) on a primary basis, subject to agreement obtained under No. S9.21. In Region 2, the band is also allocated to the radiodetermination-satellite service (space-to-Earth) on a primary basis. In Regions 1 and 3, except those countries listed in Nos. S5.369 and S5.400, the band is also allocated to the radiodetermination-satellite service (space-to-Earth) on a secondary basis. The use by the radiodetermination-satellite service is limited to feeder links in conjunction with the radiodetermination-satellite service operating in the bands 1610–1626.5 MHz and/or 2483.5–2500 MHz. The total power flux-density at the Earth's surface shall in no case exceed -159 dBW/m^2 in any 4 kHz band for all angles of arrival.

S5.449 The use of the band 5350–5470 MHz by the aeronautical radionavigation service is limited to airborne radars and associated airborne beacons.

S5.452 Between 5600 MHz and 5650 MHz, ground-based radars used for meteorological purposes are authorized to operate on a basis of equality with stations of the maritime radionavigation service.

S5.458 In the band 6425–7075 MHz, passive microwave sensor measurements are carried out over the oceans. In the band 7075–7250 MHz, passive microwave sensor measurements are carried out. Administrations should bear in mind the needs of the Earth exploration-satellite (passive) and space research (passive) services in their future planning of the bands 6425–7025 MHz and 7075–7250 MHz.

S5.472 In the bands 8850–9000 MHz and 9200–9225 MHz, the maritime radionavigation service is limited to shore-based radars.

S5.474 In the band 9200–9500 MHz, search and rescue transponders (SART) may be used, having due regard to the appropriate ITU-R Recommendation (see also Article S31).

S5.476 In the band 9300–9320 MHz in the radionavigation service, the use of shipborne radars, other than those existing on 1 January 1976, is not permitted until 1 January 2001.

S5.479 The band 9975–10,025 MHz is also allocated to the meteorological-satellite service on a secondary basis for use by weather radars.

S5.486 Different category of service: in Mexico and the United States, the allocation of the band 11.7–12.1 GHz to the fixed service is on a secondary basis (see No. S5.32).

S5.487A Additional allocation: in Region 1, the band 11.7–12.5 GHz, in Region 2, the band 12.2–12.7 GHz and, in Region 3, the band 11.7–12.2 GHz, are also allocated to the fixed-satellite service (space-to-Earth) on a primary basis, limited to non-geostationary systems and subject to application of the provisions of No. S9.12 for coordination with other non-geostationary-satellite systems in the fixed-satellite service. Non-geostationary-satellite systems in the fixed-satellite service shall not claim protection from geostationary-satellite networks in the broadcasting-satellite service operating in accordance with the Radio Regulations, irrespective of the dates of receipt by the Bureau of the complete coordination or notification information, as appropriate, for the non-GSO FSS systems and of the complete coordination or notification information, as appropriate, for the GSO networks, and No. S5.43A does not apply. Non-geostationary-satellite systems in the fixed-satellite service in the above bands shall be operated in such a way that any unacceptable interference that may occur during their operation shall be rapidly eliminated.

S5.488 The use of the band 11.7–12.2 GHz by geostationary-satellite networks in the fixed-satellite service in Region 2 is subject to the provisions of Resolution 77 (WRC-2000). For the use of the band 12.2–12.7 GHz by the broadcasting-satellite service in Region 2, see Appendix S30.

S5.490 In Region 2, in the band 12.2–12.7 GHz, existing and future terrestrial radiocommunication services shall not cause harmful interference to the space services operating in conformity with the broadcasting-satellite Plan for Region 2 contained in Appendix S30.

S5.497 The use of the band 13.25–13.4 GHz by the aeronautical radionavigation service is limited to Doppler navigation aids.

S5.502 In the band 13.75–14 GHz, the e.i.r.p. of any emission from an earth station in the fixed-satellite service shall be at least 68 dBW, and should not exceed 85 dBW, with a minimum antenna diameter of 4.5 m. In addition the e.i.r.p., averaged over one second, radiated by a station in the radiolocation or radionavigation services towards the geostationary-satellite orbit shall not exceed 59 dBW.

S5.503 In the band 13.75–14 GHz, geostationary space stations in the space research service for which information for advance publication has been received by the Bureau prior to 31 January 1992 shall operate on an equal basis with stations in the fixed-satellite service; after that date, new geostationary space stations in the space research service will operate on a secondary basis. The e.i.r.p. density of emissions from any earth station in the fixed-satellite service shall not exceed 71 dBW in any 6 MHz band in the frequency range 13.772–13.778 GHz until those geostationary space stations in the space research service for which information for advance publication has been received by the Bureau prior to 31 January 1992 cease to operate in this band. Automatic power control may be used to increase the e.i.r.p. density above 71 dBW in any 6 MHz band in this frequency range to compensate for rain attenuation, to the extent that the power-flux density at the fixed-satellite service space station does not exceed the value resulting from use of an e.i.r.p. of 71 dBW in any 6 MHz band in clear sky conditions.

S5.503A Until 1 January 2000, stations in the fixed-satellite service shall not cause harmful interference to non-geostationary space stations in the space research and Earth exploration-satellite services. After that date, these non-geostationary space stations will operate on a secondary basis in relation to the fixed-satellite service. Additionally, when planning earth stations in the fixed-satellite service to be brought into service between 1 January 2000 and 1 January 2001, in order to accommodate the needs of spaceborne precipitation radars operating in the band 13.793–13.805 GHz, advantage should be taken of the consultation process and the information given in Recommendation ITU-R SA.1071.

S5.519 Additional allocation: the band 18.1–18.3 GHz is also allocated to the meteorological-satellite service (space-to-Earth) on a primary basis. Its use is limited to geostationary satellites and shall be in accordance with the provisions of Article S21, Table S21-4.

S5.525 In order to facilitate interregional coordination between networks in the mobile-satellite and fixed-satellite services, carriers in the mobile-satellite service that are most susceptible to interference shall, to the extent practicable, be located in the higher parts of the bands 19.7–20.2 GHz and 29.5–30 GHz.

S5.526 In the bands 19.7–20.2 GHz and 29.5–30 GHz in Region 2, and in the bands 20.1–20.2 GHz and 29.9–30 GHz in Regions 1 and 3, networks which are both in the fixed-satellite service and in the mobile-satellite service may include links between earth stations at specified or unspecified points or while in motion, through one or more satellites for point-to-point and point-to-multipoint communications.

S5.527 In the bands 19.7–20.2 GHz and 29.5–30 GHz, the provisions of No. S4.10 do not apply with respect to the mobile-satellite service.

S5.528 The allocation to the mobile-satellite service is intended for use by networks which use narrow spot-beam antennas and other advanced technology at the space stations. Administrations operating systems in the mobile-satellite service in the band 19.7–20.1 GHz in Region 2 and in the band 20.1–20.2 GHz shall take all practicable steps to ensure the continued availability of these bands for administrations operating fixed and mobile systems in accordance with the provisions of No. S5.524.

S5.529 The use of the bands 19.7–20.1 GHz and 29.5–29.9 GHz by the mobile-satellite service in Region 2 is limited to satellite networks which are both in the fixed-satellite service and in the mobile-satellite service as described in No. S5.526.

S5.533 The inter-satellite service shall not claim protection from harmful interference from airport surface detection equipment stations of the radionavigation service.

S5.536 Use of the 25.25–27.5 GHz band by the inter-satellite service is limited to space research and Earth exploration-satellite applications, and also transmissions of data originating from industrial and medical activities in space.

S5.543 The band 29.95–30 GHz may be used for space-to-space links in the Earth exploration-satellite service for telemetry, tracking, and control purposes, on a secondary basis.

S5.551 Radars located on spacecraft may be operated on a primary basis in the band 35.5–35.6 GHz. (SUP–WRC–97)

S5.551A In the band 35.5–36.0 GHz, active spaceborne sensors in the earth exploration-satellite and space research services shall not cause harmful interference to, claim protection from, or otherwise impose constraints on operation or development of the radiolocation service, the meteorological aids service and other services allocated on a primary basis.

S5.553 In the bands 43.5–47 GHz, 66–71 GHz, 95–100 GHz, 134–142 GHz, 190–200 GHz and 252–265 GHz, stations in the land mobile service may be operated subject to not causing harmful interference to the space radiocommunication services to which these bands are allocated (see No. S5.43).

S5.554 In the bands 43.5–47 GHz, 66–71 GHz, 95–100 GHz, 134–142 GHz, 190–200 GHz and 252–265 GHz, satellite links connecting land stations at specified fixed points are also authorized when used in conjunction with the mobile-satellite service or the radionavigation-satellite service.

S5.555 Additional allocation: the bands 48.94–49.04 GHz, 97.88–98.08 GHz, 140.69–140.98 GHz, 144.68–144.98 GHz, 145.45–145.75 GHz, 146.82–147.12 GHz, 250–251 GHz and 262.24–262.76 GHz are also allocated to the radio astronomy service on a primary basis.

S5.556A Use of the bands 54.25–56.9 GHz, 57–58.2 GHz and 59–59.3 GHz by the inter-satellite service is limited to satellites in the geostationary-satellite orbit. The single-entry power flux-density at all altitudes from 0 km to 1 000 km above the Earth's surface produced by a station in the inter-satellite service, for all conditions and for all methods of modulation, shall not exceed -147 dB(W/m^2/100 MHz) for all angles of arrival, (WRC-97).

S5.558 In the bands 55.78–58.2 GHz, 59–64 GHz, 66–71 GHz, 116–134 GHz, 170–182 GHz and 185–190 GHz, stations in the aeronautical mobile service may be operated subject to not causing harmful interference to the inter-satellite service (see No. S5.43), (WRC-97). [Author's Note: See footnote 909 below.]

S5.559 In the bands 59–64 GHz and 126–134 GHz, airborne radars in the radiolocation service may be operated subject to not causing harmful interference to the inter-satellite service (see No. S5.43).

S5.560 In the band 78–79 GHz radars located on space stations may be operated on a primary basis in the Earth exploration-satellite service and in the space research service.

S5.561 In the band 84–86 GHz, stations in the fixed, mobile and broadcasting services shall not cause harmful interference to broadcasting-satellite stations operating in accordance with the decisions of the appropriate frequency assignment planning conference for the broadcasting-satellite service.

S5.565 The frequency band 275–400 GHz may be used by administrations for experimentation with, and development of, various active and passive services. In this band a need has been identified for the following spectral line measurements for passive services:

Radio astronomy service: 278–280 GHz and 343–348 GHz;

Earth exploration-satellite service (passive) and space research service (passive): 275–277 GHz, 300–302 GHz, 324–326 GHz, 345–347 GHz, 363–365 GHz and 379–381 GHz.

Future research in this largely unexplored spectral region may yield additional spectral lines and continuum bands of interest to the passive services. Administrations are urged to take all practicable steps to protect these passive services from harmful interference until the next competent world radiocommunication conference.

Old Numbering Scheme

510 For the use of these bands allocated to the amateur service at 3.5 MHz, 7.0 MHz, 10.1 MHz, 14.0 MHz, 18.068 MHz, 21.0 MHz, 24.89 MHz and 144 MHz in the event of natural disasters, see Resolution 640.

591 Subject to agreement obtained under the procedure set forth in Article 14, the band 117.975–137 MHz is also allocated to the aeronautical mobile-satellite (R) service on a secondary basis and on the condition that harmful interference is not caused to the aeronautical mobile (R) service.

599A The use of the band 137–138 MHz by the mobile-satellite service is subject to the application of the coordination and notification procedures set forth in Resolution 46. However, coordination of a space station of the mobile-satellite service with respect to terrestrial services is required only if the power flux-density produced by the station exceeds -125 dB(W/m^2/4 kHz) at the Earth's surface. The above power flux-density limit shall apply until such time as a competent world administrative radio conference revises it. In making assignments to the space stations in the mobile-satellite service in the above band, administrations shall take all practicable steps to protect the radio astronomy service in the 150.05–153 MHz band from harmful interference from unwanted emissions.

599B The use of the bands 137–138 MHz, 148–149.9 MHz and 400.15–401 MHz by the mobile-satellite service and the band 149.9–150.05 MHz by the land mobile-satellite service is limited to non-geostationary-satellite systems.

608A The use of the band 148–149.9 MHz by the mobile-satellite service is subject to the application of the coordination and notification procedures set forth in Resolution 46 (WARC-92). The mobile-satellite service shall not constrain the development and use of fixed, mobile and space operation services in the band 148–149.9 MHz. Mobile earth stations in the mobile-satellite service shall not produce a power flux-density in excess of -150 dB(W/m^2/4 kHz) outside national boundaries.

608B The use of the band 149.9–150.05 MHz by the land mobile-satellite service is subject to the application of the coordination and notification procedures set forth in Resolution 46 (WARC-92). The land mobile-satellite service shall not constrain the development and use of the radionavigation-satellite service in the band 149.9–150.05 MHz. Land mobile earth stations of the land mobile-satellite service shall not produce power flux-density in excess of -150 dB(W/m^2/4 kHz) outside national boundaries.

647B The use of the band 400.15–401 MHz by the mobile-satellite service is subject to the application of the coordination and notification procedures set forth in Resolution 46. However, coordination of a space station of the mobile-satellite service with respect to terrestrial services is required only if the power flux-density produced by the station exceeds -125 dB(W/m^2/4 kHz) at the Earth's surface. The above power flux-density limit shall apply until such time as a competent world administrative radio conference revises it. In making assignments to the space stations in the mobile-satellite service in the above band, administrations shall take all practicable steps to protect the radio astronomy service in the band 406.1–410 MHz from harmful interference from unwanted emissions.

669 In the maritime mobile service, the frequencies 457.525 MHz, 457.550 MHz, 457.575 MHz, 467.525 MHz, 467.550 MHz and 467.575 MHz may be used by on-board communication stations. The use of these frequencies in territorial waters may be subject to the national regulations of the administration concerned. The characteristics of the equipment used shall conform to those specified in Appendix 20.

733 The bands 1610–1626.5 MHz, 5000–5250 MHz and 15.4–15.7 GHz are also allocated to the aeronautical mobile-satellite (R) service on a primary basis. Such use is subject to agreement obtained under the procedure set forth in Article 14.

792A The use of the bands 4500–4800 MHz, 6725–7025 MHz, 10.7–10.95 GHz, 11.2–11.45 GHz and 12.75–13.25 GHz by the fixed-satellite service shall be in accordance with the provisions of Appendix 30B.

796 The band 5000–5250 MHz is to be used for the operation of the international standard system (microwave landing system) for precision approach and landing. The requirements of this system shall take precedence over other uses of this band.

797 The bands 5000–5250 MHz and 15.4–15.7 GHz are also allocated to the fixed-satellite service and the inter-satellite service, for connection between one or more earth stations at specified fixed points on the Earth and space stations, when these services are used in conjunction with the aeronautical radionavigation and/or aeronautical mobile (R) service. Such use shall be subject to agreement obtained under the procedure set forth in Article 14.

909 In the bands 54.25–58.2 GHz, 59–64 GHz, 116–134 GHz, 170–182 GHz and 185–190 GHz, stations in the aeronautical mobile service may be operated subject to not causing harmful interference to the inter-satellite service (see No. 435). [Author's Note: The 1997 World Radiocommunication Conference modified and renumbered footnote 909 as S5.558. The U.S. has not, at this writing, completely replaced footnote 909 in its domestic allocations. See Docket ET 99-261 (FCC 00-442).]

917 In the bands 140.69–140.98 GHz all emissions from airborne stations, and from space stations in the space-to-Earth direction, are prohibited.

U.S. FOOTNOTES

These Footnotes, each consisting of the letters US followed by one or more digits, denote stipulations applicable to both U.S. federal government and non-federal government stations.

Only those U.S. Footnotes that pertain to, or are proposed to pertain to, allocations in 30 MHz–300 GHz are shown. Footnotes occasionally refer to various other materials not included in this book.

US7 In the band 420–450 MHz and within the following areas, the peak envelope power output of a transmitter employed in the amateur service shall not exceed 50 watts, unless expressly authorized by the Commission after mutual agreement, on a case-by-case basis, between the Federal Communications Commission Engineer in Charge at the applicable district office and the military area frequency coordinator at the applicable military base. For areas (e) through (j), the appropriate military coordinator is located at Peterson AFB, CO.

(a) Those portions of Texas and New Mexico bounded on the south by latitude 31° 45′ North, on the east by longitude 104° 00′ West, on the north by latitude 34° 30′ North, and on the west by longitude 107° 30′ West; (b) The entire State of Florida including the Key West area and the areas enclosed within a 322-kilometer (200-mile) radius of Patrick Air Force Base, Florida (latitude 28° 21′ North, longitude 80° 43′ West), and within a 322-kilometer (200-mile) radius of Eglin Air Force Base, Florida (latitude 30° 30′ North, longitude 86° 30′ West); (c) The entire State of Arizona; (d) Those portions of California and Nevada south of latitude 37° 10′ North, and the areas enclosed within a 322-kilometer (200-mile) radius of the Pacific Missile Test Center, Point Mugu, California (latitude 34° 09′ North, longitude 119° 11′ West). (e) In the State of Massachusetts within a 160-kilometer (100-mile) radius around locations at Otis Air Force Base, Massachusetts (latitude 41° 45′ North, longitude 70° 32′ West). (f) In the State of California within a 240-kilometer (150-mile) radius around locations at Beale Air Force Base, California (latitude 39° 08′ North, longitude 121° 26′ West). (g) In the State of Alaska within a 160-kilometer (100-mile) radius of Clear, Alaska (latitude 64° 17′ North, longitude 149° 10′ West). (h) In the State of North Dakota within a 160-kilometer (100-mile) radius of Concrete, North Dakota (latitude 48° 43′ North, longitude 97° 54′ West). (i) In the States of Alabama, Georgia and South Carolina within a 200-kilometer (124-mile) radius of Warner Robins Air Force Base, Georgia (latitude 32° 38′ North, longitude 83° 35′ West). (j) In the State of Texas within a 200-kilometer (124-mile) radius of Goodfellow Air Force Base, Texas (latitude 31° 25′ North, longitude 100° 24′ West).

US8 The use of the frequencies 170.475, 171.425, 171.575, and 172.275 MHz east of the Mississippi River, and 170.425, 170.575, 171.475, 172.225

and 172.375 MHz west of the Mississippi River may be authorized to fixed, land and mobile stations operated by non-Federal forest firefighting agencies. In addition, land stations and mobile stations operated by non-Federal conservation agencies, for mobile relay operation only, may be authorized to use the frequency 172.275 MHz east of the Mississippi River and the frequency 171.475 MHz west of the Mississippi River. The use of any of the foregoing nine frequencies shall be on the condition that no harmful interference will be caused to Government stations.

US10 The use of the frequencies 26.62, 143.75, 143.90 and 148.15 MHz may be authorized to Civil Air Patrol land stations and Civil Air Patrol mobile stations.

US11 The use of the frequencies 166.250 and 170.150 MHz may be authorized to non-Government remote pickup broadcast base and land mobile stations and to non-Government base, fixed and land mobile stations in the public safety radio services (the sum of the bandwidth of emission and tolerance is not to exceed 25 kHz, except that authorizations in existence as of December 20, 1974, using a larger bandwidth are permitted to continue in operation until December 20, 1979) in the continental United States (excluding Alaska) only, except within the area bounded on the west by the Mississippi River, on the north by the parallel of latitude 37° 30' N., and on the east and south by that arc of the circle with center at Springfield, Illinois, and radius equal to the airline distance between Springfield, Illinois, and Montgomery, Alabama, subtended between the foregoing west and north boundaries, on the condition that harmful interference will not be caused to Government stations present or future in the Government band 162–174 MHz. The use of these frequencies by remote pickup broadcast stations will not be authorized for locations within 150 miles of New York City; and use of these frequencies by the public safety radio services will not be authorized except for locations within 150 miles of New York City.

US13 For the specific purpose of transmitting hydrological and meteorological data in co-operation with agencies of the Federal Government, the following frequencies may be authorized to non-Government fixed stations on the condition that harmful interference will not be caused to Government stations.

MHz 169.425 169.525 170.300 171.075 171.850 406.125 412.625 169.450 170.225 170.325 171.100 171.875 406.175 412.675 169.475 170.250 171.025 171.125 171.900 409.675 412.725 169.500 170.275 171.050 171.825 171.925 409.725 412.775

Licensees holding a valid authorization on June 11, 1962, to operate on the frequencies 169.575, 170.375 or 171.975 MHz may continue to be authorized for such operations on the condition that harmful interference will not be caused to Government stations.

US18 Navigation aids in the US and possessions in the bands 9–14 kHz, 90–110 kHz, 190–415 kHz, 510–535 kHz, 2700–2900 MHz are normally operated by the U.S. Government. However, authorizations may be made by the FCC for non-government operation in these bands subject to the conclusion of appropriate arrangements between the FCC and the Government agencies concerned and upon special showing of need for service which the Government is not yet prepared to render.

US26 The bands 117.975–121.4125 MHz, 123.5875–128.8125 MHz and 132.0125–136.0 MHz are for air traffic control communications.

US28 The band 121.5875–121.9375 MHz is for use by aeronautical utility land and mobile stations, and for air traffic control communications.

US30 The band 121.9375–123.0875 MHz is available to FAA aircraft for communications pursuant to flight inspection functions in accordance with the Federal Aviation Act of 1958.

US31 Except as provided below the band 121.9375–123.0875 MHz is for use by private aircraft stations.

The frequencies 122.700, 122.725, 122.750, 122.800, 122.950, 122.975, 123.000, 123.050 and 123.075 MHz may be assigned to aeronautical advisory stations. In addition, at landing areas having a part-time or no airdrome control tower or FAA flight service station, these frequencies may be assigned on a secondary non-interference basis to aeronautical utility mobile stations, and may be used by FAA ground vehicles for safety related communications during inspections conducted at such landing areas.

The frequencies 122.850, 122.900 and 122.925 MHz may be assigned to aeronautical multicom stations. In addition, 122.850 MHz may be assigned on a secondary non-interference basis to aeronautical utility mobile stations. In case of 122.925 MHz, US213 applies.

Air carrier aircraft stations may use 122.000 and 122.050 MHz for communication with aeronautical stations of the Federal Aviation Administration and 122.700, 122.800, 122.900 and 123.000 MHz for communications with aeronautical stations pertaining to safety of flight with and in the vicinity of landing areas not served by a control tower.

Frequencies in the band 121.9375–122.6875 MHz may be used by aeronautical stations of the Federal Aviation Administration for communication with private aircraft stations only, except that 122.000

and 122.050 MHz may also be used for communication with air carrier aircraft stations concerning weather information.

US32 Except for the frequencies 123.3 and 123.5 MHz, which are not authorized for Government use, the band 123.1125–123.5875 MHz is available for FAA communications incident to flight test and inspection activities pertinent to aircraft and facility certification on a secondary non-interference basis.

US33 The band 123.1125–123.5875 MHz is for use by flight test and aviation instructional stations. The frequency 121.950 MHz is available for aviation instructional stations.

US41 The Government radiolocation service is permitted in the band 2450–2500 MHz on condition that harmful interference is not caused to non-Government services.

US44 The non-Government radiolocation service may be authorized in the band 2900–3100 MHz on the condition that no harmful interference is caused to Government services.

US48 The non-Government radiolocation service may be authorized in the bands 5350–5460 MHz and 9000–9200 MHz on the condition that it does not cause harmful interference to the aeronautical radionavigation service or to the Government radiolocation service.

US49 The non-Government radiolocation service may be authorized in the band 5460–5470 MHz on the condition that it does not cause harmful interference to the aeronautical or maritime radionavigation services or to the Government radiolocation service.

US50 The non-Government radiolocation service may be authorized in the band 5470–5600 MHz on the condition that it does not cause harmful interference to the maritime radionavigation service or to the Government radiolocation service.

US51 In the band 5600–5650 MHz and 9300–9500 MHz, the non-Government radiolocation service shall not cause harmful interference to the Government radiolocation service.

US53 In view of the fact that the band 13.25–13.4 GHz is allocated to doppler navigation aids, Government, and non-Government airborne doppler radars in the aeronautical radionavigation service are permitted in the band 8750–8850 MHz only on the condition that they must accept any interference that may be experienced from stations in the radiolocation service in the band 8500–10000 MHz.

US54 Temporarily, and until certain operations of the radiolocation service in the band 9000–9200 MHz can be transferred to other appropriate frequency bands, the aeronautical radionavigation service may, in certain geographical areas, be subject to receiving some degree of interference from the radiolocation service.

US58 In the band 10000–10500 MHz, pulsed emissions are prohibited, except for weather radars on board meteorological satellites in the band 10000–10025 MHz. The amateur service and the non-Government radiolocation service, which shall not cause harmful interference to the Government radiolocation service, are the only non-Government services permitted in this band. The non-Government radiolocation service is limited to survey operations as specified in footnote US108.

US59 The band 10.5–10.55 GHz is restricted to systems using type N0N (A0) emission with a power not to exceed 40 watts into the antenna.

US65 The use of the band 5460–5650 MHz by the maritime radionavigation service is limited to shipborne radars.

US66 The use of the band 9300–9500 MHz by the aeronautical radionavigation service is limited to airborne radars and associated airborne beacons. In addition, ground-based radar beacons in the aeronautical radionavigation service are permitted in the band 9300–9320 MHz on the condition that harmful interference is not caused to the maritime radionavigation service.

US67 The use of the band 9300–9500 MHz by the meteorological aids service is limited to ground-based radars. Radiolocation installations will be coordinated with the meteorological aids service and, insofar as practicable, will be adjusted to meet the requirements of the meteorological aids service.

US69 In the band 31.8–33.4 GHz, ground-based radionavigation aids are not permitted except where they operate in cooperation with airborne or shipborne radionavigation devices.

US70 The meteorological aids service allocation in the band 400.15–406.0 MHz does not preclude the operation therein of associated ground transmitters.

US71 In the band 9300–9320 MHz, low-powered maritime radionavigation stations shall be protected from harmful interference caused by the operation of land-based equipment.

US74 In the bands 25.55–25.67, 73.0–74.6, 406.1–410.0, 608–614, 1400–1427, 1660.5–1670.0, 2690–2700 and 4990–5000 MHz and in the bands 10.68–10.7, 15.35–15.4, 23.6–24.0, 31.3–31.5, 86–92, 105–116 and 217–231 GHz, the radio astronomy service shall be protected from extraband radiation only to the extent that such radiation exceeds the level which would be present if the offending station were operating in compliance with the technical standards or criteria applicable to the service in which it operates.

US77 Government stations may also be authorized: (a) Port operations use on a simplex basis by coast and ship stations of the frequencies 156.6 and 156.7 MHz; (b) Duplex port operations use of the frequency 157.0 MHz for ship stations and 161.6 MHz for coast stations; (c) Inter-ship use of 156.3 MHz on a simplex basis; and (d) Vessel traffic services under the control of the U.S. Coast Guard on a simplex basis by coast and ship stations on the frequencies 156.25, 156.55, 156.6 and 156.7 MHz.

(e) Navigational bridge-to-bridge and navigational communications on a simplex basis by coast and ship stations on the frequencies 156.375 and 156.65 MHz.

US78 In the mobile service, the frequencies between 1435 and 1535 MHz will be assigned for aeronautical telemetry and associated telecommand operations for flight testing of manned or unmanned aircraft and missiles, or their major components. Permissible usage includes telemetry associated with launching and reentry into the earth's atmosphere as well as any incidental orbiting prior to reentry of manned objects undergoing flight tests. The following frequencies are shared with flight telemetry mobile stations: 1444.5, 1453.5, 1501.5, 1515.5, 1524.5 and 1525.5 MHz.

US80 Government stations may use the frequency 122.9 MHz subject to the following conditions: (a) All operations by Government stations shall be restricted to the purpose for which the frequency is authorized to non-Government stations, and shall be in accordance with the appropriate provisions of the Commission's Rules and Regulations, Part 87, Aviation Services; (b) Use of the frequency is required for coordination of activities with Commission licensees operating on this frequency; and (c) Government stations will not be authorized for operation at fixed locations.

US81 The band 38.0–38.25 MHz is used by both Government and non-government radio astronomy observatories. No new fixed or mobile assignments are to be made and Government stations in the band 38.0–38.25 MHz will be moved to other bands on a case-by-case basis, as required, to protect radio astronomy observations from harmful interference. As an exception, however, low powered military transportable and mobile stations used for tactical and training purposes will continue to use the band. To the extent practicable, the latter operations will be adjusted to relieve such interference as may be caused to radio astronomy observations. In the event of harmful interference from such local operations, radio astronomy observatories may contact local military commands directly, with a view to effecting relief. A list of military commands, areas of coordination, and points of contact for purposes of relieving interference may be obtained upon request from the Office of the Chief Engineer, Federal Communications Commission, Washington, D.C. 20554.

US87 The frequency 450 MHz, with maximum emission bandwidth of 500 kHz, may be used by Government and non-Government stations for space telecommand at specific locations, subject to such conditions as may be applied on a case-by-case basis.

US90 In the band 2025–2110 MHz, the power flux-density at the Earth's surface produced by emissions from a space station in the space operation, Earth exploration-satellite, or space research services that is transmitting in the space-to-space direction, for all conditions and all methods of modulation, shall not exceed the following values in any 4 kHz sub-band:
(a) -154 dBW/m^2 for angles of arrival above the horizontal plane (δ) of 0° to 5°,
(b) $-154 + 0.5(\delta-5)$ dBW/m^2 for δ of 5° to 25°, and
(c) -144 dBW/m^2 for δ of 25° to 90°.

US93 In the conterminous United States, the frequency 108.0 MHz may be authorized for use by VOR test facilities, the operation of which is not essential for the safety of life or property, subject to the condition that no interference is caused to the reception of FM broadcasting stations operating in the band 88–108 MHz. In the event that such interference does occur, the licensee or other agency authorized to operate the facility shall discontinue operation on 108 MHz and shall not resume operation until the interference has been eliminated or the complaint otherwise satisfied. VOR test facilities operating on 108 MHz will not be protected against interference caused by FM broadcasting stations operating in the band 88–108 MHz nor shall the authorization of a VOR test facility on 108 MHz preclude the Commission from authorizing additional FM broadcasting stations.

US99 In the band 1668.4–1670.0 MHz, the meteorological aids service (radiosonde) will avoid operations to the maximum extent practicable. Whenever it is necessary to operate radiosondes in the band 1668.4–1670 MHz within the United States, notification of the operations shall be sent as far in advance as possible to the Electromagnetic Management Unit, National Science Foundation, Washington, D.C. 20550.

US102 In Alaska only, the frequency 122.1 MHz may also be used for air carrier air traffic control purposes at locations where other frequencies are not available to air carrier aircraft stations for air traffic control.

US106 The frequency 156.75 MHz is available for assignment to non-government and Government stations for environmental communications in accordance with an agreed plan.

US107 The frequency 156.8 MHz is the national distress, safety and calling frequency for the maritime mobile VHF radiotelephone service for use by Government and non-Government ship and coast stations. Guard bands of 156.7625–156.7875 and 156.8125–156.8375 MHz are maintained.

US108 Within the bands 3300–3500 MHz and 10000–10500 MHz, survey operations, using transmitters with a peak power not to exceed five watts into the antenna, may be authorized for Government and non-Government use on a secondary basis to other Government radiolocation operations.

US110 In the bands 3100–3300 MHz, 3500–3650 MHz, 5250–5350 MHz, 8500–9000 MHz, 9200–9300 MHz, 9500–10000 MHz, 13.4–14.0 GHz, 15.7–17.3 GHz, 24.05–24.25 GHz and 33.4–36.0 GHz, the non-Government radiolocation service shall be secondary to the Government radiolocation service and to airborne doppler radars at 8800 MHz, and shall provide protection to airport surface detection equipment (ASDE) operating between 15.7–16.2 GHz.

US112 The frequency 123.1 MHz is for search and rescue communications. This frequency may be assigned for air traffic control communications at special aeronautical events on the condition that no harmful interference is caused to search and rescue communications during any period of search and rescue operations in the locale involved.

US116 In the bands 890–902 MHz and 935–941 MHz, no new assignments are to be made to Government radio stations after July 10, 1970 except on case-by case basis, to experimental stations and to additional stations of existing networks in Alaska. Government assignments existing prior to July 10, 1970 to stations in Alaska may be continued. All other existing Government assignments shall be on a secondary basis to stations in the non-Government land mobile service and shall be subject to adjustment or removal from the bands 890–902 MHz, 928–932 MHz and 935–941 MHz at the request of the FCC.

US117 In the band 406.1–410 MHz, all new authorizations will be limited to a maximum 7 watts per kHz of necessary bandwidth; existing authorizations as of November 30, 1970 exceeding this power are permitted to continue in use.

New authorizations in this band for stations, other than mobile stations, within the following areas are subject to prior coordination by the applicant through the Electromagnetic Spectrum Management Unit, National Science Foundation, Washington, D.C. 20550, (202-357-9696):

Arecibo Observatory: Rectangle between latitudes 17° 30' N. and 19° 00' N. and between longitudes 65° 10' W. and 68° 00' W.

Owens Valley Radio Observatory: Two contiguous rectangles, one between latitudes 36° N. and 37° N. and longitudes 117° 40' W. and 118° 30' W. and the second between latitudes 37° N. and 38° N. and longitudes 118° W. and 118° 50' W.

Sagamore Hill Radio Observatory: Rectangle between latitudes 42° 10' N. and 43° 00' N. and longitudes 70° 31' W. and 71° 31' W.

Table Mountain Solar Observatory (NOAA), Boulder, Colorado (407–409 MHz only): Rectangle between latitudes 39° 30' N. and 40° 30' N. and longitudes 104° 30' W. and 106° 00' W. or the Continental Divide whichever is farther east.

The non-Government use of this band is limited to the radio astronomy service and as provided by footnote US13.

US201 In the band 460–470 MHz, space stations in the earth exploration-satellite service may be authorized for space-to-Earth transmissions on a secondary basis with respect to the fixed and mobile services.

When operating in the meteorological-satellite service, such stations shall be protected from harmful interference from other applications of the earth exploration-satellite service. The power flux produced at the Earth's surface by any space station in this band shall not exceed -152 dBW/m^2/4 kHz.

US203 Radio astronomy observations of the formaldehyde line frequencies 4825–4835 MHz and 14.470–14.500 GHz may be made at certain radio astronomy observatories as indicated below:

Bands to be Observed

4 GHz	*14 GHz*	*Observatory*
X		National Astronomy and Ionosphere Center, Arecibo, Puerto Rico
X	X	National Radio Astronomy Observatory, Green Bank, W. Va.
X	X	National Radio Astronomy Observatory, Socorro, New Mexico
X	X	Hat Creek Observatory (U of Calif.), Hat Creek, Cal.

4 GHz	*14 GHz*	*Observatory*
X	X	Haystack Radio Observatory (MIT-Lincoln Lab), Tyngsboro, Mass.
X	X	Owens Valley Radio Observatory (Cal. Tech.), Big Pine, Cal.
	X	Five College Radio Astronomy Observatory, Quabbin Reservoir (near Amherst), Massachusetts

Every practicable effort will be made to avoid the assignment of frequencies to stations in the fixed or mobile services in these bands. Should such assignments result in harmful interference to these observations, the situation will be remedied to the extent practicable.

US205 Tropospheric scatter systems are prohibited in the band 2500–2690 MHz.

US208 Planning and use of the band 1559–1626.5 MHz necessitate the development of technical and/or operational sharing criteria to ensure the maximum degree of electromagnetic compatibility with existing and planned systems within the band.

US209 The use of frequencies 460.6625, 460.6875, 460.7125, 460.7375, 460.7625, 460.7875, 460.8125, 460.8375, 460.8625, 465.6625, 465.6875, 465.7125, 465.7375, 465.7625, 465.7875, 465.8125, 465.8375, and 465.8625 MHz may be authorized, with 100 mW or less output power, to Government and non-government radio stations for one-way, non-voice bio-medical telemetry operations in hospitals, or medical or convalescent centers.

US210 [As proposed in Docket ET 00-221:] In the sub-band 40.66–40.7 MHz, frequencies may be authorized to Government and non-Government stations on a secondary basis for the tracking of, and telemetering of scientific data from, ocean buoys and wildlife. Operation in this sub-band is subject to the technical standards specified in: (a) Section 8.2.42 of the NTIA Manual for Government use, or (b) 47 C.F.R. § 90.248 for non-Government use.

US211 In the bands 1670–1690, 5000–5250 MHz and 10.7–11.7, 15.1365–15.35, 15.4–15.7, 22.5–22.55, 24–24.05, 31.0–31.3, 31.8–32.0, 40.5–42.5, 84–86, 102–105, 116–126, 151–164, 176.5–182, 185–190, 231–235, 252–265 GHz, applicants for airborne or space station assignments are urged to take all practicable steps to protect radio astronomy observations in the adjacent lands from harmful interference; however, US74 applies.

US213 The frequency 122.925 MHz is for use only for communications with or between aircraft when coordinating natural resources programs of Federal or State natural resources, agencies, including forestry management and fire suppression, fish and game management and protection and environmental monitoring and protection.

US214 The frequency 157.1 MHz is the primary frequency for liaison communications between ship stations and stations of the United States Coast Guard.

US215 Emissions from microwave ovens manufactured on and after January 1, 1980, for operation on the frequency 915 MHz must be confined within the band 902–928 MHz. Emissions from microwave ovens manufactured prior to January 1, 1980, for operation on the frequency 915 MHz must be confined within the band 902–940 MHz. Radiocommunications services operating in the band 928–940 MHz must accept any harmful interference from the operation of microwave ovens manufactured before January 1, 1980.

US216 The frequencies 150.775 and 150.790, and the bands 152–152.0150, 163.2375–163.2625, 462.9375–463.1875 and 467.9375–468.1875 MHz are authorized for Government/non-Government operations in medical radio communications systems.

US217 Pulse-ranging radiolocation systems may be authorized for Government and non-Government use in the 420–450 MHz band along the shorelines of Alaska and the contiguous 48 states. Spread spectrum radiolocation systems may be authorized in the 420–435 MHz portion of the band for operation within the contiguous 48 States and Alaska. Authorizations will be granted on a case-by-case basis; however, operations proposed to be located within the zones set forth in US228 should not expect to be accommodated. All stations operating in accordance with this provision will be secondary to stations operating in accordance with the Table of Frequency Allocations.

US218 The band 902–928 MHz is available for Location and Monitoring Service (LMS) systems subject to not causing harmful interference to the operation of all Government stations authorized in these bands. These systems must tolerate interference from the operation of industrial, scientific, and medical (ISM) devices and the operation of Government stations authorized in these bands.

US220 The frequencies 36.25 and 41.71 MHz may be authorized to Government stations and non-Government stations in the petroleum radio service, for oil spill containment and cleanup operations. The use of these frequencies for oil spill containment or cleanup operations is limited to the inland and coastal waterway regions.

US222 In the band 2025–2035 MHz Geostationary Operational Environmental Satellite Earth stations in the Space Research and Earth Exploration-Satellite Services may be authorized on a coequal basis for Earth-to-space transmissions for tracking, telemetry, and telecommand at the sites listed below:

Wallops Is., Va. 37° 50' 48" N., 75° 27' 33" W.

Seattle, Wash. 47° 34' 15" N., 122° 33' 10" W.

Honolulu, Hawaii 21° 21' 12" N., 157° 52' 36" W.

US223 Within 75 miles of the United States/Canada border on the Great Lakes, the St. Lawrence Seaway, and the Puget Sound and the Strait of Juan de Fuca and its approaches, use of coast transmit frequency 162.025 MHz and ship station transmit frequency 157.425 MHz (VHF maritime mobile service Channel 88) may be authorized for use by the maritime service for public correspondence.

US224 Government systems utilizing spread spectrum techniques for terrestrial communication, navigation and identification may be authorized to operate in the band 960–1215 MHz on the condition that harmful interference will not be caused to the aeronautical radionavigation service. These systems will be handled on a case-by-case basis. Such systems shall be subject to a review at the national level for operational requirements and electromagnetic compatibility prior to development, procurement or modification.

US228 Applicants for operation in the band 420 to 450 MHz under the provisions of US217 should not expect to be accommodated if their area of service is within the following geographic areas:

(a) Those portions of Texas and New Mexico bounded on the south by latitude 31° 45' North, on the east by longitude 104° 00' West, on the north by latitude 34° 30' North, and on the West by longitude 107° 30' West.

(b) In the State of Massachusetts within a 160 kilometers (100 miles) radius around the locations of Otis Air Force Base, Massachusetts (latitude 41° 45' North, longitude 70° 32' West).

(c) In the State of California within a 240 kilometer (150 mile) radius of Beale Air Force Base, California (latitude 39° 08' North, longitude 121° 26' West).

(d) In the State of Alaska, within a 160 kilometer (100 mile) radius of Clear, Alaska (latitude 64° 17' North, longitude 149° 10' West).

(e) In the State of North Dakota, within a 160 kilometer (100 mile) radius of Concrete, North Dakota (latitude 48° 43' North, longitude 97° 54' West).

(f) Those portions of Texas and New Mexico bounded on the south by latitude 31° 45′ North, on the east by longitude 104° 00′ West, on the north by latitude 34° 30′ North, and on the West by longitude 107° 30′ West.

(g) In the state of Alaska within a 160 kilometer (100 mile) radius of Clear, Alaska (latitude 64° 17′ North, longitude 149° 10′ West).

(h) In the state of North Dakota within a 160 kilometer (100 mile) radius of Concrete, North Dakota (latitude 48° 43′ North, longitude 97° 54′ West).

(i) In the States of Alabama, Florida, Georgia and South Carolina within a 200 kilometer (124 mile) radius of Warner Robins Air Force Base, Georgia (latitude 32° 38′ North, longitude 83° 35′ West).

(j) In the State of Texas within a 200 kilometer (124 mile) radius of Goodfellow Air Force Base, Texas (latitude 31° 25′ North, longitude 100° 24′ West).

US229 [As proposed in Docket ET 00-221:] In the band 216–220 MHz, Government operations are on a non-interference basis to authorized non-Government operations and shall not hinder the implementation of any non-Government operations, except at the following space surveillance stations where Government operations are co-primary:

	Transmit Frequency of 216.98 MHz			*Receive Frequencies of 216.965–216.995 MHz*		
Location	North Latitude/ West Longitude	Protection Radius	Location	North Latitude West Longitude	Protection Radius
Lake Kickapoo, TX	33° 32′ / 098° 45′	250 km	San Diego, CA	32° 34′ / 116° 58′	50 km
Jordan Lake, AL	32° 39′ / 086° 15′	150 km	Elephant Butte, NM	33° 26′ / 106° 59′	50 km
Gila River, AZ	33° 06′ / 112° 01′	150 km	Red River, AR	33° 19′ / 093° 33′	50 km
			Silver Lake, MO	33° 08′ / 091° 01′	50 km
			Hawkinsville, GA	32° 17′ / 083° 32′	50 km
			Fort Stewart, GA	31° 58′ / 081° 30′	50 km

US230 Non-government land mobile service is allocated on a primary basis in the bands 422.1875–425.4875 and 427.1875–429.9875 MHz within 50 statute miles of Detroit, MI, and Cleveland, OH, and in the bands 423.8125–425.4875 and 428.8125–429.9875 MHz within 50 statute miles of Buffalo, NY.

US244 [Proposed to be changed in Docket WT 00-77.] The band 136.000–137.000 MHz is allocated to the non-Government aeronautical mobile (R) service on a primary basis, and is subject to pertinent international treaties and agreements. The frequencies 136.000 MHz, 136.025 MHz, 136.050 MHz, 136.075 MHz, 136.125 MHz, 136.150 MHz, 136.175 MHz, 136.225 MHz, 136.250 MHz, 136.300 MHz, 136.325 MHz, 136.350 MHz, 136.400 MHz, 136.425 MHz and 136.450 MHz are available on a shared basis to the Federal Aviation Administration for air traffic control purposes, such as automatic weather observation services (AWOS), automatic terminal information services (ATIS) and airport control tower communications. Stations licensed prior to January 2, 1990, using the 136–137 MHz band for space operation (space-to-Earth), meteorological-satellite service (space-to-Earth) and the space research service (space-to-Earth) may continue to use this band on a secondary basis to aeronautical mobile (R) service stations. No new assignments will be made to stations in the above space services.

US245 The fixed-satellite service is limited to international intercontinental systems and subject to case-by-case electromagnetic compatibility analysis.

US246 No station shall be authorized to transmit in the following bands: 608–614 MHz, except for medical telemetry equipment (Medical telemetry equipment shall not cause harmful interference to radio astronomy operations in the band 608–614 MHz and shall be coordinated under the requirements found in 47 C.F.R. §95.1119), 1400–1427 MHz, 1660.5–1668.4 MHz, 2690–2700 MHz, 4990–5000 MHz, 10.68–10.7 GHz, 15.35–15.4 GHz, 23.6–24 GHz, 31.3–31.8 GHz, 50.2–50.4 GHz, 52.6–54.25 GHz, 86–92 GHz, 100–102 GHz, 105–116 GHz, 164–168 GHz, 182–185 GHz and 217–231 GHz.

US251 The band 12.75–13.25 GHz is also allocated to the space research (deep space) (space-to-Earth) service for reception only at Goldstone, California, 35° 18' N. 116° 54' W.

US252 The bands 2110–2120, 7145–7190 MHz, and 34.2–34.7 GHz are also allocated for Earth-to-space transmissions in the space research service, limited to deep space communications at Goldstone, California.

US254 In the band 18.6–18.8 GHz the fixed and mobile services shall be limited to a maximum equivalent isotropically radiated power of +35 dBW and the power delivered to the antenna shall not exceed −3 dBW.

US255 In addition to any other applicable limits, the power flux-density across the 200 MHz band 18.6–18.8 GHz produced at the surface of the Earth by emissions from a space station under assumed free-space propagation conditions shall not exceed -95 dB(W/m²) for all angles of arrival. This limit may be exceeded by up to 3 dB for no more than 5% of the time.

US256 Radio astronomy observations may be made in the band 1718.8–1722.2 MHz on an unprotected basis. Agencies providing other services in this band in the geographic areas listed below should bear in mind that their operations may affect those observations, and those agencies are encouraged to minimize potential interference to the observations insofar as it is practicable.

Hat Creek Observatory, Hat Creek, California: Rectangle between latitudes 40° 00′ N. and 42° 00′ N. and between longitudes 120° 15′ W. and 122° 15′ W.

Owens Valley Radio Observatory, Big Pine, California: Two contiguous rectangles, one between latitudes 36° 00′ N. and 37° 00′ N. and between longitudes 117° 40′ W. and 118° 30′ W. and the second between latitudes 37° 00′ N. and 38° 00′ N. and between longitudes 118° 00′ W and 118° 50′ W.

Haystack Radio Observatory, Tyngsboro, Massachusetts: Rectangle between latitudes 41° 00′ N. and 43° 00′ N. and between longitudes 71° 00′ W. and 73° 00′ W.

National Astronomy and Ionosphere Center, Arecibo, Puerto Rico: Rectangle between latitudes 17° 30′ N. and 19° 00′ N. and between longitudes 65° 10′ W. and 68° 00′ W.

National Radio Astronomy Observatory, Green Bank, West Virginia: Rectangle between latitudes 37° 30′ N. and 39° 15′ N. and longitudes 78° 30′ W. and 80° 30′ W.

National Radio Astronomy Observatory, Socorro, New Mexico: Rectangle between latitudes 32° 30′ N. and 35° 30′ N. and between longitudes 106° 00′ W. and 109° 00′ W.

US257 [Docket WT 00-32 proposes to replace US257 with US311.] Radio astronomy observations may be made in the 4950–4990 MHz band at certain Radio Astronomy Observatories indicated below: National Astronomy and Ionosphere Center, Arecibo, Puerto Rico: Rectangle between latitudes 17° 30′ N. and 19° 00′ N. and between latitudes (sic) 65° 10′ W. and 68° 00′ W.

Haystack Radio Observatory, Tyngsboro, Massachusetts: Rectangle between latitudes 41° 00′ N. and 43° 00′ N. and between longitudes 71° 00′ W. and 73° 00′ W.

National Radio Astronomy Observatory, Green Bank, West Virginia: Rectangle between latitudes 37° 30′ N. and 39° 15′ N. and longitudes 78° 30′ W. and 80° 30′ W.

National Radio Astronomy Observatory, Socorro, New Mexico: Rectangle between latitudes 32° 30′ N. and 35° 30′ N. and longitudes 106° 00′ W. and 109° 00′ W.

Owens Valley Radio Observatory, Big Pine, California: Two contiguous rectangles, one between latitudes 36° 00′ N. and 37° 00′ N. and between longitudes 117° 40′ W. and 118° 30′ W. and the second between latitudes 37° 00′ N. and 38° 00′ N. and between longitudes 118° 00′ W. and 118° 50′ W.

Hat Creek Observatory, Hat Creek, California: Rectangle between latitudes 40° 00′ N. and 42° 00′ N. and between longitudes 120° 15′ W. and 122° 15′ W.

Every practicable effort will be made to avoid the assignment of frequencies in the band 4950–4990 MHz to stations in the fixed and mobile services within the geographic areas given above. In addition, every practicable effort will be made to avoid the assignment of frequencies in this band to stations in the aeronautical mobile service which operate outside of those geographic areas, but which may cause harmful interference to the listed observatories.

Should such assignments result in harmful interference to these observatories, the situation will be remedied to the extent practicable.

US258 In the band 8025–8400 MHz, the non-Government earth exploration-satellite service (space-to-Earth) is allocated on a primary basis. Authorizations are subject to a case-by-case electromagnetic compatibility analysis.

US259 Stations in the radiolocation service in the band 17.3–17.7 GHz, shall be restricted to operating powers of less than 51 dBW eirp after feeder link stations for the broadcasting-satellite service are authorized and brought into use.

US260 Aeronautical mobile communications which are an integral part of aeronautical radionavigation systems may be satisfied in the bands 1559–1626.5 MHz, 5000–5250 MHz and 15.4–15.7 GHz.

US261 The use of the band 4200–4400 MHz by the aeronautical radionavigation service is reserved exclusively for airborne radio altimeters. Experimental stations will not be authorized to develop equipment for operational use in this band other than equipment related to altimeter stations. However, passive sensing in the earth exploration-satellite and space research services may be authorized in this band on a secondary basis (no protection is provided from the radio altimeters).

US262 The use of the band 31.8–32.3 GHz by the space research service (deep space) (space-to-Earth) is limited to Goldstone, California.

US263 In the bands 21.2–21.4 GHz, 22.21–22.5 GHz, 36–37 GHz, 56.26–58.2 GHz, 116–126 GHz, 150–151 GHz, 174.5–176.5 GHz, 200–202 GHz, and 235–238 GHz, the space research and the Earth exploration-satellite services shall not receive protection from the fixed and mobile services operating in accordance with the Table of Frequency Allocations.

US264 In the band 48.94–49.04 GHz, airborne stations shall not be authorized.

US265 In the band 10.6–10.68 GHz, the fixed service shall be limited to a maximum equivalent isotropically radiated power of 40 dBW and the power delivered to the antenna shall not exceed −3dBW per 250 kHz.

US266 Licensees in the public safety radio services holding a valid authorization on June 30, 1958, to operate in the frequency band 156.27–157.47 MHz or on the frequencies 161.85, 161.91 or 161.97 MHz may, upon proper application, continue to be authorized for such operation, including expansion of existing systems, until such time as harmful interference is caused to the operation of any authorized station other than those licensed in the public safety radio service.

US267 In the band 902–928 MHz, amateur radio stations shall not operate within the States of Colorado and Wyoming, bounded by the area of: latitude 39° N. to 42° N. and longitude 103° W. to 108° W.

US268 The bands 890–902 MHz and 928–942 MHz are also allocated to the radiolocation service for Government ship stations (off-shore ocean areas) on the condition that harmful interference is not caused to non-Government land mobile stations. The provisions of footnote US116 apply.

US269 In the band 2500–2690 MHz, applicants for space station assignments are urged to take all practicable steps to protect radio astronomy observations in the adjacent band, 2690–2700 MHz, from harmful interference. Further, all applicants are urged to coordinate their proposed system through the Electromagnetic Management Unit, National Science Foundation, Washington, D.C. 20550, prior to system development.

US270 The band 72.77–72.91 GHz is also allocated to the radio astronomy service. Applicants for frequency assignments in this band are urged to take all practicable steps to protect radio astronomy observations from harmful interference.

US271 The use of the band 17.3–17.8 GHz by the fixed-satellite service (Earth-to-space) is limited to feeder links for broadcasting-satellite service.

U.S. Footnotes

US273 In the 74.6–74.8 MHz and 75.2–75.4 MHz bands stations in the fixed and mobile services are limited to a maximum power of 1 watt from the transmitter into the antenna transmission line.

US274 [Proposed to be removed in Docket ET 00-221:] In the 216–220 MHz band fixed, aeronautical mobile and land mobile stations are limited to telemetering and associated telecommand operations.

US275 The band 902–928 MHz is allocated on a secondary basis to the amateur service subject to not causing harmful interference to the operations of Government stations authorized in this band or to Location and Monitoring Service (LMS) systems. Stations in the Amateur service must tolerate any interference from the operations of Industrial, Scientific and Medical (ISM) devices, LMS systems and the operations of Government stations authorized in this band. Further, the Amateur Service is prohibited in those portions of Texas and New Mexico bounded on the south by latitude 31° 41' North, on the east by longitude 104° 11' West, on the north by latitude 34° 30' North, and on the west by longitude 107° 30' West; in addition, outside this area but within 150 miles of these boundaries of White Sands Missile Range the service is restricted to a maximum transmitter peak envelope power output of 50 watts.

US276 [As proposed in Docket ET 00-221:] Except as otherwise provided for herein, use of the bands 2320–2345 MHz and 2360–2385 MHz by the mobile service is limited to aeronautical telemetering and associated telecommand operations for flight testing of manned or unmanned aircraft, missiles or major components thereof. The following four frequencies are shared on a co-equal basis by Government and non-Government stations for telemetering and associated telecommand operations of expendable and reusable launch vehicles whether or not such operations involve flight testing: 2332.5 MHz, 2364.5 MHz, 2370.5 MHz, and 2382.5 MHz. All other mobile telemetering uses shall be secondary to the above uses.

US277 The band 10.6–10.68 GHz is also allocated on a primary basis to the radio astronomy service. However, the radio astronomy service shall not receive protection from stations in the fixed service which are licensed to operate in the one hundred most populous urbanized areas as defined by the U.S. Census Bureau.

The following radio astronomy sites have been coordinated for observations in this band: National Radio Astronomy Observatory, Green Bank, West Virginia; (38° 26' 08" N.; 79° 49' 42" W.), National Radio Astronomy Observatory, Socorro, New Mexico; (34° 04' 43" N.; 107° 37' 04" W.), Harvard Radio Astronomy Station, Fort Davis, Texas; (30° 38' 08" N.; 103° 56' 42" W.), Hat Creek Observatory, Hat Creek, California; (40° 49' 03" N.; 121° 28' 24" W.), Owens Valley Radio Observatory, Big Pine,

California; (37° 13′ 54″ N.; 118° 17′ 36″ W.), Naval Research Laboratory, Maryland Point, Maryland (38° 22′ 26″ N.; 77° 14′ 00″ W.).

US278 In the 22.55–23.55 and 32–33 GHz bands non-geostationary intersatellite links may operate on a secondary basis to geostationary intersatellite links.

US291 Television pickup stations in the mobile service may be authorized to use frequencies in the band 38.6–40 GHz on a secondary basis to stations operating in accordance with the Table of Frequency Allocations.

US292 In the band 14.0–14.2 GHz stations in the radionavigation service shall operate on a secondary basis to the fixed-satellite service.

US297 The bands 47.2–49.2 GHz and 74.0–75.5 GHz are also available for feeder links for the broadcasting-satellite service.

US300 The frequencies 169.445, 169.505, 170.245, 170.305, 171.045, 171.105, 171.845 and 171.905 MHz are available for wireless microphone operations on a secondary basis to Government and non-Government operations.

US301 Except as provided in US302, broadcast auxiliary stations licensed as of November 21, 1984, to operate in the band 942–944 MHz may continue to operate on a co-equal primary basis to other stations and services operating in the band in accordance with the Table of Frequency Allocations.

US302 The band 942–944 MHz in Puerto Rico is allocated as an alternative allocation to the fixed service for broadcast auxiliary stations only.

US303 In the band 2285–2290 MHz, non-Federal government space stations in the space research, space operations and earth exploration-satellite services may be authorized to transmit to the Tracking and Data Relay Satellite System subject to such conditions as may be applied on a case-by-case basis. Such transmissions shall not cause harmful interference to authorized Federal government stations. The power flux density at the Earth's surface from such non-Federal government stations shall not exceed -144 to -154 dBW/m^2/4 kHz, depending on angle of arrival, in accordance with ITU Radio Regulation S21.16.

US307 The sub-band 5150–5216 MHz is also allocated for space-to-Earth transmissions in the fixed satellite service for feeder links in conjunction with the radiodetermination satellite service operating in the bands 1610–1626.5 MHz and 2483.5–2500 MHz. The total power flux density at the Earth's surface shall in no case exceed -159 dBW/m per 4 kHz for all angles of arrival.

US308 In the frequency bands 1549.5–1558.5 MHz and 1651–1660 MHz, the Aeronautical-Mobile-Satellite (R) requirements that cannot be accommodated in the 1545–1549.5 MHz, 1558.5–1559 MHz, 1646.5–1651 MHz and 1660–1660.5 MHz bands shall have priority access with real-time preemptive capability for communications in the mobile-satellite service. Systems not interoperable with the aeronautical mobile-satellite (R) service shall operate on a secondary basis. Account shall be taken of the priority of safety-related communications in the mobile-satellite service.

US309 Transmissions in the bands 1545–1559 MHz from terrestrial aeronautical stations directly to aircraft stations, or between aircraft stations, in the aeronautical mobile (R) service are also authorized when such transmissions are used to extend or supplement the satellite-to-aircraft links. Transmissions in the band 1646.5–1660.5 MHz from aircraft stations in the aeronautical mobile (R) service directly to terrestrial aeronautical stations, or between aircraft stations, are also authorized when such transmissions are used to extend or supplement the aircraft-to-satellite links.

US310 In the band 14.896–15.121 GHz, non-Government space stations in the space research service may be authorized on a secondary basis to transmit to Tracking and Data Relay Satellites subject to such conditions as may be applied on a case-by-case basis. Such transmissions shall not cause harmful interference to authorized Government stations. The power flux density at the Earth's surface from such non-Government stations shall not exceed -138 to -148 dBW/m^2/4 kHz, depending on the angle of arrival, in accordance with CCIR Recommendation 510-1.

US311 [As proposed in Docket ET 00-221:] Radio astronomy observations may be made in the bands 1350–1400 MHz and 4950–4990 MHz on an unprotected basis at certain radio astronomy observatories indicated below:

National Astronomy and Ionosphere Center, Arecibo, Puerto Rico	Rectangle between latitudes 17° 30' N and 19° 00' N and between longitudes 65° 10' W and 68° 00' W.
National Radio Astronomy Observatory, Socorro, New Mexico	Rectangle between latitudes 32° 30' N and 35° 30' N and between longitudes 106° 00' W and 109° 00' W.
National Radio Astronomy Observatory, Green Bank, West Virginia	Rectangle between latitudes 37° 30' N and 39° 15' N and between longitudes 78° 30' W and 80° 30' W.

National Radio Astronomy Observatory, Very Long Baseline Array Stations	80 kilometers (50 mile) radius centered on:	
	Latitude (North)	Longitude (West)
Pie Town, NM	34° 18'	108° 07'
Kitt Peak, AZ	31° 57'	111° 37'
Los Alamos, NM	35° 47'	106° 15'
Fort Davis, TX	30° 38'	103° 57'
North Liberty, IA	41° 46'	91° 34'
Brewster, WA	48° 08'	119° 41'
Owens Valley, CA	37° 14'	118° 17'
Saint Croix, VI	17° 46'	64° 35'
Mauna Kea, HI	19° 48'	155° 27'
Hancock, NH	42° 56'	71° 59'

Every practicable effort will be made to avoid the assignment of frequencies in the bands 1350–1400 MHz and 4950–4990 MHz to stations in the fixed and mobile services that could interfere with radio astronomy observations within the geographic areas given above. In addition, every practicable effort will be made to avoid assignment of frequencies in these bands to stations in the aeronautical mobile service which operate outside of those geographic areas, but which may cause harmful interference to the listed observatories. Should such assignments result in harmful interference to these observatories, the situation will be remedied to the extent practicable.

US312 The frequency 173.075 MHz may also be authorized on a primary basis to non-Government stations in the Police Radio Service (with a maximum authorized bandwidth of 20 kHz) for stolen vehicle recovery systems.

US315 In the frequency bands 1530–1544 MHz and 1626.5–1645.5 MHz maritime mobile-satellite distress and safety communications, e.g., GMDSS, shall have priority access with real-time preemptive capability in the mobile-satellite service. Communications of mobile-satellite system stations not participating in the GMDSS shall operate on a secondary basis to distress and safety communications of stations operating in the GMDSS. Account shall be taken of the priority of safety-related communications in the mobile-satellite service.

US316 The band 2900–3000 MHz is also allocated on a primary basis to the Meteorological Aids Service. Operations in this service are limited to Government Next Generation Weather Radar (NEXRAD) systems where accommodation in the 2700–2900 MHz band is not technically practical and are subject to coordination with existing authorized stations.

US317 [Proposed to be removed in Docket ET 00-221:] The band 218.0–219.0 MHz is allocated on a primary basis to the Interactive Video and Data operations.

US318 Until January 1, 2000, the use of the 137–138 MHz band by the mobile-satellite service will be secondary to Government satellite operations in the subbands: 137.333–137.367, 137.485–137.515, 137.605–137.635 and 137.753–137.787 MHz.

US319 In the bands 137–138 MHz, 148–149.9 MHz, 149.9–150.05 MHz, 399.9–400.05 MHz, 400.15–401 MHz, 1610–1626.5 MHz, and 2483.5–2500 MHz, Federal government stations in the mobile-satellite service shall be limited to earth stations operating with non-Federal government space stations.

US320 Use of the 137–138, 148–149.9, and 400.15–401 MHz bands by the mobile-satellite service is limited to non-voice, non-geostationary satellite systems and may include satellite links between land earth stations at fixed locations.

US322 Use of the bands 149.9–150.05 MHz and 399.9–400.05 MHz by the mobile-satellite service (Earth-to-space) is limited to non-voice, non-geostationary satellite systems, including satellite links between land earth stations.

US323 In the 148–149.9 MHz band, no individual mobile earth station shall transmit, on the same frequency being actively used by fixed and mobile stations and shall transmit no more than 1% of the time during any 15 minute period; except, individual mobile earth stations in this band that do not avoid frequencies actively being used by the fixed and mobile services shall not exceed a power density of −16 dBW/4 kHz and shall transmit no more than 0.25% of the time during any 15 minute period. Any single transmission from any individual mobile earth station operating in this band shall not exceed 450 ms in duration and consecutive transmissions from a single mobile Earth station on the same frequency shall be separated by at least 15 seconds. Land earth stations in this band shall be subject to electromagnetic compatibility analysis and coordination with terrestrial fixed and mobile stations.

US324 Government and non-Government satellite systems in the 400.15–401 MHz band shall be subject to electromagnetic compatibility analysis and coordination.

US325 In the band 148–149.9 MHz fixed and mobile stations shall not claim protection from land earth stations in the mobile-satellite service that have been previously coordinated; Government fixed and mobile stations exceeding 27 dBW EIRP, or an emission bandwidth greater than 38 kHz, will be coordinated with existing mobile-satellite service space stations.

US327 The 2310–2360 MHz is allocated to the broadcasting-satellite service (sound) and complementary terrestrial broadcasting service on a primary basis. Such use is limited to digital audio broadcasting and is subject to the provisions of Resolution 528.

US328 In the band 2320–2345 MHz, the mobile and radiolocation services are allocated on a primary basis until a broadcasting-satellite (sound) service has been brought into use in such a manner as to affect or be affected by the mobile and radiolocation services in those service areas. The broadcasting-satellite (sound) service during implementation should also take cognizance of the expendable and reusable launch vehicle frequency 2332.5 MHz, to minimize the impact on this mobile service use to the extent possible.

US334 In the band 17.8–20.2 GHz, Government space stations in both geostationary (GSO) and non-geostationary satellite orbits (NGSO) and associated earth stations in the fixed-satellite service (space-to-Earth) may be authorized on a primary basis. For a Government geostationary satellite network to operate on a primary basis, the space station shall be located outside the arc, measured from East to West, 70 W Longitude to 120 West Longitude. Coordination between Government fixed-satellite systems and non-Government space and terrestrial systems operating in accordance with the United States Table of Frequency Allocations is required.

(a) In the sub-band 17.8–19.7 GHz, the power flux-density at the surface of the Earth produced by emissions from a Government GSO space station or from a Government space station in a NGSO constellation of 50 or fewer satellites, for all conditions and for all methods of modulation, shall not exceed the following values in any 1 MHz band:

(1) -115 dB(W/m^2) for angles of arrival above the horizontal plane (δ) between 0° and 5°,
(2) $-115 + 0.5(\delta-5)$ dB(W/m^2) for δ between 5° and 25°, and
(3) -105 dB(W/m^2) for δ between 25° and 90°.

(b) In the sub-band 17.8–19.3 GHz, the power flux-density at the surface of the Earth produced by emissions from a Government space station in an NGSO constellation of 51 or more satellites, for all conditions and for all methods of modulation, shall not exceed the following values in any 1 MHz band:

(1) $-115 - X$ dB(W/m^2) for δ between 0° and 5°,
(2) $-115 - X + ((10 + X)/20(\delta - 5)$ dB(W/m^2) for δ between 5° and 25°, and

(3) -105 dB(W/m²) for δ between 25° and 90°; where X is defined as a function of the number of satellites, n, in an NGSO constellation as follows:

For $n \leq 288$, $X = (5/119)(n - 50)$ dB; and
For $n > 288$, $X = (1/69)(n + 402)$ dB.

US335 The primary Government and non-Government allocations for the various segments of the 220–222 MHz band are divided as follows: (1) the 220.0–220.55 / 221.0–221.55, 220.6–220.8 / 221.6–221.8, 220.85–220.90 / 221.85–221.90 and 220.925–221.0 / 221.925–222.0 MHz bands (Channels 1–110, 121–160, 171–180 and 186–200, respectively) are available for exclusive non-Government use; (2) the 220.55–220.60 / 221.55–221.60 MHz bands (Channels 111–120) are available for exclusive Government use; and (3) the 220.80–220.85 / 221.80–221.85 and 220.900–220.925 / 221.900–221.925 MHz bands (Channels 161–170 and 181–185, respectively) are available for shared Government and non-Government use. The exclusive non-Government band segments are also available for temporary fixed geophysical telemetry operations on a secondary basis to the fixed and mobile services.

US337 In the band 13.75–13.80 GHz, earth stations in the fixed-satellite service shall be coordinated on a case-by-case basis through the frequency assignment subcommittee in order to minimize harmful interference to the Tracking and Data Relay Satellite System's forward space-to-space link (TDRSS forward link-to-LEO).

US338 In the 2305–2310 MHz band, space-to-Earth operations are prohibited. Additionally, in the 2305–2320 MHz band, all Wireless Communications Service (WCS) operations within 50 kilometers of 35° 20′ North Latitude and 116° 53′ West Longitude shall be coordinated through the Frequency Assignment Subcommittee of the Interdepartment Radio Advisory Committee in order to minimize harmful interference to NASA's Goldstone Deep Space facility.

US339 The bands 2310–2320 and 2345–2360 MHz are also available for aeronautical telemetering and associated telecommand operations for flight testing of manned or unmanned aircraft, missiles or major components thereof on a secondary basis to the Wireless Communications Service. The following two frequencies are shared on a co-equal basis by Government and non-Government stations for telemetering and associated telecommand operations of expendable and re-usable launch vehicles whether or not such operations involve flight testing: 2312.5 and 2352.5 MHz. Other mobile telemetering uses may be provided on a non-interference basis to the above uses. The broadcasting-satellite (sound) service during implementation should also take cognizance of the expendable and reusable launch vehicle frequencies 2312.5 and 2352.5 MHz, to minimize the impact on this mobile service use to the extent possible.

US342 In making assignments to stations of other services to which the bands:

13360–13410 kHz	4825–4835 MHz	48.94–49.04 GHz*
37.5–38.25 MHz	14.47–14.5 GHz*	97.88–98.08 GHz*
322–328.6 MHz*	22.01–22.21 GHz*	140.69–140.98 GHz*
1330–1400 MHz*	22.21–22.5 GHz	144.68–144.98 GHz*
1610.6–1613.8 MHz*	22.81–22.86 GHz*	145.45–145.75 GHz*
1660–1670 MHz	23.07–23.12 GHz*	146.82–147.12 GHz*
3260–3267 MHz*	31.2–31.3 GHz	262.24–262.76 GHz*
3332–3339 MHz*	36.43–36.5 GHz*	265–275 GHz
3345.8–3352.5 MHz*	42.5–43.5 GHz	

are allocated (*indicates radio astronomy use for spectral line observations), all practicable steps shall be taken to protect the radio astronomy service from harmful interference. Emissions from spaceborne or airborne stations can be particularly serious sources of interference to the radio astronomy service (see Nos. 343/S4.5 and 344/S4.6 and Article 36/S29 of the ITU Radio Regulations).

US345 In the 402–405 MHz band the mobile, except mobile aeronautical, service is allocated on a secondary basis and is limited to, with the exception of military tactical mobile stations, Medical Implant Communications Service (MICS) operations. MICS stations are authorized by rule on the condition that harmful interference is not caused to stations in the Meteorological Aids, Meteorological Satellite, and Earth Exploration Satellite Services, and that MICS stations accept interference from stations in the Meteorological Aids, Meteorological Satellite, and Earth Exploration Satellite Services. Certain MICS stations are subject to the registration requirements set forth in Section 95.1215 of this chapter.

US346 Except as provided by footnote US222, the use of the band 2025–2110 MHz by the Government space operation service (Earth-to-space), Earth-exploration-satellite service (Earth-to-space), and space research service (Earth-to-space) shall not constrain the deployment of the Television Broadcast Auxiliary Service, the Cable Television Relay Service, or the Local Television Transmission Service. To facilitate compatible operations between non-Government terrestrial receiving stations at fixed sites and Government earth station transmitters, coordination is required. To facilitate compatible operations between non-government terrestrial transmitting stations and Government spacecraft receivers, the terrestrial transmitters shall not be high-density systems (see Recommendations ITU-R SA.1154 and ITU-R F.1247).

US347 In the band 2025–2110 MHz, non-Government Earth-to-space and space-to-space transmissions may be authorized in the space research and Earth exploration-satellite services subject to such conditions as may be applied on a case-by-case basis. Such transmissions shall not cause harmful interference to Government and non-Government stations operating in accordance with the Table of Frequency Allocations.

US348 The band 3650–3700 MHz is also allocated to the Government radiolocation service on a primary basis at the following sites: St. Inigoes, MD (38° 10′ N, 76° 23′ W); Pascagoula, MS (30° 22′ N, 88° 29′ W); and Pensacola, FL (30° 21′ 28″ N, 87° 16′ 26″ W). All fixed and fixed satellite operations within 80 kilometers of these sites shall be coordinated through the Frequency Assignment Subcommittee of the Interdepartmental Radio Advisory Committee on a case-by-case basis.

US349 The band 3650–3700 MHz is also allocated to the Government radiolocation service on a non-interference basis for use by ship stations located at least 44 nautical miles in off-shore ocean areas on the condition that harmful interference is not caused to non-Government operations.

US350 In the bands 608–614 MHz, 1395–1400 MHz, and 1429–1432 [MHz], the land mobile service is limited to medical telemetry and telecommand operations. Additionally, the band 1429–1432 MHz may be used on [a] secondary basis for non-Government land mobile telemetry and telecommand and fixed telemetry.

US351 In the band 1390–1400 MHz, Government operations, except for medical telemetry operations in the sub-band 1395–1400 MHz, are on a non-interference basis to authorized non-Government operations and shall not hinder implementation of any non-Government operations. However, Government operations authorized as of March 22, 1995 at 17 sites identified below will be continued on a fully protected basis until January 1, 2009.

Sites	Lat/Long	Radius	Sites	Lat/Long	Radius
Eglin AFB, FL	30°28′N/ 086°31′W	80 km	Ft. Greely, AK	63°47′N/ 145°52′W	80 km
Dugway PG, UT	40°11′N/ 112°53′W	80	Ft. Rucker, AL	31°13′N/ 085°49′W	80
China Lake, CA	35°41′N/ 117°41′W	80	Redstone, AL	34°35′N/ 086°35′W	80

Sites	Lat/Long	Radius	Sites	Lat/Long	Radius
Ft. Huachuca, AZ	31°33'N/ 110°18'W	80	Utah Test Range, UT	40°57'N/ 113°05'W	80
Cherry Point, NC	34°57'N/ 076°56'W	80	WSM Range, NM	32°10'N/ 106°21'W	80
Patuxent River, MD	38°17'N/ 076°25'W	80	Holloman AFB, NM	33°29'N/ 106°50'W	80
Aberdeen PG, MD	39°29'N/ 076°08'W	80	Yuma, AZ	32°29'N/ 114°20'W	80
Wright-Patterson AFB, OH	39°50'N/ 084°03'W	80	Pacific Missile Range, CA	34°07'N/ 119°30'W	80
Edwards AFB, CA	34°54'N/ 117°53'W	80			

US352 [As proposed in Docket ET 00-221:] In the band 1427–1432 MHz, Government operations, except for medical telemetry operations in the sub-band 1429–1432 MHz, are on a non-interference basis to authorized non-Government operations and shall not hinder the implementation of any non-Government operations, except at the sites identified below where Government operations are co-primary until January 1, 2004.

Location	North Latitude/ West Longitude	Radius	Location	North Latitude/ West Longitude	Radius
Patuxent River, MD	38° 17' / 076° 25'	70 km	Mountain Home AFB, ID	43° 01' / 115° 50'	160 km
NAS Oceana, VA	36° 49' / 076° 02'	100 km	NAS Fallon, NV	39° 24' / 118° 43'	100 km
MCAS Cherry Point, NC	34° 54' / 076° 52'	100 km	Nellis AFB, NV	36° 14' / 115° 02'	100 km
Beaufort MCAS, SC	32° 26' / 080° 40'	160 km	NAS Lemore, CA	36° 18' / 119° 47'	120 km
NAS Cecil Field, FL	30° 13' / 081° 52'	160 km	Yuma MCAS, AZ	32° 39' / 114° 35'	160 km
NAS Whidbey IS., WA	48° 19' / 122° 24'	70 km	China Lake, CA	35° 29' / 117° 16'	80 km
Yakima Firing Ctr AAF, WA	46° 40' / 120° 15'	70 km	MCAS Twenty Nine Palms, CA	34° 15' / 116° 03'	80 km

US353 In the sub-bands 56.24–56.29 GHz, 58.422–58.472 GHz, 59.139–59.189 GHz, 59.566–59.616 GHz, 60.281–60.331 GHz, 60.41–60.46 GHz, and 62.461–62.511 GHz, space-based radio astronomy observations may be made on an unprotected basis.

US354 In the sub-band 58.422–58.472 GHz, airborne stations and space stations in the space-to-Earth direction shall not be authorized.

US355 In the band 10.7–11.7 GHz, non-geostationary satellite orbit licensees in the fixed-satellite service (space-to-Earth), prior to commencing operations, shall coordinate with the following radio astronomy observatories to achieve a mutually acceptable agreement regarding the protection of the radio telescope facilities operating in the band 10.6–10.7 GHz:

Observatory	*West Longitude*	*North Latitude*	*Elevation*
Arecibo Obs.	66° 45' 11"	18° 20' 46"	496 m
Green Bank Telescope (GBT)	79° 50' 24"	38° 25' 59"	825 m
Very Large Array (VLA)	107° 37' 04"	34° 04' 44"	2126 m
Very Long Baseline Array (VLBA) Stations:			
Pie Town, NM	108° 07' 07"	34° 18' 04"	2371 m
Kitt Peak, AZ	111° 36' 42"	31° 57' 22"	1916 m
Los Alamos, NM	106° 14' 42"	35° 46' 30"	1967 m
Ft Davis, TX	103° 56' 39"	30° 38' 06"	1615 m
N Liberty, IA	91° 34' 26"	41° 46' 17"	241 m
Brewster, WA	119° 40' 55"	48° 07' 53"	255 m
Owens Valley, CA	118° 16' 34"	37° 13' 54"	1207 m
St Croix, VI	64° 35' 03"	17° 45' 31"	16 m
Hancock, NH	71° 59' 12"	42° 56' 01"	309 m
Mauna Kea, HI	155° 27' 29"	19° 48' 16"	3720 m

US356 In the band 13.75–14 GHz, an earth station in the fixed-satellite service shall have a minimum antenna diameter of 4.5 m and the e.i.r.p. of any emission should be at least 68 dBW and should not exceed 85 dBW. In addition the e.i.r.p., averaged over one second, radiated by a station in the radiolocation service towards the geostationary-satellite orbit shall not exceed 59 dBW. Receiving space stations in the fixed-satellite service shall not claim protection from radiolocation transmitting stations operating in accordance with the United States Table of Frequency Allocations. ITU Radio Regulation No. S5.43A does not apply.

US357 In the band 13.75–14 GHz, geostationary space stations in the space research service for which information for advance publication has been received by the ITU Radiocommunication Bureau (Bureau) prior to

31 January 1992 shall operate on an equal basis with stations in the fixed-satellite service; after that date, new geostationary space stations in the space research service will operate on a secondary basis. Until those geostationary space stations in the space research service for which information for advance publication has been received by the Bureau prior to 31 January 1992 cease to operate in this band:

a) the e.i.r.p. density of emissions from any earth station in the fixed-satellite service operating with a space station in geostationary-satellite orbit shall not exceed 71 dBW in any 6 MHz band from 13.77 to 13.78 GHz;

b) the e.i.r.p. density of emissions from any earth station in the fixed-satellite service operating with a space station in non-geostationary-satellite orbit shall not exceed 51 dBW in any 6 MHz band from 13.77 to 13.78 GHz.

Automatic power control may be used to increase the e.i.r.p. density in any 6 MHz band in this frequency range to compensate for rain attenuation, to the extent that the power flux-density at the fixed-satellite service space station does not exceed the value resulting from use by an earth station of an e.i.r.p. of 71 dBW or 51 dBW, as appropriate, in any 6 MHz band in clear-sky conditions.

USxxx [As proposed in Docket ET 00-221:] In the band 1432–1435 MHz, Government operations are on a non-interference basis to authorized non-Government operations and shall not hinder the implementation of any non-Government operations, except at the sites identified below where Government operations are co-primary:

Location	North Latitude/ West Longitude	Protection Radius	Location	North Latitude/ West Longitude	Protection Radius
China Lake/Edwards AFB, CA	35° 29' / 117° 16'	100 km	AUTEC	24° 30' / 078° 00'	80 km
White Sands Missile Range/ Holloman AFB, NM	32° 11' / 106° 20'	160 km	Beaufort MCAS, SC	32° 26' / 080° 40'	160 km
Utah Test and Training/ Range Dugway ProvingGround, Hill AFB, UT	40° 57' / 113° 05'	160 km	MCAS Cherry Point, NC	34° 54' / 076° 53'	100 km

Location	North Latitude/ West Longitude	Protection Radius	Location	North Latitude/ West Longitude	Protection Radius
Patuxent River, MD	38° 17' / 076° 24'	70 km	NAS Cecil Field, FL	30° 13' / 081° 52'	160 km
Nellis AFB, NV	37° 29' / 114° 14'	130 km	NAS Fallon, NV	39° 30' / 118° 46'	100 km
Fort Huachuca, AZ	31° 33' / 110° 18'	80 km	NAS Oceana, VA	36° 49' / 076° 01'	100 km
Eglin AFB/ Gulfport ANG Range, MS/ Fort Rucker, AL	30° 28' / 086° 31'	140 km	NAS Whidbey Island, WA	48° 21' / 122° 39'	70 km
Yuma Proving Ground, AZ	32° 29' / 114° 20'	160 km	NCTAMS, GUM	13° 35' / 144° 51' East	80 km
Fort Greely, AK	63° 47' / 145° 52'	80 km	Lemoore, CA	36° 20' / 119° 57'	120 km
Redstone Arsenal, AL	34° 35' / 086° 35'	80 km	Savannah River, SC	33° 15' / 081° 39'	3 km
Alpene Range, MI	44° 23' / 083° 20'	80 km	Naval Space Operations Center, ME	44° 24' / 068° 01'	80 km
Camp Shelby, MS	31° 20' / 089° 18'	80 km			

Usyyy [As proposed in Docket ET 00-221:] In the band 1670–1675 MHz, Government operations are on a non-interference basis to authorized non-Government operations and shall not hinder the implementation of any non-Government operations, except that the Geostationary Orbit Environmental Satellite receiving earth station at Wallops Island, VA (37° 56' 47" N, 75° 27' 37" W) operates on a co-primary basis.

USzzz [As proposed in Docket ET 00-221:] Until January 1, 2005, the band 2385–2390 MHz is also allocated to the Government mobile and radiolocation services on a co-primary basis and to the Government fixed service on a secondary basis. Use of the mobile service is limited to aeronautical telemetry and associated telecommand operations for flight testing of manned or unmanned aircraft, missiles or major components thereof. Use of the radiolocation service is limited to the military services. On January 1, 2005, Government operations in the band 2385–2390 MHz shall be on a non-interference basis to authorized non-Government operations and shall not hinder the implementation of any non-Government operations, except at the sites identified below where Government operations are co-primary until January 1, 2007.

Location	North Latitude/ West Longitude	Protection Radius	Location	North Latitude/ West Longitude	Protection Radius
Yuma Proving Ground, AZ	32° 54' / 114° 20'	160 km	Palm Beach County, FL	26° 54' / 0 80° 19'	160 km
Nellis AFB, NV	37° 48' / 116° 28'	160 km	Barking Sands, HI	22° 07' / 159° 40'	160 km
White Sands Missile Range, NM	32° 58' / 106° 23'	160 km	Roosevelt Roads, PR	18° 14' / 065° 38'	160 km
Utah Test Range, UT	40° 12' / 112° 54'	160 km	Glasgow, MT	48° 25' / 106° 32'	160 km
China Lake, CA	35° 40' / 117° 41'	160 km	Edwards AFB, CA	34° 54' / 117° 53'	100 km
Eglin AFB, FL	30° 30' / 086° 30'	160 km	Patuxent River, MD	38° 17' / 076° 25'	100 km
Cape Canaveral, FL	28° 33' / 080° 34'	160 km	Wichita, KS	37° 40' / 097° 26'	160 km
Seattle, WA	47° 32' / 122° 18'	160 km	Roswell, NM	33° 18' / 104° 32'	160 km
St Louis, MO	38° 45' / 090° 22'	160 km			

GOVERNMENT FOOTNOTES

These Footnotes, each consisting of the letter G followed by one or more digits, apply only to stations of the U.S. government.

Only those Government Footnotes that pertain to, or are proposed to pertain to, allocations in 30 MHz–300 GHz are shown. Footnotes occasionally refer to various other materials not included in this book.

G2 [As proposed in Docket ET 00-221:] In the bands 220–225 MHz, 420–450 MHz (except as provided by US217), 890–902 MHz, 928–942 MHz, 1300–1390 MHz, 2310–2385 MHz, 2417–2450 MHz, 2700–2900 MHz, 5650–5925 MHz, and 9000–9200 MHz, the Government radiolocation service is limited to the military services.

G5 In the bands 162.0125–173.2, 173.4–174, 406.1–410 and 410–420 MHz, the fixed and mobile services are all allocated on a primary basis to the Government non-military agencies.

G6 Military tactical fixed and mobile operations may be conducted nationally on a secondary basis: (1) To the meteorological aids service in the band 403–406 MHz; and (2) to the radio astronomy service in the band 406.1–410 MHz. Such fixed and mobile operations are subject to local coordination to ensure that harmful interference will not be caused to the services to which the bands are allocated.

G8 Low power Government radio control operations are permitted in the band 420–450 MHz.

G11 Government fixed and mobile radio services, including low power radio control operations, are permitted in the band 902–928 MHz on a secondary basis.

G15 Use of the band 2700–2900 MHz by the military fixed and shipborne air defense radiolocation installations will be fully coordinated with the meteorological aids and aeronautical radionavigation services. The military air defense installations will be moved from the band 2700–2900 MHz at the earliest practicable date. Until such time as military air defense installations can be accommodated satisfactorily elsewhere in the spectrum, such operations will, insofar as practicable, be adjusted to meet the requirements of the aeronautical radionavigation service.

G19 Use of the band 9000–9200 MHz by military fixed and shipborne air defense radiolocation installations will be fully coordinated with the aeronautical radionavigation service, recognizing fully the safety aspects of the latter. Military air defense installations will be accommodated ultimately outside this band. Until such time as military defense installations can be accommodated satisfactorily elsewhere in the spectrum such operations will, insofar as practicable, be adjusted to meet the requirements of the aeronautical radionavigation services.

G27 [As proposed in Docket ET 00-221:] In the bands 255–328.6 MHz, 335.4–399.9 MHz, and 1350–1390 MHz, the fixed and mobile services are limited to the military services.

G30 In the bands 138–144, 148–149.9, 150.05–150.8, 1427–1429, and 1432–1435 MHz, the fixed and mobile services are limited primarily to operations by the military services.

G31 In the bands 3300–3500 MHz, the Government radiolocation is limited to the military services, except as provided by footnote US108.

G32 Except for weather radars on meteorological satellites in the band 9975–10025 MHz and for Government survey operations (see footnote US108), Government radiolocation in the band 10000–10500 MHz is limited to the military services.

G34 In the band 34.4–34.5 GHz, weather radars on board meteorological satellites for cloud detection are authorized to operate on the basis of equality with military radiolocation devices. All other non-military radiolocation in the band 33.4–36.0 GHz shall be secondary to the military services.

G42 Space command, control, range and range rate systems for earth station transmission only (including installations on certain Navy ships) may be accommodated on a co-equal basis with the fixed and mobile services in the band 1761–1842 MHz. Specific frequencies required to be used at any location will be satisfied on a coordinated case-by-case basis.

G56 Government radiolocation in the bands 1215–1300, 2900–3100, 5350–5650 and 9300–9500 MHz is primarily for the military services; however, limited secondary use is permitted by other Government agencies in support of experimentation and research programs. In addition, limited secondary use is permitted for survey operations in the band 2900–3100 MHz.

G59 In the bands 902–928 MHz, 3100–3300 MHz, 3500–3650 MHz, 5250–5350 MHz, 8500–9000 MHz, 9200–9300 MHz, 13.4–14.0 GHz, 15.7–17.7 GHz and 24.05–24.25 GHz, all Government non-military radiolocation shall be secondary to military radiolocation, except in the sub-band 15.7–16.2 GHz airport surface detection equipment (ASDE) is permitted on a co-equal basis subject to coordination with the military departments.

G100 The bands 235–322 MHz and 335.4–399.9 MHz are also allocated on a primary basis to the mobile-satellite service, limited to military operations.

G104 In the bands 7450–7550 and 8175–8215 MHz, it is agreed that although the military space radio communication systems, which include earth stations near the proposed meteorological-satellite installations will precede the meteorological-satellite installations, engineering adjustments to either the military or the meteorological-satellite systems or both will be made as mutually required to assure compatible operations of the systems concerned.

G106 The bands 2501–2502 kHz, 5003–5005 kHz, 10003–10005 kHz, 15005–15010 kHz, 19990–19995 kHz, 20005–20010 kHz and 25005–25010 kHz are also allocated, on a secondary basis, to the space research service. The space research transmissions are subject to immediate temporary or permanent shutdown in the event of interference to the reception of the standard frequency and time broadcasts.

G109 All assignments in the band 157.0375–157.1875 MHz are subject to adjustment to other frequencies in this band as long term U.S. maritime VHF planning develops, particularly that planning incident to support of the National VHF-FM Radiotelephone Safety and Distress System (See Doc. 15624/1–1.9.111/1.9.125).

G110 Government ground-based stations in the aeronautical radionavigation service may be authorized between 3500–3650 MHz when accommodation in the band 2700–2900 MHz is not technically and/or economically feasible.

G114 [As proposed in Docket ET 00-221:] The band 1369.05–1390 MHz is also allocated to the fixed-satellite service (space-to-Earth) and to the mobile-satellite service (space-to-Earth) on a primary basis for the relay of nuclear burst data.

G116 The band 7125–7155 MHz is also allocated for Earth-to-space transmissions in the Space Operations Service at a limited number of sites (not to exceed two), subject to established coordination procedures.

G117 In the bands 7.25–7.75 GHz, 7.9–8.4 GHz, 17.8–21.2 GHz, 30–31 GHz, 39.5–40.5 GHz, 43.5–45.5 GHz, and 50.4–51.4 GHz the Government fixed-satellite and mobile-satellite services are limited to military systems.

G118 Government fixed stations may be authorized in the band 1700–1710 MHz only if spectrum is not available in the band 1710–1850 MHz.

G120 [As proposed in Docket ET 00-221:] Development of airborne primary radars in the band 2310–2385 MHz with peak transmitter power in excess of 250 watts for use in the United States is not permitted.

G122 The bands 2390–2400, 2402–2417 and 4660–4685 MHz were identified for immediate reallocation, effective August 10, 1994, for exclusive non-Government use under Title VI of the Omnibus Budget Reconciliation Act of 1993. Effective August 10, 1994, any Government operations in these bands are on a non-interference basis to authorized non-Government operations and shall not hinder the implementation of any non-Government operations.

G123 [Proposed to be removed in Docket ET 00-221:] The bands 2300–2310 and 2400–2402 MHz were identified for reallocation, effective August 10, 1995, for exclusive non-Government use under Title VI of the Omnibus Budget Reconciliation Act of 1993.

Effective August 10, 1995, any Government operations in these bands are on a non-interference basis to authorized non-Government operations and shall not hinder the implementation of any non-Government operations.

G124 The band 2417–2450 MHz was identified for reallocation, effective August 10, 1995, for mixed Government and non-Government use under Title VI of the Omnibus Reconciliation Act of 1993.

G126 [Author's Note: The FCC proposed to add this footnote in Docket ET 98-142. It already is in the NTIA Table. The FCC stated (DA 99-2743) that it intends to fold consideration of adoption of this footnote in 108–117.975 MHz and 1559–1610 MHz into a proceeding that will deal primarily with implementation of World Radio Conference decisions.] Differential-Global-Positioning-System (DGPS) Stations may be authorized on a primary basis in the bands 108–117.975 MHz, 1559–1610 MHz, and 5000–5150 MHz for the specific purpose of transmitting DGPS information intended for aircraft navigation.

G128 Use of the band 56.9–57 GHz by inter-satellite systems is limited to transmissions between satellites in geostationary orbit, to transmissions between satellites in geostationary satellite orbit and those in high-Earth orbit, to transmissions from satellites in geostationary satellite orbit to those in low-Earth orbit, and to transmissions from non-geostationary satellites in high-Earth orbit to those in low-Earth orbit. For links between satellites in the geostationary satellite orbit, the single entry power flux-density at all altitudes from 0 km to 1000 km above the Earth's surface, for all conditions and for all methods of modulation, shall not exceed -147 dB (W/m^2/100 MHz) for all angles of arrival.

NON-GOVERNMENT FOOTNOTES

These Footnotes, each consisting of the letters NG followed by one or more digits, apply only to non-Federal Government stations under U.S. jurisdiction.

Only those non-Government Footnotes that pertain to, or are proposed to pertain to, allocations in 30 MHz–300 GHz are shown. Footnotes occasionally refer to various other materials not included in this book.

NG2 Facsimile broadcasting stations may be authorized in the band 88–108 MHz.

NG3 Control stations in the domestic public mobile radio service may be authorized frequencies in the band 72–73 and 75.4–76 MHz on the condition that harmful interference will not be caused to operational fixed stations.

NG4 The use of the frequencies in the band 152.84–153.38 MHz may be authorized, in any area, to remote pickup broadcast base and mobile stations on the condition that harmful interference will not be caused to stations operating in accordance with the Table of Frequency Allocations.

NG6 Stations in the public safety radio services authorized as of June 30, 1958, to use frequencies in the band 159.51–161.79 MHz in areas other than Puerto Rico and the Virgin Islands may continue such operation, including expansion of existing systems, on the condition that harmful interference will not be caused to stations in the services to which these bands are allocated. In Puerto Rico and the Virgin Islands this authority is limited to frequencies in the band 160.05–161.37 MHz. No new public radio service system will be authorized to operate on these frequencies.

NG12 Frequencies in the bands 454.40–455 MHz and 459.40–460 MHz may be assigned to domestic public land and mobile stations to provide a two-way air-ground public radiotelephone service.

NG17 Stations in the land transportation radio services authorized as of May 15, 1958 to operate on the frequency 161.61 MHz may, upon proper application, continue to be authorized for such operation, including expansion of existing systems, on the condition that harmful interference will not be caused to the operation of any authorized station in the maritime mobile service. No new land transportation radio service system will be authorized to operate on 161.61 MHz.

NG23 Frequencies in the band 2100–2200 MHz may also be assigned to stations in the International Fixed Public Radiocommunication Services located south of 25° 30′ North Latitude in the State of Florida and in U.S. insular areas in the Caribbean, except that no new assignments in the band 2150–2162 MHz will be made to such stations after February 25,

1974 and no new assignments in the band 2165-2200 MHz will be made to such stations after June 27, 2000.

NG28 The frequency band 160.86-161.40 MHz is available for assignment to remote pickup base and remote pickup mobile stations in Puerto Rico and the Virgin Islands only on a shared basis with the land transportation radio service.

NG30 Stations in the international fixed public radiocommunication service in Florida, south of 25° 30′ north latitude, may be authorized to use frequencies in the band 716-890 MHz on the condition that harmful interference will not be caused to the broadcasting service of any country. This is an interim allocation the termination of which will later be specified by the Commission when it is determined that equipments are generally available for use in bands allocated internationally to the fixed services.

NG31 Stations in the Rural Radio Service licensed for Basic Exchange Telecommunications Radio Service may be authorized to use some frequencies in the bands 816-820 MHz (fixed subscriber) and 861-865 MHz (central office or base), on a co-primary basis with private land mobile radio licensees, pursuant to Part 22 Subpart H.

NG41 Frequencies in the bands 3700-4200 MHz, 5925-6425 MHz, and 10.7-11.7 GHz may also be assigned to stations in the international fixed public and international control services located in U.S. Possessions in the Caribbean area.

NG42 Non-Government stations in the radiolocation service shall not cause harmful interference to the amateur service.

NG47 In Alaska, frequencies within the band 2655-2690 MHz are not available for assignment to terrestrial stations.

NG49 The following frequencies may be authorized for mobile operations in the Manufacturers Radio Service subject to the condition that no interference is caused to the reception of television stations operating on channels 4 and 5 and that their use is limited to a manufacturing facility (MHz) 72.02 72.06 72.10 72.14 72.18 72.22 72.26 72.30 72.34 72.38 72.04 72.08 72.12 72.16 72.20 72.24 72.28 72.32 72.36 72.40.

Further, the following frequencies may be authorized for mobile operations in the Special Industrial Radio Service, Manufacturers Radio Service, Railroad Radio Service and Forest Products Radio Service subject to the condition that no interference is caused to the reception of television stations operating on channels 4 and 5; and that their use is limited to a railroad yard, manufacturing plant, logging site, mill, or similar industrial facility. (MHz) 72.44 72.48 72.52 72.56 72.60 75.44 75.48 75.52 75.56 75.60.

NG51 In Puerto Rico and the Virgin Islands only, the bands 150.8–150.98 MHz and 150.98–151.49 MHz are allocated exclusively to the business radio service.

NG53 In the band 12.7–13.15 GHz, television pickup stations and CARS pickup stations shall be assigned channels on a co-equal basis and shall operate on a secondary basis to fixed stations operating in accordance with the Table of Frequency Allocations. In the 13.15–13.20 GHz band television pickup stations and CARS pickup stations shall be assigned on an exclusive basis in the top one hundred markets, as set out in Section 76.51.

NG56 In the bands 72.0–73.0 and 75.4–76.0 MHz, the use of mobile radio remote control of models is on a secondary basis to all other fixed and mobile operations.

Such operations are subject to the condition that interference will not be caused to common carrier domestic public stations, to remote control of industrial equipment operating in the 72–76 MHz band, or to the reception of television signal on channels 4 (66–72 MHz) or 5 (76–82 MHz). Television interference shall be considered to occur whenever reception of regularly used television signals is impaired or destroyed, regardless of the strength of the television signal or the distance to the television station.

NG59 The frequencies 37.60 and 37.85 MHz may be authorized only for use by base, mobile, and operational fixed stations participating in an interconnected or coordinated power service utility system.

NG60 In the 5850–5925 MHz band, the use of the non-Federal government mobile service is limited to Dedicated Short Range Communications operating in the Intelligent Transportation System radio service.

NG63 Television Broadcast translator stations holding valid licenses on November 15, 1971, to operate in the frequency band 806–890 MHz (channels 70–83), may continue to operate in this band, pursuant to periodic license renewals, on a secondary basis to the land mobile radio service.

NG66 The frequency band 470–512 MHz is allocated for use in the broadcasting and land mobile radio services. In the land mobile services it is available for assignment in the domestic public, public safety, industrial, and land transportation radio services at, or in the vicinity of 13 urbanized areas of the United States, as set forth in the table below.

Additionally, in the land mobile services, TV Channel 16 is available for assignment in the public safety radio services at, or in the vicinity of Los Angeles. Such use in the land mobile services is subject to the conditions set forth in Parts 22 and 90 of this chapter, CFR 47.

Urbanized area / TV channel
New York-Northeastern New Jersey / 14, 15
Los Angeles /14, 20
Chicago-Northwestern Indiana / 14, 15
Philadelphia, PA-New Jersey / 19, 20
Detroit, Michigan / 15, 16
San Francisco-Oakland, California / 16, 17
Boston, Mass / 14, 16
Washington D.C.,-Maryland-Virginia / 17, 18
Pittsburgh, PA / 14, 18
Cleveland, Ohio / 14, 15
Miami, Florida / 14
Houston, Texas / 17
Dallas, Texas / 16

NG70 In Puerto Rico and the Virgin Islands only, the bands 159.240–159.435 and 160.410–160.620 MHz are also available for assignment to base stations and mobile stations in the special industrial radio service.

NG101 The use of the band 2500–2690 MHz by the broadcasting-satellite service is limited to domestic and regional systems for community reception of educational television programming and public service information. Such use is subject to agreement among administrations concerned and those having services operating in accordance with the table, which may be affected. Unless such agreement includes the use of higher values, the power flux density at the Earth's surface produced by emissions from a space station in this service shall not exceed those values set forth in Part 73 of the rules for this frequency band.

NG102 Use of the fixed-satellite service in the bands 2500–2655 MHz (space-to-Earth) and 2655–2690 MHz (Earth-to-space) is limited as follows: (a) For common carrier use in Alaska, for intra-Alaska service only, and in the mid- and western Pacific areas including American Samoa, Guam, the Northern Mariana Islands, and Hawaii, and under the Compacts of Free Association with the Federated States of Micronesia and the Republic of the Marshall Islands.

(b) For educational use in the contiguous United States, Alaska, and the mid- and western Pacific areas including American Samoa, Guam, the Northern Mariana Islands, and Hawaii.

Such use is subject to agreement with administrations having services operating in accordance with the Table, which may be affected. In the band 2500–2655 MHz, unless such agreement includes the use of higher values, the power flux density at the Earth's surface produced by emissions from a space station in this service shall not exceed the values set forth in Part 25 of the Rules for this frequency band.

NG104 The use of the bands 10.7–11.7 GHz (space-to-Earth) and 12.75–13.25 GHz (Earth-to-space) by the fixed-satellite service in the geostationary-satellite orbit shall be limited to international systems, i.e., other than domestic systems.

NG111 The band 157.4375–157.4625 MHz may be used for one way paging operations in the special emergency radio service.

NG112 The frequencies 25.04, 25.08, 150.980, 154.585, 158.445, 159.480, 454.000 and 459.000 MHz may be authorized to stations in the petroleum radio service for use primarily in oil spill containment and cleanup operations and secondarily in regular land mobile communication.

NG114 In the Gulf of Mexico offshore from the Louisiana-Texas coast, the frequency band 476–494 MHz (TV channels 15, 16 and 17) is allocated to the Domestic Public and Private Land Mobile Radio Services in accordance with the regulations set forth in Parts 22 and 90 respectively.

NG115 In the 174 to 216 MHz band wireless microphones may be authorized to operate on a secondary, non-interfering basis, subject to terms and conditions set forth in Part 74 of these Rules and Regulations.

NG117 The frequency 156.050 and 156.175 MHz may be assigned to stations in the maritime mobile service for commercial and port operations in the New Orleans Vessel Traffic Service (VTS) area and the frequency 156.250 MHz may be assigned to stations in the maritime mobile service for port operating in the New Orleans and Houston VTS areas.

NG118 In the band 2025–2110 MHz, television translator relay stations may be authorized to use frequencies on a secondary basis to other stations in the Television Broadcast Auxiliary Service that are operating in accordance with the Table of Frequency Allocations.

NG120 Frequencies in the band 928–960 MHz may be assigned for multiple address systems and mobile operations on a primary basis as specified in 47 C.F.R. Part 101.

NG122 Television Pickup stations may be authorized under Part 74 in the 6425–6525 MHz band on a secondary basis to stations operating in accordance with the Table of Frequency Allocations.

NG124 Within designated segments of the bands that comprise 30.85–47.41 MHz, 150.8–159.465 MHz, and 453.0125–467.9875 MHz, police licensees are authorized to operate low power radio transmitters on a secondary, non-interference basis in accordance with the provisions of 47 C.F.R. §§ 2.803 and 90.20(e)(5).

NG127 In Hawaii, the frequency band 488–494 MHz is allocated exclusively to the fixed service for use by common carrier control and repeater stations for point-to-point inter-island communications only.

NG128 In the band 535–1705 kHz, AM broadcast licensees or permittees may use their AM carrier on a secondary basis to transmit signals intended for both broadcast and non-broadcast purposes. In the band 88–108 MHz, FM broadcast licensees or permittees are permitted to use subcarriers on a secondary basis to transmit signals intended for both broadcast and non-broadcast purposes. In the bands 54–72, 76–88, 174–216, 470–608 and 614–806 MHz, TV broadcast licensees or permittees are permitted to use subcarriers on a secondary basis for both broadcast and non-broadcast purposes.

NG129 In Alaska, the bands 76–88 MHz and 88–100 MHz are also allocated to the Fixed service on a secondary basis. Broadcast stations operating in these bands shall not cause interference to non-Government fixed operations authorized prior to January 1, 1982.

NG134 In the band 10.45–10.5 GHz non-Government stations in the radiolocation service shall not cause harmful interference to the amateur and amateur-satellite services.

NG135 In the 420–430 MHz band the amateur service is not allocated north of line A (def. § 2.1).

NG141 The frequencies 42.40 MHz and 44.10 MHz are authorized on a primary basis in the State of Alaska for meteor burst communications by fixed stations in the Rural Radio Service operating under the provisions of Part 22 of this Chapter. The frequencies 44.20 MHz and 45.90 MHz are authorized on a primary basis in Alaska for meteor burst communications by fixed private radio stations operating under the provisions of Part 90 of the Chapter. The private radio station frequencies may be used by Common Carrier stations on a secondary, non-interference basis and the Common Carrier frequencies may be used by private radio stations for meteor burst communications on a secondary, non-interference basis. Users shall cooperate to the extent practical to minimize potential interference. Stations utilizing meteor burst communications shall not cause harmful interference to stations of other radio services operating in accordance with the Table of Frequency Allocations.

NG143 In the band 11.7–12.2 GHz, protection from harmful interference shall be afforded to transmissions from space stations not in conformance with ITU Radio Regulation S5.488 only if the operations of such space stations impose no unacceptable constraints on operations or orbit locations of space stations in conformance with S5.488.

NG144 Stations authorized as of September 9, 1983 to use frequencies in the band 17.7–18.58 GHz and 19.3–19.7 GHz may, upon proper application, continue operations. Fixed stations authorized in the band 18.58–19.3 GHz that remain co-primary under the provisions of §§21.901(e), 74.502(c), 74.602(g), 78.18(a)(4), and 101.174(r) may continue operations consistent with the provisions of those sections.

NG145 In the band 11.7–12.2 GHz, transponders on space stations in the fixed-satellite service may be used additionally for transmissions in the broadcasting-satellite service, provided that such transmissions do not have a maximum e.i.r.p. greater than 53 dBW per television channel and do not cause greater interference or require more protection from interference than the coordinated fixed-satellite service frequency assignments. With respect to the space services, this band shall be used principally for the fixed satellite service.

NG147 Stations in the broadcast auxiliary service and private radio services licensed as of July 25, 1985, or on a subsequent date following as a result of submitting an application for license on or before July 25, 1985, may continue to operate on a primary basis with the mobile-satellite service and the radiodetermination satellite service.

NG148 The frequencies 154.585 MHz, 159.480 MHz, 160.725 MHz, 160.785 MHz, 454.000 MHz and 459.000 MHz may be authorized to maritime mobile stations for offshore radiolocation and associated telecommand operations.

NG149 The frequency bands 54–72 MHz, 76–88 MHz, 174–216 MHz, 470–512 MHz, 512–608 MHz, and 614–746 MHz are also allocated to the Fixed Service to permit subscription television operations in accordance with Part 73 of the rules.

NG151 In the frequency bands 824–849 MHz and 869–894 MHz, cellular land mobile licensees are permitted to offer auxiliary services on a secondary basis subject to the provisions of Part 22.

NG152 The band 219–220 MHz is also allocated to the amateur service on a secondary basis for stations participating, as forwarding stations, in point-to-point fixed digital message forwarding systems, including intercity packet backbone networks.

NG153 The bands 2110–2150 MHz and 2160–2165 MHz are reserved for future emerging technologies on a co-primary basis with the fixed and mobile services. Allocations to specific services will be made in future proceedings.

NG155 The bands 159.500–159.675 MHz and 161.375–161.550 MHz are allocated to the maritime service as described in Part 80 of this chapter. Additionally, the frequencies 159.550, 159.575 and 159.600 MHz are used for low-power intership communications.

NG156 The band 1990–2025 MHz is also allocated to the fixed and mobile services on a primary basis for facilities where the receipt date of the initial application was prior to June 27, 2000, and on a secondary basis for all other initial applications. Not later than September 6, 2010, the band 1990–2025 MHz is allocated to the fixed and mobile services on a secondary basis.

NG158 The frequency bands 764–776 MHz and 794–806 MHz are available for assignment exclusively to the public safety radio services, to be defined in Docket No. WT 96-86.

NG159 Full power analog television stations licensed pursuant to applications filed before January 2, 2001, and new digital television (DTV) broadcasting operations in the 746–806 MHz band will be entitled to protection from harmful interference until the end of the DTV transition period. After the end of the DTV transition period, the Commission may assign licenses in the 746–806 MHz band without regard to existing television and DTV operations. Low power television and television translators in the 746–806 MHz band must cease operations in the band at the end of the DTV transition period.

NG160 In the 5850–5925 MHz band, the use of the non-Federal government mobile service is limited to Dedicated Short Range Communications operating in the Intelligent Transportation System radio service.

NG163 The allocation to the broadcasting-satellite service in the band 17.3–17.7 GHz shall come into effect on 1 April 2007.

NG164 The use of the band 18.3–18.8 GHz by the fixed-satellite service (space-to-Earth) is limited to systems in the geostationary-satellite orbit.

NG165 The use of the band 18.8–19.3 GHz by the fixed-satellite service (space-to-Earth) is limited to systems in non-geostationary-satellite orbits.

NG166 The use of the band 19.3–19.7 GHz by the fixed-satellite service (space-to-Earth) is limited to feeder links for the mobile-satellite service.

NG167 The use of the fixed-satellite service (Earth-to-space) in the band 24.75–25.25 GHz is limited to feeder links for the broadcasting-satellite service operating in the band 17.3–17.7 GHz. The allocation to the fixed-satellite service (Earth-to-space) in the band 24.75–25.25 [GHz] shall come into effect on 1 April 2007.

NG168 The band 2165–2200 MHz is also allocated to the fixed and mobile services on a primary basis for facilities where the receipt date of the initial application was prior to January 16, 1992, and on a secondary basis for all other initial applications. Not later than September 6, 2010, the band 2165–2200 MHz is allocated to the fixed and mobile services on a secondary basis.

NG169 After December 1, 2000, operations on a primary basis by the fixed-satellite service (space-to-Earth) in the band 3650–3700 MHz shall be limited to grandfathered earth stations. All other fixed-satellite service earth station operations in the band 3650–3700 MHz shall be on a secondary basis. Grandfathered earth stations are those authorized prior to December 1, 2000, or granted as a result of an application filed prior to December 1, 2000, and constructed within 12 months of initial authorization. license applications for primary operations for new earth stations, major amendments to pending earth station applications, or applications for major modifications to earth station facilities filed on or after December 18, 1998, and prior to December 1, 2000, shall not be accepted unless the proposed facilities are in the vicinity (*i.e.* within 10 miles) of an authorized primary earth station operating in the band 3650–3700 MHz. License applications for primary operations by new earth stations, major amendments to pending earth station applications, and applications for major modifications to earth station facilities, filed after December 1, 2000, shall not be accepted, except for changes in polarization, antenna orientation or ownership of a grandfathered earth station.

NG170 In the band 3650–3700 MHz, the mobile except aeronautical mobile service is limited to base station operations. These base stations are subject to the same coordination procedures as fixed service operations in the band 3650–3700 MHz.

NGXXX [As proposed in Docket WT 00-32:] Fixed-satellite service systems that operate primarily outside the 3650–3700 MHz band may be authorized to perform space operations, such as, telemetry, tracking and telecommand operations in the band 3650–3700 MHz, provided the requirement in §25.202(g)(1) of this chapter is satisfied.

Subject Index

Symbols

218–219 MHz Service, 82
24 GHz Service, see TFGS, 349
3G wireless services (Third Generation), 263
802.11 standard, 253
802.11b standard, 253
802.15 standard, 253

A

A segment, cellular phone systems, 144
ABMs (anti-ballistic missiles), 75
ACARS (Aircraft Communications Addressing and Reporting System), 52
ACAS (Airborne Collision Avoidance System), 171
ACATS (Advisory Committee on Advanced Television Service), 120
active sensors, NASA, 318
ACTS (Air Combat Training Systems), 212
ADS (Automatic Dependent Surveillance), 68, 175, 281
Advanced Air Defense System, 323
Advanced EHF Satellite System, 341
Advanced Mobile and Fixed Communications Service (AMFCS), 210, 219
Aegis system, 268
AeroAstro, SENS satellite system, 206, 236
aeronautical advisory stations, see unicoms
Aeronautical Enroute Service, 52
Aeronautical Mobile (Route) Service, 49
aeronautics
  flight testing, 51
  search and rescue, 51
  telemetry, 186, 191, 246
aerostats, 185, 212, 275
AERS (Automobile Emergency Radio Service), 16
AfriSpace, 187
AfriStar, 187
AFTRCC (Aerospace and Flight Test Radio Coordinating Council), 186
AGRAS (Air-Ground Radiotelephone Automated Service), 109
Air Force
  aeronautical telemetry, 191
  aircraft stationkeeping systems, 269
  ARSR, 176
  BMEWS, 107
  electronic warfare training, 236
  general purpose mobile usage, 105
  Global Weather Central system, 208
  GPS aircraft positioning, 179
  hazardous material communications, 81
  JSS, 172
  navigation system certification, 247
  Radio Solar Telescope Network, 265

Satellite Control Network, 212
SBIRS, 296
SBWAS, 233
SCN, 233
SGLS, 212, 233
Space Command, GPS satellites, 173
test range radars, 284
tropospheric scatter communications, 277
UAVs, 301
WBS, 352
air radionavigation, 364
air traffic control, 49, 51-52, 270
airborne precision ground mapping, 363
AirCell, 109, 147
aircraft
  search and rescue, 51
  stationkeeping systems, 269
aircraft carrier landing systems, 363
Airfone, 146
AirPort (Apple Computer), 222, 248
Airport radar, 323
AIRSAR (airborne synthetic aperture radar), 323
airship platforms, 375
Allen Telescope Array, 183
allotment, FM radio broadcasting, 37
ALMA (Atacama Large Millimeter Array), 360
ALS (Aircraft Landing Systems), 322
Altair (Motorola), 337
altimeters, 274
Amateur Radio Service, 31-33, 247, 270, 375, 387-388, 393-395, 401-402
  digital network links, 85
  EME (Earth-Moon-Earth), 59, 236, 291
  EME operations, 59

licenses, 59
NVNG MSS, 54
Phase 3D amateur satellite, 347
primary allocation status, 252
repeaters, 60
tropospheric ducting, 303
Amateur Satellite Service, 177, 251-252
American Personal Communications, PCS, 216
AmeriStar, 187
AMFCS, 210, 219
AMPS (Advanced Mobile Phone Service), 144
AMS(R)S (Aeronautical Mobile Satellite (Route) Service), 189, 193-194
AMSAT (Radio Amateur Satellite Corporation), 59, 177
AMSC-1 (Motient Corp), 189
AMST (Association for Maximum Service Television), 278
AMTS (Automated Maritime Telecommunication System), 66, 80-81
anti-collision vehicle radar, 374, 387
APA (Administrative Procedures Act), 336
Apple Computer
  AirPort, 222, 248
  Data-PCS, 221, 247
  NII, 282
Archives of Humanity (SpaceArc), 310
Arecibo Observatory, 107, 245
ARGOS data collection system, 99
ARINC (Aeronautical Radio Inc.), 52
ARISS (Amateur Radio International Space Station), 59
Army
  Corps of Engineers, hydropower station control, 213

general purpose mobile usage, 105
MSE, 212
Patriot system radar, 284
SINCGARS, 12
tactical radio relay systems, 183
THAAD, 303
unmanned vehicle control, 179
ArrayComm, i-Burst, 181, 237
ARRL (American Radio Relay
   League), 7, 28, 252
ARSR (Air Route Surveillance
   Radar), 172, 176
ART wireless services, 368
Ascom Wireless Solutions, UPCS
   phones, 223
ASDE (Airport Surface Detection
   Equipment), 300, 323
AsiaStar, 187
ASRs (Airport Surveillance
   Radars), 266
Astrolink (Lockheed Martin), 331
astronomy
   radio, see radio astronomy
   star formation research, 402
AstroVision, 297
asynchronous UPCS, 223
AT&T
   Digital Broadband, 215
   Inflight air phone, 147
ATCRBS (Air Traffic Control Radar
   Beacon Systems), 170
ATF (Bureau of Alcohol, Tobacco
   and Firearms), 72
ATG (Air-to-Ground), 146–148
ATIS
   Automatic Terminal Information
      Systems, 53
   Automatic Transmitter
      Identification System, 65,
      384
atmospheric temperature profiles,
   385
ATSC (Advanced Television
   Systems Committee), 120

ATV (Amateur TV), 108
auditory assistance devices, 34, 86
auroral propagation, 32
Automated Highway Systems, 290
automated unicoms, 50
automatic vehicle identification
   tags, 287
aviation
   inclement weather guidance
      systems, 386
   radio, 50
avionics, see aircraft
AWACS (Airborne Warning and
   Control System), 268
AWOS (Automated Weather
   Observation Systems), 53

B

B segment, cellular phone
   systems, 145
B-GAN (Broadband Global Area
   Network), INMARSAT, 191
baby monitors, 29
background radiation, Big
   Bang, 396
backscatter, 325
balloon platforms, 375
band managers, MWCS, 132
Band, of spectrum, xi
bar code readers, 158
BAS (Broadcast Auxiliary
   Services), 108, 291, 327
base stations, BAS, 108
BATF, see ATF
BBA (Balanced Budget Act of
   1997), spectrum
   reallocations, 249
Beacon bucks, 85
beacons, Instrument Landing
   System, 36
Bentspace.com, 167

BETRS (Basic Exchange Telecommunications Radio Service), 22, 70
bhangmeters, 179
Big Bang background radiation, 396
Big LEOs, 196
  Eagle Eye Tag, 199
  ECCO, 199
  Ellipso 2G, 199
  Globalstar, 199
  interference, 258
  Iridium, 198
  Odyssey, 199
birds
  migration radar, 301
  tracking backpacks, 99
BLAST (Bell Labs Layered Space-Time), 133, 136, 211
blimp platforms, 375
Blossom Point Tracking Facility, 233
Bluetooth, 253
BMEWS (Ballistic Missile Early Warning System), 107
Boeing
  AMS(R)S system, 191
  GMPCS system, 227
bomb disposal systems, 257
border interdiction, Treasury Dept, 277
BRAN (Broadband Radio Access Networks), 281
broadband PCS, 217
broadcast licensing, 38
Broadcasting Service, 371
BSS (Broadcast Satellite Service), 325, 367
BTAs (Basic Trading Areas), 369
  LMDS, 354
  N-PCS, 152
  PCS, 216–217

C

C-band
  cable TV, 273
  GSO FSS, 273
  TTC, 274
Cable TV, C-band usage, 273
Calling Communications, *see* Teledesic
CAPs (Competitive Access Providers), wireless services, 368
CaribStar, 187
CARS (Cable Television Relay Service), 293, 313, 327
Cassini, 235
CB (Citizens Band), 113
CBR (Cosmic Background Radiation), radio astronomy, 396, 399
CDPD (Cellular Digital Packet Data), 145
CEC (Cooperative Engagement Capability), 276
Celestri (Motorola), 330, 367
cell splitting, cellular phone systems, 145
cellular phone systems
  A segment, 144
  B segment, 145
  CDPD, 145
  cell-splitting, 145
  MSAs, 144
  RSAs, 144
CellularVision, 354
Celsat America, GMPCS system, 226
Citizens Band Radio Services, MICS, 100
Civil Air Patrol
  search and rescue, 58–60
Civil air traffic control radar, 270

E

E-band, 32, 387
E-bombs, 46
E-SAT, 56
E911 (Enhanced 911), 220
Eagle Eye Tag, 199
EAN (Emergency Action Notification), 44
Earth Exploration-Satellite Service (EESS), 370
Earth observation radar, 285
EarthWatch, 297
EAs (Economic Areas)
    LMS, 158
    MAS, 163
    wireless services, 369
EAS (Emergency Alert System), 44
eavesdropping satellites, 154, 328, 335
EBS (Emergency Broadcast System), 44
ECCO, 199
Echelon, 335
EchoStar (DISH Network), 311
EESS (Earth Exploration-Satellite Service), 277, 285, 297, 325, 341, 352, 378
electromagnetic energy spectrum, vi
electromagnetic warfare, 46
electronic eavesdropping, 154
electronic warfare training, Air Force, 236
Ellipso, 199
Ellipso Virgo, 309
ELTs (Emergency Locator Transmitters), 49–50, 92, 102
emergency beacons, 99
Emergency Crew Return Vehicle, 173

Emergency Managers Weather Information Network, 62
Emergency Medical Radio Service, 62
Emerging Technologies, 215
EMRS, 116
EMWIN (Emergency Managers Weather Information Network), 209
ENG (Electronic News Gathering), 224, 278, 293
Enhanced GPWS, 200
environmental remote sensing band, 380
environmental services, 366
EON Corp., IVDS, 82–84
EOS (Earth Observation System), 297
EOSAT (Earth Observation Satellite Corporation), Landsat, 234
ephemeris data, 98
EPIRBs (Emergency Position Indicating Radio Beacons), 49, 69, 92, 102
ERDS (Emergency Radio Data System), 39, 349
ESMR (Enhanced SMR), 139–141
    legal restrictions, 140
    Nextel origins, 139
ET communications, 245
ETS (Electronic Tracking), 85
ETSI (European Telecommunications Standards Institute), Hiperlan, 281
Eureka, 40
European Union, Galileo, 176
Experimental Radio Service and Experimental Broadcast Stations, viii
extended C-band, 271–272, 289

F

F-band, 390, 392
FAA (Federal Aviation
    Administration)
  ADS, 281
  air traffic control, 49
  ARSR, 176
  facility microwave links, 319–320
  GPS, 281
  JSS, 172
  RCL, 168, 293
  TDWR, 281
  UWB opposition, 5
  WAAS, 175
FACS (Final Analysis
    Communication Services), 55
Fantasma Networks, UWB home
    networking, 5
fax broadcasting, 45
FCC (Federal Communications
    Commission)
  Amateur Radio Service
    licenses, 59
  broadcast licensing, 38
  CARS, 314
  DEMS, 304
  DSRC, 290
  FSS, extended C-band
    application freezes, 272
  MVPDs, 314
  NGSO MSS, 292
  PCOs, 314
  PURAC, 114
  SDR, 2, 6
  secondary markets, 2, 8
  TT&C, 306
  U-NII, 282–283
  UWB, 2–3
  Watch Officer, viii
  WCS
    licensing, 278
    spectrum reallocation, 271
  Wireless Radio Services, 13

FEMA (Federal Emergency
    Management Agency), 76
fence, 79
FHWA (Federal Highway
    Administration), 89
field disturbance sensors, 287, 304
fire radio, 142
FIS (Flight Information
    Services), 53
fish tracking, 99
fixed microwave, 215
Fixed Satellite Service, 370
Fixed Services, 368–369
  NATO usage, 275
Fleet Call, *see* Nextel
fleet tracking, 156
FLEWUG (Federal Law
    Enforcement Wireless Users
    Group), 135
flight test telemetry, 246
float free EPIRBs, 203
FLTSATCOM (Fleet Satellite
    Communications System),
    Navy, 92
Flumpits, *see* IMT-2000
FM radio broadcasting, 37–39
FPLMTS (Future Public Land
    Mobile Telecommunications
    System), 218
Free Flight, Data Link, 175
FreePage Corp., PCS, 14
FreeSpace, fixed cellular Internet
    service, 131
frequency allocation, vi
frequency reuse, 144
frequency-hopping radar, 172
FRS (Family Radio Service), 63,
    112, 115–116
FSS (Fixed Satellite Service), 272,
    280, 367, 378
  Intelsat/Inmarsat, 270
  Ku-band, 307
FWA (Fixed Wireless Access), 269

G

G-band, 394
Gabriel, Peter, Archives of Humanity, 310
Galileo, 176
GE American Communications, GE StarPlus, 367
General Aviation Air-Ground Radiotelephone Service, 109
General pool, PMRS, 16
geophysical telemetry, 81
Geosat, Navy, 316
Geostar, 196
GESN (Global EHF Satellite Network), 367
GFO (Geosat Follow-On), 316
Gigalink (Harmonix Corp.), 382, 385
Global Verification and Location System, 179
Global Weather Central system, Air Force, 208
Globalstar, 199, 367
GLONASS (Russian Federation Global Orbital Navigation Satellite System), 176, 195, 200
GMDSS (Global Maritime Distress and Safety System), 49, 67, 203
GMPCS (Global Mobile Personal Communications by Satellite), 226
GMRS (General Mobile Radio Service), 112–114, 361
GOES (Geostationary Operational Environmental Satellites), 98, 117, 206–207
Goldstone
  Deep Space Network, 315, 363
  Solar System Radar, 244, 300
GPMRS (General Purpose Mobile Radio Service), 151
GPR (Ground Penetrating Radar), 2, 6
GPS (Global Positioning System), 48, 280
  DGPS, 175
  dropsondes, 98
  FRS usage, 115–116
  health messages, 174
  NUDET, 179
  receivers, 173
  SA, 174
  tracking implants, 176
  UWB objections, 4
GPWS (Ground Proximity Warning System), 200
grandfathering FSS stations, 272
Green Bank U.S. Naval Observatory Interferometer, 265
Gross, Al, 112
GS-40, Globalstar, GMPCS system, 227, 367
GSM (Global System for Mobile Communications), 145
GSO FSS (Geostationary Fixed Satellite Service), 273, 291
GSO MSS (Geostationary Mobile Satellite Service), 188, 202
GSOs (Geostationary Orbit Satellites), 367
GSTS (Global Stratospheric Telecommunications Service), Sky Station International, 375–376
guard bands, MWCS, 131
GWCS (General Wireless Communication Service), 276–278

H

HALO aircraft (High Altitude Long Operation), LMDS, 355
Ham radio, 31–33
HAPS (High Altitude Platform Stations), 375–377
Harmonix Corp., GigaLink, 385
Hat Creek Observatory, 183
HDFS (High Density Fixed Services), 372
HDTV (High Definition TV), 118
health care assistance devices, 86
health messages, GPS, 174
hearing assistance devices, 34, 289
heart monitors, 77
Hendrix, Jimi, Archives of Humanity, 310
HEOs (Highly Elliptical Orbit Satellites), 367
HiperAccess, 282
Hiperlan, 281
HiperLink, 282
home networking, UWB, 5
HomeRF Working Group, SWAP, 254–255
homing beacons, 104
HPM weapons (High Power Microwave), 46
HST (Hubble Space Telescope), 228
Hughes Communications
  Hughes Expressway, 367
  HughesLINK, 308
  SpaceCast system, 367
  Spaceway EXP, 331
  StarLynx system, 368
Hughes Expressway (Hughes Communications), 367
HughesLINK (Hughes Communications), 308

I

i-Burst (ArrayComm), 237
Ibiquity Digital Corp., 40
IBOC (in-band and on-channel), 40
ICAO (International Civil Aviation Organization), 281
ICR (Inter-City Relay links), radio/TV, 169
IEEE (Institute of Electrical and Electronics Engineers)
  U-NII standards, 283
  wireless LAN standards, 252
IFF radar (Identification Friend or Foe), 171, 285
illegal eavesdropping, 154
illegal radiotelephones, 96
ILS (Instrument Landing System), 47, 95, 281
imaging radiometry, 366
IMPI (International Microwave Power Institute), 385
implanted GPS devices, 176
IMT-2000 (International Mobile Telecommunications-2000), 142, 210–213, 218–220, 263
Incidental radiators, xii
inclement weather guidance systems, 386
industrial microwave stations, 209
industrial RF devices, 255
industrial safety and control sensors, 385
Industrial/Business pool, PMRS, 16–17
Inflight air phone, 146
INMARSAT (International Maritime Satellite Organization), 49, 103
  B-GAN, 190
  FSS, 270
insect tracking radar, 302

Subject Index

Intentional radiators, xii
Instrument Landing System, 36
intelligence-collection satellites, 332–333
Intelsat, FSS, 270
Interdepartmental Radio Advisory Committee (IRAC), viii
interference
  Big LEOs, 258
  ISM devices, 255
  MDS, 260
  satellite systems, 258
  spread spectrum, 254
internal paging, 167
International Broadcasting Bureau, 124
International Space Station, 228
  Amateur Radio Service operations, 59
  radar systems, 302
  spacecraft communications, 319
International Telecommunication Union (ITU), vi
international unlicensed spectrum applications, 382
Internet in the Sky (Teledesic), 333
interrogating transponders, 287
Iridium, 198, 227, 308
ISERS (Indoor Sports and Entertainment Radio Service), 44
ISM (Industrial, Scientific and Medical), 30, 236, 288, 345, 400
  applications, 385
  hearing aids, 289
  interference, 255
isochronous UPCS, 222
ISS (Intersatellite Service), 380
ITFS (Instructional Television Fixed Service), 258–261, 264, 314

ITS (Intelligent Transportation Systems), 89, 289, 378, 386, 396
  DSRC, 289–290
  LMS, 156
ITU (International Telecommunication Union), vi
IVDS (Interactive Video and Data Service), 80–84

J

jamming radar, aircraft stationkeeping systems, 269
Jet Propulsion Laboratory, DSN, 234
joint frequency agreements, NATO, 365–370
JSS (Joint Surveillance System), 172
JTCTS, 212
JTIDS (Joint Tactical Information Distribution System), 171
Justice Department video surveillance, 277

K

Ka-band, 291, 364–365
Kitchen sink band, 153
Ku-band, 306
  DTH FSS, 273
  FSS, 307, 317
  VSAT, 307
KuBS (Ku-Band Supplement), 308

L

L-band, 186
LAAS (Local Area Augmentation System), 175

LAMPS (Light Airborne Multipurpose System), 277
land remote-sensing satellites, 297
Landsat, 228, 234, 297
LANs, wireless, *see* wireless LANs
Large Millimeter Wave Telescope, 360
LECs (Local Exchange Carriers), wireless services, 368
licenses
  Amateur Radio Service, 59
  FCC, 38
  MWCS, 130
light tap video assist devices, 77
liquid level UWB sensors, 2
Little LEO satellite systems, 110
LMCC (Land Mobile Communications Council), 108, 171, 206
LMCS (Land Mobile Communications Service), 18, 180, 184
LMDS (Local Multipoint Distribution Service), 353, 361–362
  BTAs, 354
  CellularVision, 354
  HALO operations, 355
  mobile services, 354
  VisionStar Inc., 330
LMS (Location and Monitoring Service), 74, 155–156, 238
  EAs, 158
  interference, 160
  non-multilateration, 157–158
local homing systems, 50
local positioning systems, 256
Lockheed Martin
  Astrolink, 331
  GSOs, 367
  UWB cautions, 5
LOIs (Letters of Intent), 226
LoJack, 73

long range search radar, 142–143, 148, 164
Loral Space and Communications
  CyberPath system, 368, 371
  Globalstar, 199
low VHF UWB systems, 12
LPAS (Low-Power Auxiliary Stations), 133
LPDS (Limited Program Distribution Service), 14
LPFM (Low Power FM), 41–43
LPRS (Low Power Radio Service), 35, 100
LPTV (Low Power TV), 123
LTTS (Local Television Transmission Service), 224, 229, 291, 306
LUTs (Local User Terminals), EPIRB/ELT tracking, 103

M

M-Star (Motorola), 330, 367
Magellan, mobile vehicle tracking units, 55
Magic Band, 32
manned space exploration missions, 366, 370
MAP (Microwave Anisotropy Probe), 342, 399
maritime bandwidths
  distress signaling, 203
  navigation radar, 301
  safety broadcasts, 69
  telephone service, 70
  VHF, 66
Mars Pathfinder, 111
MAS (Multiple Address Systems), 162–164, 168–170
MAVs (Micro Air Vehicles), 301
MDS (Multipoint Distribution Service), 230, 259, 264

Subject Index

data transmission, 261
interference, 260
response channels, 261
response station hubs, 262
WCA, 262
MEAs (Major Economic Areas)
Paging Service, 22
PCP, 165
WCS, 240
medical ISM devices, 255
medical telemetry, 116, 124–126
medical uses of UWB, 2
MEOs (Medium Earth Orbit Satellites), 368
Mercury radar imaging, 245
Messaging Services, CMRS, 13
Met Aids (meteorological aids), 97, 101
meteor burst tracking, 26
Meteorological-Satellite Service, 295
meteorology radar, 365
Metricom, 159, 239
micro radio, 41–42
micrometeor tracking, 26
Microsoft, Data-PCS, 248
MicroTrax, PLMS, 206, 237
microwave lighting, 255
microwave oven interference, 255
microwave radiometry, 276
microwave road condition sensors, 256
MICS (Medical Implant Communications Service), 100
Millenium (Motorola), 329
Millimeter Array, 402
millimeter wavelength uses, 360
Milstar (Military, Strategic and Tactical Relay system), 339–340, 373

Miscellaneous Wireless Communications Services, see WCS
missile launch tracking systems, 268
MLS (Microwave Landing System), 281
MLUs (Mt. Hood Locator Units), 90
MMBS (Mobile Multimedia Broadcast Service), 129
Mobile Communications Holdings
Ellipso, 199
Ellipso2G, 227
mobile LMDS, 354
Mobile Services
CMRS, 13
NATO usage, 275
mobile units, BAS, 108
mobile VHF services, 12
model airplane/boat/car remote controls, 34–35
moonbounce, see EME
Morse Code, 32, 59
Motient Corp., AMSC-1, 189
Motorola
Altair, 337
Celestri system, 367
Iridium, 198
M-Star, 367
Millenium, 329
Piano system, 385
MoU (Memorandum of Understanding), GMPCS, 226
MPRs (multipurpose radars), 322
MRI (Magnetic Resonance Imaging), 33
MSAs (Metropolitan Statistical Areas), cellular phone systems, 144
MSAT-1, 190
MSAT-2, 190
MSS (Mobile Satellite Service), x, 224, 367, 386

MTAs (Metropolitan Trading Areas), 369
N-PCS, 152, 166
PCS, 216
Mtel, N-PCS, 151
MTSO (Mobile Telephone Switching Office), 144
MURS (Multi-Use Radio Service), 63–64
MVDDS (Multichannel Video Distribution and Data Service), 311–312
MVPDs (Multichannel Video Programming Distributors), 121, 314
MWCS, 128–129, 136
  band managers, 132
  guard bands, 131
  licensing, 130
MWCWG (Millimeter Wave Communications Working Group), 383

N

nanosatellites, 56
N-PCS (Narrowband Personal Communications Services), 13, 151–152, 166, 213
NASA (National Aeronautics and Space Administration)
  active sensors, 318
  DSN, 230, 234–236, 315, 363
  EOS, 297
  Goldstone Solar System Radar, 244, 300
  MAP, 342
  Mars Pathfinder, 111
  National Scientific Balloon Facility, 108
  ozone measurements, 154
  POCS, 352
  range safety surveillance radar, 266
  Space Network, 234
  Space Shuttle communications, 91
  TDRSS, 317, 320–321, 344, 352
National Air and Space Museum, microwave lighting, 255
National Astronomy and Ionosphere Center, Arecibo Observatory, 245
National Guard, tactical radio relay systems, 183
National Missile Defense System, radar system upgrades, 269
National Scientific Balloon Facility, 108
national security satellites, 335
National Telecommunications and Information Administration, see NTIA
NATO
  joint frequency agreements, 365–370
  JTIDS, 171
  military fixed and mobile services, 275
  SATCOM, 298
Naval Research Laboratory, Blossom Point Tracking Facility, 234
navigation system certification, Navy, 247
NAVSPACECOM (Naval Space Command), 79
Navstar Global Positioning System, 173
NAVTEX, 68
Navy
  Aegis system, 268
  CEC, 276
  Cobra Judy, 268
  DWTS, 212

Subject Index

FLTSATCOM, 92
general purpose mobile
   usage, 105
Geosat, 316
LAMPS, 277
long range search radar, 142–143, 148, 164
navigation system certification, 247
NAVSPACECOM, 79
RACONs, 268
SBWAS, 233
sea skimmer detection, 154
shipboard radar, 316
Space Operations Center, 80
SPASUR, 79
surface search radar, 284
Transit satellites, 61
video interception transmissions, 212
NAWAS (National Warning System), 76
NCE FM stations (Non-Commercial Educational), 37
NDS (National Distress System), 66
neighborhood Watch groups, 134
Nellis Air Force Base, NEST location, 275
NESDIS (National Environmental Satellite, Data, and Information Service), 206
NEST (Nuclear Emergency Search Team), 275
NEXRAD weather radar, 266–270
Nextel, ESMR, 139
NGSO FSS (Non-Geostationary Fixed Satellite Service), 307–308
NGSO MSS (Non-Geostationary Mobile Satellite Service), 196, 224, 258, 280, 292

NGSOs (Non-Geostationary Orbit Satellites), 367
NII (National Information Infrastructure), 282
NMR (Nuclear Magnetic Resonance), 33
NOAA (National Oceanic and Atmospheric Administration), 56
  GOES, 98, 206
  NWS, 74
  POES, 98, 208
  wind profilers, 107
NOI (Notice of Inquiry), 3, UWB usage, 3
Nokia, SDR, 7
non-multilateration LMS, 157–158
Northpoint Technology, MVDDS, 311
nosebleed bands, 371
NOSS (Naval Ocean Surveillance System), 233
NOTAMS (Notices to Airmen), 53
NPOESS (National Polar-Orbiting Operational Environmental Satellite), 209
NPR (National Public Radio), LPFM objections, 43
NPRMs (Notices of Proposed Rulemaking), 3, 375
NPSPAC (National Public Safety Planning Advisory Committee), 142
NQR (Nuclear Quadrupole Resonance), 34
NRAO (National Radio Astronomy Observatory), 360
  Millimeter Array, 402
  VLBA, 126
NRO (National Reconnaissance Office), intelligence-gathering spacecraft, 233

NSA (National Security Agency), intelligence-gathering spacecraft, 233
NTIA (National Telecommunications and Information Administration), vi
  HDFS, 372
  UWB, 3
NTSC standard (National Television Systems Committee), 118
Nucentrix Broadband Networks, MDS, 259
NUDET (Nuclear Detonation Detection System), 179
NVNG MSS (Non-Voice Non-Geostationary Mobile Satellite Service), 54–57, 97
NWS (National Weather Service), 74, 208

O

OBRA (Omnibus Budget Reconciliation Act), 249, 271
ocean buoy telemetry, 81
ocean temperature measurements, passive sensors, 292
Odyssey, 199
OFDM (Orthogonal Frequency Division Multiplexing), 253, 281
OMB (Office of Management and Budget), extended C-band auction proceeds, 273
Omnipoint Corp., PCS, 216
OmniTracs, Qualcomm, 318
Orbcomm, NVNG MSS, 55
Orbimage, 297
Orbit Act (Open-Market Reorganization for the Betterment of International Telecommunications), 272
Orbital Sciences OrbLink, 367
origins of astronomy, 24
ORS (Offshore Radiotelephone Service), 119
overscan, TV data transmission, 123
ozone measurements, 154

P

packet radio networks, 59
PACS (Personal Access Communications System), 239
paging, internal organizational, 167
Paging and Radiotelephone Service, 110
Paging Service, 20–22
PanAmSat, V-Stream system, 368
PANDORA (Personal Alerting Network Deployed Only in Remote Areas), 105
PANs (Personal Area Networks), 253, 284
Parcae, 233
passive imaging, 390
passive radiometers, 264
passive sensors, 275
  earth exploration services, 378
  ocean temperature measurements, 292
Patriot system radar, Army, 284
PAVE PAWS (Precision Acquisition Vehicle Entry Phased Array Warning System), 107
PCN (Personal Communications Network), 216
PCOs (Private Cable Operators), 314, 327

Subject Index

PCP (Private Carrier Paging), 21, 165
PCS (Personal Communications Service), 213–215
   BTAs, 216–217
   E911, 220
   FreePage Corp., 14
   MTAs, 216
   origins, 215
   technical standards, 218
PCSAT (Personal Communications Satellite Corporation), GMPCS system, 226
PELTS (Personal Emergency Locator Transmitter Service), 89, 104
Pentriad system, Denali Telecom, 367
PER (Public Emergency Radio), 75
PERKI, 75
Personal Location and Monitoring Service (MicroTrax), 206
Personal Radio Services, 112
phase 3D amateur satellite, 347
phased-array radar, 172
phono oscillator, 27
PHS (Personal Handyphone System), 216
Piano system (Motorola), 385
Pioneer, 235, 245
Pioneer's Preferences, 151
PLBs (Personal Locating Beacons), 103
PLMRS (Private Land Mobile Radio Services), 12
PLMS (Personal Location and Monitoring Service), 158, 237
PMRS (Private Mobile Radio Services), 12–13, 65, 137, 141, 206
   General pool, 16
   Industrial/Business pool, 16–17
   Public Safety pool, 16–19

refarming channels, 18
SMR pool, 16
POCS (Proximity Operations Communication System), 352
POES (Polar-orbiting Operational Environmental Satellites), 98, 208
POFS (Private Operational Fixed Point-to-Point Microwave Service), 213
police radio, 23–25, 142
police speed radar, 303, 365
police surveillance systems, 257
POV cameras (Point of View), 224
power utility services, 23
Power, Petroleum, and Railroad Radio Services, 16
Prayers Heavenbound, 167
PRCS, 149–150
Primary service, ix
private coast stations, Coast Guard, 68
Private Operational Fixed Point-to-Point Microwave Service, 230, 334
ProNet Inc., TracPacs, 85
prostate disease microwave treatments, 177
protospace-based services, 375
PRSG (Personal Radio Steering Group), 64
PSAPs (Public Service Answering Points), E911, 220
pseudolites, 175
psuedosats, 175
PSWAC (Public Safety Wireless Advisory Committee), 143
PTTs (Platform Transmitter Terminals), 99
public coast stations, Coast Guard, 68
Public Mobile Rural Radiotelephone Service, 110

public safety channels, 134, 143, 147
public safety digital radio, 135
Public Safety pool, PMRS, 16–19
Public Safety Services, 137
pulsar observations, 95
pulsed altimeters, 200, 275
PURAC (Personal Use Radio Advisory Committee), 114

Q

Q-band, 352, 364, 370
Qualcomm
  Globalstar, 199
  OmniTracs, 318
  PCS, 217
quasars, 306
quasi-optical gyrotrons, 389

R

R/C Radio Service (Radio Control), 35
R/L DBS, 311
RACONs (Radar Transponder Beacons), 268
radar
  ASDE, 323
  AWACS, 268
  bird migration, 301
  civil air traffic control, 270
  Coast Guard shipboard, 316
  coastal, 300
  Earth observation, 285
  frequency-hopping, 172
  Goldstone Solar System Radar, 244
  IFF, 285
  insect tracking, 302
  International Space Station, 302
  interrogating transponders, 287
  long range search, 142–143, 148, 164
  maritime navigation, 301
  Navy shipboard, 316
  NEXRAD, 266–270
  PAVE PAWS, 107
  phased-array, 172
  RACONs, 268
  SARs, 269
  TDWR, 281, 286
  UWB, 3, 6
  weather, 301
  wind profilers, 154
radar astronomy, observation data return links, 366
radar cruise control, 374
Radar Services, 364
radio
  altimeters, 200, 274
  trunked systems, 138
radio astronomy, 362, 390, 394
  ammonia spectral line, 344
  brightness distribution studies, 279
  CBR reasearch, 399
  CO spectral line imaging systems, 392
  continuum emission observations, 106, 264, 321, 362, 373
  formaldehyde spectral line, 320
  galactic distance measurements, 343
  galactic molecular material studies, 378
  hydrogen spectral line emissions, 179
  hydroxyl radical spectral line observations, 201, 205, 211
  mass distribution/motion observations, 182
  methanol spectral line, 292

millimeter wavlength observations, 360
objections to expansion of frequency allocation, 24
origins, 24
pulsar observations, 95
quasars, 306
radiation observations, 24
receive-only bands, 279
SETI, 183
solar wind readings, 35
space-based research, 380–382
spectral emission observations, 320, 393
supernovas, 277
VLBA, 373
VLBI, 125
radio paging, 20–22
radio services, vi
Radio Solar Telescope Network, 265
radio spectrum, vi
radiodetermination, 196
radiolocation, 154
Radiolocation Service, 17
radios
  spread spectrum, 252
  stations, 169
radiosondes, 98, 208
Radiotelephone Service, 20
railroad communications, 71
range safety surveillance radar, NASA, 266
RAS (Radio Astronomy Service), 276–277
RCCs (Radio Common Carriers), 14, 21
RCLs (Radio Communication Links), 168, 293
RDS (Radio Data System), 39
RDSS (Radiodetermination Satellite Service), 196, 292
REAGs (Regional Economic Area Groupings), 240, 376

reallocating spectrum, 249
rear-looking vehicle radar, 374
receive-only scientific bands, 362
receivers, GPS, 173
Red Bat Communication, secondary markets, 9
Red Cross frequency, 31
refarming, PMRS, 18
remote control model aircraft, 34
remote rainfall sensing, 306
remote satellite sensing systems, *see* satellite remote sensing systems
rendezvous radar, Space Shuttle, 318
repeaters, 60
reserved spectrum bands, 251, 257
response channels, MDS, 261
response station hubs, MDS, 262
reverse band working, 326
RF (Radio Frequency), 2
RF weapons, 46
RNSS (Radio Navigation Satellite Service), 61
road condition sensors, 256
robotic horse jockeys, 159
rocketsondes, 98
RRS (Radio Reading Services), 38
RSAs (Rural Statistical Areas), cellular phone systems, 144
RTUs (Response Transmitter Units), IVDS, 83
Rural Radiotelephone Service, 22

S

SA (Selective Availability), GPS, 174
safe harbor policy, spread spectrum, 160
safety issues, HAPS, 376
SAME (Specific Area Message Encoding), 45

Subject Index

SAR (Search And Rescue), 102–104
SARs (Synthetic Aperture Radars), 172, 269, 302, 320
SARSAT (Search and Rescue Satellite-Aided Tracking), 102, 193
SARTs (Search and Rescue Transponders), 301
SATCOM, 298
satellite altimeters, 316
Satellite Control Network, Air Force, 212
Satellite DARS, 243–244
satellite engineering beacons, 251
satellite remote sensing devices, 389–401
satellite services, 360
satellite systems, extended C-band, 271
Satellite TV, C-band usage, 273
Sativod, see SkyBridge
SBC Wireless, SDR, 7
SBIRS (Space Based Infrared System), 296
SBRs (Surface Based Radars), 322
SBWAS (Space Based Wide Area Surveillance System), 233
SCADA (System Control And Data Acquisition), 162, 296
scanning receivers, 148
scatterometers, 319
scientific data collection, 61
scintillation, 339
SDARS, 243–244
SDR (Software-Defined Radio), 2, 6–7
SDR Forum, 7
SDTV (multiple Standard Definition TV), 118
sea skimmer detection, 154
search and rescue, aircraft, 51
secondary markets, 2, leases, 8
Secondary service, ix
Segments, of spectrum, xi

seismic telemetry, 81
SENS (Satellite Enabled Notification System), 206, 236
sensors, industrial safety and control, 385
SETI (Search for Extraterrestrial Intelligence), 182–183
SGLS (Space-Ground Link Subsystem), 212, 233, 296
SHUCS (Spacehab Universal Communications Systems), 202
side-looking vehicle radar, 374
SIGINT satellites (signals intelligence), 335
SINCGARS (Single Channel Ground and Airborne Radio System), 12
Sirius Satellite Radio, SDARS, 243
Sky Station International, GSTS, 375–376
SkyBridge, 308, 331
SkyTel, 152
SMR (Specialized Mobile Radio), 16, 137–141, 149
Snotel, 26
SNRs (Signal-to-Noise Ratios), BLAST, 133
Solar System Radar, NASA, 244
solar wind readings, 35
soundings, 98, 208
Space Command, Air Force, 173
space exploration, 366
Space Imaging, 297
Space Network, NASA, 234
Space Operation Service, 106
Space Research Service, 370
Space Science Services, 106
Space Shot service, 166
Space Shuttle
  communications options, 91
  rendezvous radar, 318
  SARs, 302

space-based radio astronomy, 380–382
SpaceArc Archives of Humanity, 245, 310
spaceborne cloud radars, 391
SpaceCast system, Hughes Communications, 367
spacecraft communications, International Space Station, 319
spacecraft control, 228
Spaceway EXP (Hughes Communications), 331
SPASUR (Naval Space Surveillance System), 79
spectral line imaging systems, 392
spectrum allocation, vi
spectrum etiquette, Data-PCS, 252
spectrum reallocation, 249
   civil air traffic control radar, 270
   WCS, 271
speed detectors, 348
SpeedUS.com, 355
sporadic E-band, 32
spread spectrum, 4, 159–161, 254, 287
   radio, 252
   robotic horse jockeys, 159
Sprint, MDS, 259
SRS (Space Research Service), 106, 285, 315, 324, 352, 363
SSR (Secondary Surveillance Radar), 170
star formation research, 402
StarBand, 311
StarLynx system, Hughes Communications, 368
state highway maintenance services, 31
stationkeeping systems for aircraft, 269
STLs (Studio-Transmitter Links), 169, 293
stolen item tracking, 85

sub-video method, TV data transmission, 123
submillimeter wavelength
   uses, 360
sufferance, unlicensed devices, 28
SUPERNet (Shared Unlicensed Personal Radio Network), 282
surface search radar, Navy, 284
surveillance, 154
SVRS (Stolen Vehicle Recovery Systems), 73
SWAP (Shared Wireless Access Protocol), 254–255
SWAS (Submillimeter Wave Astronomy Satellite), 402
sweeping altimeters, 200
SWS (Safety Warning System), 39, 347–348
synthetic vision systems, 365, 390

T

Table of Frequency Allocations, vi
TACAN (Tactical Air Navigation), 171
TACSRAD (Tactical Satellite Radar), 173
TAPR (Tucson Amateur Packet Radio Corporation), 59, 155
TCAS (Traffic Alert and Collision Avoidance System), 171
TDRSS (Tracking and Data Relay Satellite System), 228, 317, 320–321, 344, 352
TDWR (Terminal Doppler Weather Radar), 281, 286
technical standards, PCS, 218
TEDs (Transient Electromagnetic Devices), 46
Teledesic Corp.
   Internet in the Sky, 333
   KuBS, 308
   NGSO system, 368

telemetry, 81
Telepoint phones, 215
Teletrac, 156
television
   data transmission, 123
   NTSC standard, 118
   station links, 169
   VBI, 123
Teligent L.L.C., DEMS licenses, 335
Tennessee Valley Authority, SCADA, 296
terminal radars, 266
terrorist electromagnetic weapons, 47
test range radars, Air Force, 284
TFGS, 325, 349–351
THAAD (Theater High-Altitude Area Defense), 246, 303
threat simulation, 320
thunderstorm physics, 365
TIA (Telecommunications Industry Association), HDFS, 372
Time Domain Corp., UWB applications, 3
time signal satellite services, 360
timed traffic signal controls, 361
TIROS (Television Infrared Observation Satellite), 208
tracking implants, GPS, 176
tracking objects in Earth orbit, 107
TracPacs (ProNet Inc.), 85
traffic light control systems, 112, 346, 361
transmitter ID systems, 384
traveling wave tubes, 389
Treasury Department, border interdiction, 277
Tropical Rainfall Measurement Mission, 365
troposphere
   bending, 32
   ducting, 303
   scatter communications, 277

trunked radio systems, 138–139
TRW
   GESN, 367
   Odyssey, 199
   passive millimeter wave video cameras, 390
TSYKADA system, 61
TT&C (Telemetry, Tracking and Control), 228, 274, 306
TV Marti, 124

U

U-band, 364, 370
U-NII (Unlicensed NII), 282–284, 346
U.S. Department of Agriculture, Snotel, 26
U.S. FM radio broadcasting, 37–39
U.S. International Broadcasting Bureau, 124
U.S. Naval Observatory Interferometer, 265
UAVs (Unmanned Aerial Vehicles)
   MAVs, 301
   radar control systems, 289
   SARs, 320
UFO satellites (UHF Follow-On), 92
UFOs (Unidentified Flying Objects)
   misobservations of NOSS objects, 233
   SPASUR detection, 80
Ulysses, 235
Unicoms, 50
Unintentional radiators, xii
Unmanned military vehicle control, 179
UPCS (Unlicensed PCS), 161, 221–223, 247
UPS (United Parcel Service), 87
USADR (USA Digital Radio Partners), 40

USGS (US Geological Survey), seismic telemetry, 81
USSB (U.S. Satellite Broadcasting), 310
UTAM
  Data-PCS, 248
  UPCS, 222
Utility meter readings, 184
UWB (Ultra Wideband Radio), 2
  FAA opposition, 5
  FCC actions, 3
  GPR, 2, 6
  GPS objections, 4
  home networking, 5
  liquid level sensors, 2
  Lockheed Martin cautions, 5
  low VHF experimental systems, 12
  medical uses, 2
  microphones, 2
  radar, 3
  RF phasing, 2
  XtremeSpectrum, 4

V

V-Stream system, PanAmSat, 368
VBI (Vertical Blanking Interval), TV, 123
VDL (VHF Digital Link), 54
Vehicle collision-avoidance radars, 348, 374, 391
Vehicle identification tags, 287
Venus, radar mapping, 245
VHF (Very High Frequency), 12
VHF cordless phones, 29
Video eavesdropping, 154
Video surveillance, 277
Virgo, 309
VisionStar Inc., LMDS, 330
VITA (Volunteers in Technical Assistance), 55
VITASAT satellite, 55
VLBA (Very Large Baseline Array), 126, 373, 378
VLBI (Very Long Baseline Interferometry), 125, 235, 279, 343
Voice eavesdropping, 154
VoiceStream (Western Wireless Corp.), 217
VOR (VHF Omnidirectional Range), 47
Voyager spacecraft, 235, 245
VPC areas (VHF Public Coast), 68
VSAT (Very Small Aperture Terminal), 307, 318
VTS (Vessel Traffic Services), 68, 116

W

W-band, 387, 390
WAAS (Wide Area Augmentation System), 175
WCA (Wireless Communications Association), 262, 362
WCS (Wireless Communications Service), 238–240, 244, 271, 278
WCS Radio Inc., SDARS, 241
Weather radar, 266, 301
Weather satellites, 205
WECA (Wireless Ethernet Compatibility Alliance), 253
Western Wireless Corp, VoiceStream, 217
WGU-20, 75
White Cloud, 233
Wi-Fi products, 253
Wildlife telemetry, 61, 72, 81, 99
Wind profilers, 107, 154
WINForum (Wireless Information Networks Forum), 221, 282
WinStar, 368
Wireless cable video, 259

Wireless LANs, 252–253
Wireless Medical Telemetry
    Service, 181, 185
Wireless microphones, 77
Wireless Radio Services
  FCC classifications, 13
  secondary market leases, 8
Wireless services, 366–369
Wireless utility meter
    readings, 184
WMTS (Wireless Medical
    Telemetry Service),
    125–126, 185

Worldcom, MDS, 259
WorldSpace, 187, 244

X

X-band, 303
XM Satellite Radio, SDARS, 243
XtremeSpectrum, UWB uses, 4

Band Index

30—30.56 MHz 12	123.5875—128.8125 MHz..... 51
30.56—32 MHz 12	128.8125—132.0125 MHz..... 51
32—33 MHz................ 19	132.0125—136 MHz 52
33—34 MHz................ 19	136—137 MHz.............. 53
34—35 MHz................ 19	137—137.025 MHz 54
35—36 MHz................ 20	137.025—137.175 MHz...... 56
36—37 MHz................ 23	137.175—137.825 MHz...... 57
37—37.5 MHz.............. 23	137.825—138 MHz 57
37.5—38 MHz.............. 23	138—144 MHz.............. 58
38—38.25 MHz 25	144—146 MHz.............. 58
38.25—39 MHz 25	146—148 MHz.............. 60
39—40 MHz................ 25	148—149.9 MHz 60
40—40.98 MHz 26	149.9—150.05 MHz......... 61
40.98—42 MHz 26	150.05—150.8 MHz......... 62
42—43.69 MHz 26	150.8—152.855 MHz........ 62
43.69—46.6 MHz........... 27	152.855—154 MHz 65
46.6—47 MHz 30	154—156.2475 MHz 65
47—49.6 MHz 30	156.2475—157.0375 MHz.... 69
49.6—50 MHz 31	157.0375—157.1875 MHz.... 69
50—54 MHz................ 31	157.1875—157.45 MHz...... 70
54—72 MHz................ 33	157.45—161.575 MHz....... 70
72—73 MHz................ 34	161.575—161.625 MHz...... 71
73—74.6 MHz 35	161.625—161.775 MHz...... 71
74.6—74.8 MHz............ 35	161.775—162.0125 MHz..... 72
74.8—75.2 MHz............ 36	162.0125—173.2 MHz....... 72
75.2—75.4 MHz............ 36	173.2—173.4 MHz.......... 76
75.4—76 MHz 36	173.4—174 MHz 77
76—88 MHz................ 37	174—216 MHz.............. 77
88—108 MHz............... 37	216—220 MHz.............. 78
108—117.975 MHz 47	220—222 MHz.............. 87
117.975—121.9375 MHz..... 49	222—225 MHz.............. 90
121.9375—123.0875 MHz.... 50	225—235 MHz.............. 91
123.0875—123.5875 MHz.... 51	235—267 MHz.............. 91

Band	Page	Band	Page
267—322 MHz	94	869—894 MHz	148
322—328.6 MHz	95	894—896 MHz	148
328.6—335.4 MHz	95	896—901 MHz	149
335.4—399.9 MHz	96	901—902 MHz	151
399.9—400.05 MHz	96	902—928 MHz	153
400.05—400.15 MHz	97	928—929 MHz	162
400.15—401 MHz	97	929—930 MHz	164
401—402 MHz	98	930—931 MHz	165
402—403 MHz	100	931—932 MHz	167
403—406 MHz	101	932—935 MHz	168
406—406.1 MHz	101	935—940 MHz	168
406.1—410 MHz	105	940—941 MHz	169
410—420 MHz	106	941—944 MHz	169
420—450 MHz	107	944—960 MHz	169
450—454 MHz	108	960—1215 MHz	170
454—455 MHz	109	1215—1240 MHz	172
455—456 MHz	110	1240—1300 MHz	176
456—460 MHz	111	1300—1350 MHz	178
460—462.5375 MHz	111	1350—1390 MHz	178
462.5375—462.7375 MHz	112	1390—1395 MHz	180
462.7375—467.5375 MHz	116	1395—1400 MHz	181
467.5375—467.7375 MHz	117	1400—1427 MHz	182
467.7375—470 MHz	117	1427—1429 MHz	183
470—512 MHz	117	1429—1432 MHz	185
512—608 MHz	125	1432—1435 MHz	185
608—614 MHz	125	1435—1525 MHz	186
614—698 MHz	127	1525—1530 MHz	188
698—746 MHz	127	1530—1535 MHz	191
746—764 MHz	128	1535—1544 MHz	192
764—776 MHz	134	1544—1545 MHz	192
776—794 MHz	136	1545—1549.5 MHz	193
794—806 MHz	137	1549.5—1558.5 MHz	194
806—821 MHz	137	1558.5—1559 MHz	194
821—824 MHz	142	1559—1610 MHz	195
824—849 MHz	143	1610—1610.6 MHz	195
849—851 MHz	146	1610.6—1613.8 MHz	201
851—866 MHz	147	1613.8—1626.5 MHz	201
866—869 MHz	147	1626.5—1645.5 MHz	202

1645.5—1646.5 MHz........ 203	3.1—3.3 GHz.............. 268
1646.5—1651 MHz......... 203	3.3—3.5 GHz.............. 269
1651—1660 MHz........... 204	3.5—3.6 GHz.............. 270
1660—1660.5 MHz......... 204	3.6—3.65 GHz............. 270
1660.5—1668.4 MHz....... 204	3.65—3.7 GHz............. 271
1668.4—1670 MHz......... 205	3.7—4.2 GHz.............. 273
1670—1675 MHz........... 205	4.2—4.4 GHz.............. 274
1675—1700 MHz........... 207	4.4—4.5 GHz.............. 275
1700—1710 MHz........... 209	4.5—4.8 GHz.............. 276
1710—1755 MHz........... 210	4.8—4.94 GHz............. 276
1755—1850 MHz........... 211	4.94—4.99 GHz............ 277
1850—1990 MHz........... 213	4.99—5 GHz 279
1990—2025 MHz........... 224	5—5.25 GHz 280
2025—2110 MHz........... 227	5.25—5.35 GHz............ 284
2110—2150 MHz........... 229	5.35—5.46 GHz............ 285
2150—2160 MHz........... 230	5.46—5.47 GHz............ 285
2160—2165 MHz........... 231	5.47—5.6 GHz............. 286
2165—2200 MHz........... 231	5.6—5.65 GHz............. 286
2200—2290 MHz........... 232	5.65—5.83 GHz............ 287
2290—2300 MHz........... 234	5.83—5.85 GHz............ 288
2300—2305 MHz........... 235	5.85—5.925 GHz........... 288
2305—2310 MHz........... 238	5.925—6.425 GHz.......... 291
2310—2320 MHz........... 242	6.425—6.525 GHz.......... 291
2320—2345 MHz........... 242	6.525—6.875 GHz.......... 292
2345—2360 MHz........... 244	6.875—7.075 GHz.......... 292
2360—2385 MHz........... 245	7.075—7.125 GHz.......... 293
2385—2390 MHz........... 246	7.125—7.19 GHz........... 293
2390—2400 MHz........... 247	7.19—7.235 GHz........... 294
2400—2402 MHz........... 250	7.235—7.25 GHz........... 294
2402—2417 MHz........... 256	7.25—7.3 GHz............. 294
2417—2450 MHz........... 256	7.3—7.45 GHz............. 294
2450—2483.5 MHz......... 257	7.45—7.55 GHz............ 295
2483.5—2500 MHz......... 258	7.55—7.75 GHz............ 295
2500—2655 MHz........... 259	7.75—7.9 GHz............. 296
2655—2690 MHz........... 264	7.9—8.025 GHz............ 296
2690—2700 MHz........... 264	8.025—8.175 GHz.......... 297
2700—2900 MHz........... 265	8.175—8.215 GHz.......... 298
2900—3100 MHz........... 268	8.215—8.4 GHz............ 298

Band	Page	Band	Page
8.4—8.45 GHz	299	18.58—18.6 GHz	332
8.45—8.5 GHz	299	18.6—18.8 GHz	332
8.5—9 GHz	299	18.8—19.3 GHz	333
9—9.2 GHz	300	19.3—19.7 GHz	337
9.2—9.3 GHz	300	19.7—20.1 GHz	338
9.3—9.5 GHz	301	20.1—20.2 GHz	338
9.5—10 GHz	302	20.2—21.2 GHz	339
10—10.45 GHz	302	21.2—21.4 GHz	341
10.45—10.5 GHz	303	21.4—22 GHz	342
10.5—10.55 GHz	303	22—22.21 GHz	342
10.55—10.6 GHz	304	22.21—22.5 GHz	342
10.6—10.68 GHz	305	22.5—22.55 GHz	343
10.68—10.7 GHz	305	22.55—23.55 GHz	343
10.7—11.7 GHz	306	23.55—23.6 GHz	344
11.7—12.2 GHz	307	23.6—24 GHz	344
12.2—12.7 GHz	309	24—24.05 GHz	345
12.7—12.75 GHz	313	24.05—24.25 GHz	346
12.75—13.25 GHz	315	24.25—24.45 GHz	349
13.25—13.4 GHz	315	24.45—24.65 GHz	350
13.4—13.75 GHz	316	24.65—24.75 GHz	350
13.75—14 GHz	317	24.75—25.05 GHz	350
14—14.2 GHz	318	25.05—25.25 GHz	351
14.2—14.4 GHz	319	25.25—27 GHz	351
14.4—14.47 GHz	319	27—27.5 GHz	353
14.47—14.5 GHz	320	27.5—29.5 GHz	353
14.5—14.7145 GHz	320	29.5—29.9 GHz	356
14.7145—15.1365 GHz	321	29.9—30 GHz	357
15.1365—15.35 GHz	321	30—31 GHz	360
15.35—15.4 GHz	321	31—31.3 GHz	360
15.4—15.7 GHz	322	31.3—31.8 GHz	362
15.7—16.6 GHz	322	31.8—32 GHz	363
16.6—17.1 GHz	324	32—32.3 GHz	364
17.1—17.2 GHz	324	32.3—33 GHz	364
17.2—17.3 GHz	324	33—33.4 GHz	364
17.3—17.7 GHz	325	33.4—36 GHz	365
17.7—17.8 GHz	326	36—37 GHz	365
17.8—18.3 GHz	328	37—37.6 GHz	366
18.3—18.58 GHz	328	37.6—38.6 GHz	366

Band Index

38.6—39.5 GHz 368	92—95 GHz 390
39.5—40 GHz 370	95—100 GHz 391
40—40.5 GHz 370	100—102 GHz............. 391
40.5—41 GHz 371	102—105 GHz............. 392
41—42.5 GHz 372	105—116 GHz............. 392
42.5—43.5 GHz............ 373	116—119.98 GHz 392
43.5—45.5 GHz............ 373	119.98—120.02 GHz....... 393
45.5—46.9 GHz............ 374	120.02—126 GHz 393
46.9—47 GHz 374	126—134 GHz............. 394
47—47.2 GHz 375	134—142 GHz............. 394
47.2—48.2 GHz............ 375	142—144 GHz............. 394
48.2—50.2 GHz............ 377	144—149 GHz............. 395
50.2—50.4 GHz............ 378	149—150 GHz............. 395
50.4—51.4 GHz............ 379	150—151 GHz............. 395
51.4—52.6 GHz............ 379	151—164 GHz............. 396
52.6—54.25 GHz........... 379	164—168 GHz............. 396
54.25—55.78 GHz 380	168—170 GHz............. 396
55.78—56.9 GHz........... 380	170—174.5 GHz 397
56.9—57 GHz 381	174.5—176.5 GHz......... 397
57—58.2 GHz 381	176.5—182 GHz 397
58.2—59 GHz 382	182—185 GHz............. 397
59—59.3 GHz 383	185—190 GHz............. 398
59.3—64 GHz 383	190—200 GHz............. 398
64—65 GHz 386	200—202 GHz............. 398
65—66 GHz 386	202—217 GHz............. 399
66—71 GHz 386	217—231 GHz............. 399
71—74 GHz 387	231—235 GHz............. 399
74—75.5 GHz 387	235—238 GHz............. 400
75.5—76 GHz 387	238—241 GHz............. 400
76—77 GHz 387	241—248 GHz............. 400
77—77.5 GHz 388	248—250 GHz............. 400
77.5—78 GHz 388	250—252 GHz............. 401
78—81 GHz 389	252—265 GHz............. 401
81—84 GHz 389	265—275 GHz............. 401
84—86 GHz 389	275—300 GHz............. 402
86—92 GHz 390	

About the Author

Bennett Z. Kobb is a Washington, D.C.-based technology writer and consultant.

As a trade journalist, he has covered policy and regulation at the Federal Communications Commission since 1983. He founded *Federal Communications TechNews* (now TR Wireless) and was editor of *Cellular Radio News* and *Personal Communications Magazine*, the first periodicals for the cellular telephone industry.

He co-founded (with Apple Computer) and served as the first executive director of the Wireless Information Networks Forum, a trade association of telephone and computer manufacturers.

He is a member of the Institute of Electrical and Electronics Engineers, a licensed radio amateur, and a contributor to *Communications World* on the Voice of America.

Bennett received a master's degree in telecommunications from the University of Colorado and a bachelor's degree in radio-television-film from the University of Texas.

After hours, he may be found amid flying objects as a devotee of juggling, the happiest art.